Mathematicians Fleeing
from Nazi Germany

Mathematicians Fleeing from Nazi Germany

Individual Fates and Global Impact

Reinhard Siegmund-Schultze

PRINCETON UNIVERSITY PRESS

PRINCETON AND OXFORD

Copyright 2009 © by Princeton University Press
Published by Princeton University Press, 41 William Street, Princeton, New Jersey 08540
In the United Kingdom: Princeton University Press, 6 Oxford Street, Woodstock,
Oxfordshire OX20 1TW

Library of Congress Cataloging-in-Publication Data

Siegmund-Schultze, R. (Reinhard)
   Mathematicians fleeing from Nazi Germany: individual fates and global impact /
Reinhard Siegmund-Schultze.
       p.   cm.
   Includes bibliographical references and index.
   ISBN 978-0-691-12593-0 (cloth) — ISBN 978-0-691-14041-4 (pbk.)
1. Mathematicians—Germany—History—20th century.   2. Mathematicians—
United States—History—20th century.   3. Mathematicians—Germany—Biography.
4. Mathematicians—United States—Biography.   5. World War, 1939–1945—
Refuges—Germany.   6. Germany—Emigration and immigration—History—1933–1945.
7. Germans—United States—History—20th century.   8. Immigrants—United
States—History—20th century.   9. Mathematics—Germany—History—20th century.
10. Mathematics—United States—History—20th century.   I. Title.
   QA27.G4S53 2008
   510.09'04—dc22   2008048855

British Library Cataloging-in-Publication Data is available

This book has been composed in Sabon
Printed on acid-free paper. ∞
press.princeton.edu
Printed in the United States of America
10 9 8 7 6 5 4 3 2 1

Contents

Chapter 8
The American Reaction to Immigration: Help and Xenophobia   186

CHAPTER 9
Acculturation, Political Adaptation, and the American Entrance
into the War   230

CHAPTER 10
The Impact of Immigration on American Mathematics   267

CHAPTER 11
Epilogue: The Postwar Relationship of German
and American Mathematicians   319

Figures and Tables

Figures

Tables

Preface

IN NOVEMBER 1934, Richard Courant, a German mathematician driven out of Göttingen and now living in New York City, commented on his long-time friend Wolfgang Sternberg's book project with the following words:

> An extensive manuscript by Sternberg on the calculus of probability is already at the printing-house of Vieweg's and should have appeared a long time ago. However, in spite of an existing contract, the publisher Vieweg decided to stop the printing at the last moment, pointing to problems of publishing a book by a non-Aryan author at this time. I believe, that for that reason the mathematical public has lost a useful and valuable work.[1]

This incident encapsulates the political background and various events to be described in this book.

This book shows the prominent role played by Courant—former organizer of the mathematical institute in Göttingen—as an emigrant in the United States, in particular his efforts in reinstalling German mathematicians who had been dismissed from academic positions. The present publication also deals with the fate of the victims who, like Sternberg, never regained a position commensurate with their abilities,[2] or who, as the Prague mathematician Berwald, the addressee of Courant's letter, were murdered by the Nazis. The incident with Sternberg's manuscript says something indirectly about the specificity of the various waves of purges, since Berwald, being in Prague, was, in 1934, still safe.[3] The book also considers the more general sociological consequences of Nazi rule for science and mathematics. In fact, the censorship and the dismissal of the Jewish author Sternberg exemplify the losses for German mathematics due to Nazi interference.

---

[1]Courant to Ludwig Berwald, CPP (T), November 10, 1934.

[2]In an interview with the *Sources of History of Quantum Physics* in 1962, Courant describes the promising beginnings of Sternberg's career in Breslau (Kuhn et al. 1967), transcript of interview with Courant, May 9, 1962, p. 3. The Sternberg files in the Oswald Veblen Papers at the Library of Congress in Washington, DC give a deeply distressing picture of Sternberg's miserable living conditions after his immigration to the United States in 1939.

[3]Prague was occupied by Germany in 1939. For Austria, however, one also has to consider the years between 1933 and 1938 as causing forced emigration, because of the various pre-Fascist regimes at the time. This led for example to the gradual emigration of the members of the Vienna Circle of neopositivist philosophy and to the assassination of its leader, Moritz Schlick, in 1936.

Breslau, the place of Sternberg's dismissal, became a part of Poland as a consequence of the war.[4] It had not only been home to many German mathematicians;[5] it also had broader cultural importance. For example, Edith Stein (1891–1942) the Jewish-Catholic martyr (incidentally a close cousin of Richard Courant's) came from there, too.[6] Thus the losses to Germany as a whole, not restricted to only its mathematics, are also exemplified by examining the fate of Sternberg and of his birthplace Breslau.

However, this book is also a description of the *bridge to the present*, constituted by the often very successful and, for today's mathematics, very significant emigration of so many important European mathematicians.

Finally, this book is about the responsibility of the living to keep the memory of these historical events alive. The publisher of an earlier (1998) though much changed German version of this book, Vieweg, the very same publisher who once felt forced to stop the publication of Sternberg's work, took this responsibility.[7]

In order to make the discussion more systematic, the following general limitations have been imposed: The discussion is principally restricted to mathematicians who had completed a full university education in a German-speaking environment[8] and were persecuted in a Nazi-dominated territory between 1933 and 1945 while still working as mathematicians. Within this population the focus is on the emigrants rather than on those victims remaining in Germany. Motivations for this restriction and a discussion of the relevant concepts such as "German-speaking," "mathematician," "emigration," and "persecution" will follow in chapters 1 and 2.

Major changes in the communication structure of science, such as emigration, are generally of great interest and importance to historians. The

[4]Breslau is today the Polish town Wrocław.

[5]Beside Courant and the friend of his youth, Sternberg, further prominent émigrés in mathematics (E. Hellinger, O. Toeplitz, and less known H. Kober and his wife K. Silberberg) grew up in the Breslau area. Others were purged from there later in their lives (F. Noether, H. Rademacher). Also, Courant's first wife, Nelly Neumann, came from this city. As a mathematics teacher she had fewer chances for emigration than the others and was murdered by the Nazis.

[6]See Stein (1986), where Edith Stein says: "He was bound to me by our close relationship" (p. 262). She also explains the reasons for Courant's and Neumann's divorce.

[7]In the archives of Vieweg no correspondence or other documentation could be traced for Sternberg's planned publication.

[8]With respect to this point the present English version tries to be more consistent than the German book of 1998. It included for instance Kurt Bing (1914–1997), born in Köln, but who received his mathematical training only after his emigration in Palestine (Letter by Mrs. P. Bing to the author, March 6, 1998). Also Horst Tietz (born 1921), who has done so much to bring about a critical discussion of the Nazi past, is for systematic reasons no longer included, since he, too, had not completed a full study of mathematics before persecution. See the printed manuscript (Tietz 1998) with interesting information on the situation of mathematics in Hamburg during the 1940s.

actions of the scientists in these periods have an enormous impact on the future path of their disciplines. The discussion will show that it needed both the human qualities and scientific competence of mathematicians such as Richard Courant, Hermann Weyl, and Oswald Veblen to be effective in saving and shaping much of the potential of their science for future generations. Indeed, only a few of the emigrants would have accepted a conspicuous (if fortunately often only temporary) downgrading of their social status, if they had not been able to trust in the scientific competence, even superiority, of the organizers of emigration to secure them fair treatment under very restricted conditions. It also took the scientific intuition and instinct of Courant, Weyl,[9] and others to foresee future scientific trends, as well as recognizing the potential of the young refugees and the best places for them. Thus this book also makes a principled case for the engagement of the highly competent and prominent scientist in the social and organizational affairs of his or her discipline.

Motives for dealing with the particular social and historical problem of scientific emigration in this book are manifold, recent political events being among them.

Given the increased globalization of international scientific, economic, and cultural relations, problems of acculturation in foreign societies now play an ever-increasing role. Reforms of archaic, ethnically based naturalization laws such as the German one, which in the 1920s impeded acculturation and later on, in the 1930s, emigration, have to be based on historical experiences.[10]

Another more purely scientific motive for dealing with the emigration problem is the need to supplement the existing, predominantly popular or anecdotal treatment even in meritorious publications such as those by Constance Reid and Max Pinl,[11] by a presentation based on verifiable sources. Unfortunately, I have not been able to study the recollections of emigrants, collected by Pinl for his publication of 1969, since the present owners of the material have not made them accessible to me.[12] The still existing political sensitivity of the topic has discouraged several other historians of the period in their efforts. Public financial support for historical research on mathematics under the Third Reich has not been consistent in the past either; several projects to explore the newly opened files of the Deutsche Mathematiker-Vereinigung in Freiburg have been turned down.

[9] Weyl's competence as a mathematician enabled him "instinctively" to judge the mathematical potential in some younger refugees and see the exceptional ability in them, even if he was no specialist in the respective field.

[10] The German naturalization laws were maintained long after World War II and have only recently been changed to reflect more modern criteria, among them geographical ones.

[11] For instance, Reid (1976) and Pinl (1969–72).

[12] This was already criticized in Schaper (1992).

I have principally tried to use, and prefer, *new* archival sources, in particular reports by the German-speaking mathematicians involved. They serve to illustrate and back up general claims on scientific emigration, some of which have been made by other authors already. Since those reports were for the most part written long after the events and because both the sorrow felt by the protagonists and the self-interest that is inevitably present in biographical recollections play a role, the historian has to judge and to select with care. Particularly valuable recollections are those by Frucht (1982), Hirsch (1986), Menger (1994, edited by L. Golland et al.), Mahler (1991, in Poorten), Thullen (2000), and Wasow (1986), the latter only printed for private circulation. The autobiography by Fraenkel (1967), though invaluable as a source of information on religious Jews in the Republic of Weimar and on the first years of the university in Jerusalem, has to be treated with particular care because it contains a number of inadvertent inaccuracies. Reid's biography (1976) of Courant remains valuable because of its many hints and superb literary style in spite of a lack of documentation of the sources. With regard to the general conditions of scientific emigration I refer to the rather broad literature available. In particular there will be no new description of the network of support organizations for emigration. The microlevel of the politics of mathematicians has priority over the general policies of international relations among states, the strong impact of these general conditions on the fates of the scholars notwithstanding. Since this book focuses on the processes of emigration and immigration, the mechanisms of the purges from Germany are not given priority here. These mechanisms have most recently been described by Remmert (2004) and Segal (2003). Nevertheless, some typical examples of the purges and of traces of resistance are given in chapter 4.

The most important historical source used in this book is the private correspondence of Richard Courant from 1933 to 1936, henceforth cited as CPP.[13] Courant, the former director of the Göttingen Mathematical Institute and later the leader of a mathematical center (named after him today) with a strong leaning toward applications at New York University, was without any doubt a key figure within emigration in mathematics. Further important historical sources are the papers of mathematicians

---

[13]The correspondence CPP that was in the possession of Courant's son Ernest is now deposited at New York University (Elmer Holmes Bobst Library). This correspondence, which has been partially used already by Reid (1976) and Beyerchen (1982), is mostly pre-1936. No further private correspondence by Courant has so far come to light. However, the Bobst Library contains further correspondence by Courant from the files of the Courant Institute of New York University, which is partly private in character. As of today (2008), the Courant Papers at the Bobst Library (abbreviated as CP, which now include CPP) have not been fully catalogued.

such as von Mises, von Kármán, Birkhoff, Richardson, Weyl, Wiener, and Veblen. Among them the Oswald Veblen Papers at the Library of Congress, which include extensive correspondence of the refugees with Hermann Weyl, stand out in importance. The above papers, the Courant correspondence, as well as the files of the New York Emergency Committee in Aid of Displaced (later: Foreign) Scholars (S. Duggan) and the Harlow Shapley Refugee Files (Harvard University), which have also been used, are far from exhausted in their value as historical sources in the present investigation. Further unpublished sources are given in Spalek (1978) and in *American Council for Émigrés in the Professions: Records 1941–1974* (AC), also referred to in the appendices.

The present book is deliberately unconventional in its structure and its purpose. It focuses on one particular and important aspect of the influence of Nazi rule on mathematics, namely, emigration, relying on a three-part division into analysis, documentation, and case studies, supplemented with extensive appendices. While it hopefully presents a study of the varying conditions and motives constructing historical "objectivity" and provides a critical evaluation of various myths that have arisen in the work of historians and in the commentary of contemporaries alike, the book is deliberately not adopting the rather extremist position in some modern methodological work which stipulates that "there are no facts." Thus about half of the book is primarily documentation, allowing mathematicians of today to find information about the lives, policies, and, not least, sufferings of their predecessors. Hopefully this information can be gathered easily through the indices and the table of contents. Since different individuals sometimes faced similar dilemmas, such a documentary approach necessarily involves some repetition of arguments.

The individual chapters of this book (beginning with chapter 3) will be divided under two headings: first, a discussion of the most relevant problems of emigration relating to the topic of the chapter, and second, documents (D) and case studies (S), which illustrate the preceding claims. The latter follow many biographical sidelines, relevant to the topic under discussion but which, for reasons of space, cannot always be fully explored. For example, the study in chapter 8 on anti-Semitic remarks by the prominent American mathematician George David Birkhoff (1884–1944) would benefit from being embedded within a detailed biographical investigation exploring Birkhoff's motivations more thoroughly.

The introductory parts of the chapters are devoted to an analysis of the crucial aspects of emigration such as "voluntary" and "enforced emigration"; "losses," "gains," and "impact" of emigration; "aftereffects of the war"; and similar problems. Readers primarily interested in the general and theoretical questions of science and mathematics under the Third Reich are advised to focus on this part of the book. This will hopefully encourage further studies of the history of emigration in other disciplines during the period of the Third Reich as well, a history still largely unexplored.

Since the German edition of this book (1998), the Freiburg files focusing on inner-German developments have been opened and have been the subject of studies by V. Remmert. Other major topical work published since 1998 is by Litten, Menzler-Trott, Segal,[14] Dawson, and the present author. The publications of the last two mentioned are closely linked to emigration.

In addition to updating the literature, and correcting typographical or minor errors, the present English edition considerably changes and extends the German original in various directions, essentially making it a new book. The most important addition concerns the files of the British Society for the Protection of Science and Learning (SPSL), kept at the Bodleian Library in Oxford. However, the extensive correspondence contained in these files should be the basis for a separate study of emigration to Britain, which is not the main focus of this book. Some new information was gathered from the Louis Joel Mordell Papers at St John's College Library, Cambridge, UK. As to emigration to the United States, several archives, in particular the Oswald Veblen Papers in Washington, DC, the Files of the Emergency Committee at the New York Public Library, and the Courant Papers of the New York University Archives have been revisited and researched more completely,[15] and the archives of the Institute for Advanced Study in Princeton have been visited for the first time. New material has been added, for instance, on immigration to and through Norway (my home since 2000). Another addition in this book concerns the fate of two particular émigrés, Hilda Geiringer and her husband since 1943, Richard von Mises, on whom I have worked extensively. The two new appendices 3.3 and 4.3 contain interesting excerpts from von Mises's diary concerning his emigration to Turkey (1933) and to the United States (1939). Another major change is the inclusion of an autobiographical report by the function theorist Peter Thullen on the circumstances of his flight from Germany as a Catholic dissenter (Appendix 6). The last-mentioned document widens the perspective of the reasons for emigration, while this book, for the most part, retains a natural focus on the dominating anti-Semitic policies of the Nazis. Particularly helpful was a list of doctoral students with bio-bibliographical information recently published by Renate Tobies, showing especially that the fate of mathematics schoolteachers under the Nazis is notoriously difficult to trace.[16]

---

[14]Segal (2003) does not yet use the Freiburg files.
[15]Appendix 5.1 contains a previously unpublished letter from the Courant Papers.
[16]Tobies (2006).

Although the present book aims at understanding emigration and acculturation of both mathematics and mathematicians as a broader process, in its sources and in its mode of presentation it proceeds primarily from the individual biographical perspective. This choice of presentation is in part determined by my conviction that in a book on emigration of mathematicians, the destruction of lives should receive as much attention as the development of mathematics, which is the usual focus of books on the history of the discipline.[17]

Finally, a word on the terminology used in this book as well as in the German original. I am convinced that the "National Socialism" movement had neither socialist aims nor acted in the interest of the "nation." Therefore the term "National Socialism" has been largely avoided here. Instead the abbreviation "Nazi," applied disparagingly already at the time, has been used. Another word applied to the regime by contemporaries (and not just communists)—"Fascism"—is still used by some scholars in order to stress common traits (social demagogy, economy, antimodernism, and militarism) between the Italian and German regimes in the 1930s. I do not use that term, even in contexts where the indisputable differences between the two regimes (differences in implementing anti-Semitism) do *not* matter, in order to prevent any possible misunderstanding. In contrast to Pinl's publications of 1969–72 on mathematicians in Hitler's Germany, the "reasons" (or pretexts) for dismissal and emigration will be named in each case. Nonscientific and ideological Nazi jargon such as "Aryan" and "Third Reich" will generally be put in quotation marks in order to express distance. The same will be done with words such as "Jewish," if they do not clearly reflect self-assessment of the respective mathematician and if they, instead, express the Nazi-construction of the notion in question.

As far as technical details are concerned, quotations are always in English, either original or translated. My own translations from German into English are marked by (T) in parentheses; translations from other languages or by other translators are noted accordingly. Occasional commentary by myself within quotations, in particular original German words, will be added in square brackets. Emphasis within the text as well as titles of journals and books are marked in italics. The first time a person is mentioned, their given names and dates are included, as far as known, but only if they are *not* included in Appendices 1, which list emigrants and victims.

---

[17]On a more theoretical level, Mehrtens (1990a/b) critically discusses traditional historiography of mathematics in which the emergence of collective memories in the sciences is prioritized, and political and moral categories usually do not play any role.

I have a great many individuals and institutions to thank for their support. First and foremost the mathematicians described in the book and their relatives, and also colleagues, libraries, and archives, as well as publishing houses that granted publication rights for various illustrations. Representative for all sources mentioned in the appendix (archives, correspondence, illustrations) I name the private correspondence by Richard Courant (CPP) from the early 1930s that Courant's son, Professor Ernest Courant (Bayport, New York) kindly made accessible to me when I prepared the German book of 1998. For comments and advice I have to thank the following individuals: Gerard Alberts (Amsterdam), Manfred Agethen (St. Augustin near Bonn), Joseph-James Ahern (Philadelphia), Tom Archibald (Vancouver), Wolfgang Arnold (Berlin), Liliane Beaulieu (Nancy), Christa Binder (Vienna), Clare Bunce (Cold Spring Harbor), Janet Bunde (New York City), Leo Corry (Tel Aviv), Jonathan Coss (New York), Jeremy Gray (London), Peter Gruber (Vienna), Heinrich Guggenheimer (Farmingdale, New York), Per Christian Hemmer (Trondheim), Claus-Dieter Krohn (Hamburg), Hans Lausch (Clayton, Australia), Elisabeth Lebensaft (Vienna), Freddy Litten (Munich), Ralf Lohan (St. Augustin), Herbert Mehrtens (Braunschweig), Erica Mosner (Princeton), Volker Peckhaus (Erlangen), Allan Pinkus (Haifa), Claudia Pinl (Köln), Susann Puchta (Hof), Volker Remmert (Mainz), Hans Romberg (Heidelberg), Peter Roquette (Heidelberg), Karl-Heinz Schlote (Leipzig), Friedrich Schreiber (Aachen), Gert Schubring (Bielefeld), Wolfgang Schwarz (Frankfurt am Main), Christoph J. Scriba (Hamburg), Sanford L. Segal (Rochester), Karl Sigmund (Vienna), Silke Slembek (Strasbourg), Alexander Soifer (Colorado Springs), Georg and Sylvia Thullen (Genthod), Jennifer Ulrich (New York City), Marianne Wenger (Vienna), and Günther Wirth (Berlin). To the foundation Stiftung Centrum Judaicum Berlin (Sabine Hank) and the editorial staff of the Poggendorff Biographical Dictionary in Leipzig now dissolved (Mss. Köstler, Kühn, and Marschallek) I am indebted for individual biographical information. I am particularly grateful to the following colleagues and friends who read the original German manuscript critically and made many valuable suggestions: Heinrich Begehr, Kurt R. Biermann (now deceased), and Katrin Liebich (all Berlin), David E. Rowe (Mainz), Norbert Schappacher (Strasbourg), Winfried Scharlau (Münster), Skúli Sigurdsson (Seltjarnarnesi/Berlin), Heinrich Wefelscheid (Essen), and Dirk Werner (Berlin). I owe much to discussion with Renate Tobies (Berlin), profiting from her comprehensive knowledge of the social conditions for mathematics in Germany in the 1920s and 1930s. Invaluable help with respect to language and content, as well as personal encouragement, came from June Barrow-Green (London), and my daughter Ulrike Romberg (Berlin). Kathleen P. Nordgarden (Kristiansand), and Dawn Hall (Albuquerque, New Mexico) gave great help in copyediting the manuscript.

Much of the material on which this book is based was collected during several stays in the United States (1991–97), financed by the Alexander von Humboldt Foundation (Bonn) within its Feodor-Lynen Program. The work on the material for the book of 1998 was done mainly within a project of the Deutsche Forschungsgemeinschaft (DFG, 1994–96), directed by Roswitha März (Humboldt Universität Berlin). The work on the present book was largely done during a sabbatical year in the United States in 2005/6, financed by my home institution, the University of Agder (Kristiansand, Norway), and supported by the History of Science Department at Harvard University, in particular by Professor Gerald Holton. The publication through Princeton University Press owes much to the initiative of senior editor Vickie Kearn, who was untiring in her support during the project, to Heath Renfroe, the production editor at Princeton, and to Alexander Soifer (Colorado Springs).

The incentive for publishing the German book was the International Congress of Mathematicians,[18] held in Berlin in 1998, for the first time in Germany since Heidelberg 1904. In the intervening years Germany had lost its international supremacy in the field of mathematics mainly due to the events described in this book. It is to be hoped that the publication of this book will be taken as a sign that German mathematicians are prepared to face the problems and responsibilities of the past after the successful reemergence of German research in international mathematics.

—Reinhard Siegmund-Schultze
—Kristiansand, Norway, August 2008

[18]See also Brüning et al. (1998), the English catalogue of an exhibition on dismissed Berlin mathematicians organized for that occasion. A somewhat later exhibition on émigré-mathematicians from Vienna was organized by Karl Sigmund. See Sigmund (2001). A special exhibition to commemorate Gödel's centenary in 2006 was organized by Sigmund and others. See the related volume, Sigmund/Dawson/Mühlberger (2006).

Terror and Exile

Persecution and Expulsion
of Mathematicians from Berlin
between 1933 and 1945

An Exhibition
on the Occasion of the
International Congress of Mathematicians 1998

Deutsche Mathematiker–Vereinigung

Figure 1   *Berlin Exhibit. The title page of the catalogue for the exhibit on dismissed and persecuted Berlin mathematicians, organized by the Deutsche Mathematiker-Vereinigung in 1998.*

Mathematicians Fleeing
from Nazi Germany

Figure 2 *Richard Courant (1888–1972). The student of David Hilbert (1862–1943) and Felix Klein (1849–1925) had been the organizer of the flourishing mathematics in Göttingen of the 1920s. From 1935 on Courant built a graduate school for mathematics at New York University, which became a center for research in applied mathematics during the war and later on a pioneer institution for the incorporation of computing techniques into mathematics. Courant helped many immigrants adapt to American society. His personal correspondence from the first part of the 1930s is an important source for the present book.*

ONE

The Terms "German-Speaking Mathematician," "Forced," and "Voluntary Emigration"

THIS CHAPTER tries to settle some fundamental concepts to be used in the book concerning the overall process of expulsion of scientists by the Nazi regime and which are not specific to "mathematics," although the concrete examples are from that particular field. In addition, this chapter outlines the structure of argumentation and the mode of presentation used in the book.

The expulsion of many European mathematicians from their jobs and from their home countries between 1933 and the early 1940s forced upon them by Hitler's regime is undoubtedly the central event of the social history of mathematics between the two world wars.

That momentous event has to be put, on the one hand, into a broader historical perspective and to be treated with some claim of historical completeness. On the other hand, however, the discussion has to be appropriately restricted to exemplary case studies that can be dealt with in a limited volume.

The restrictions concern basically an emphasis on the special process of "emigration" within the overall "expulsion,"[1] a focus on German-speaking émigrés,[2] and an appropriate delimitation of the notion of a "mathematician."[3] The demand for completeness and broader perspective implies a concern for as detailed data as possible with respect to the group of mathematicians in mind (as mainly reflected in the appendices). It also implies an embodiment of Nazi-enforced emigration into broader processes of cultural and scientific "emigration," regarding both the change in historical conditions and the motives.

These restrictions enable a consistency of historical method, since the persons described were united by common traits of scientific education and socialization and by a common language, even if they in many cases

---

[1] "Expulsion" and "persecution"—the latter notion including more than "emigration"—will be discussed for the example of mathematics in chapters 2, 4, and 5.

[2] This category is also the basis for a recent comprehensive German dictionary of emigrants (Krohn et al., eds. 1998).

[3] For the latter delimitation see the next chapter.

had their origins in peripheral countries[4] and entered the German-Austrian[5] system in order to undertake their university education or to work as mathematicians there. Thus "German-speaking" as used in this book means more than just fluency in the German language. It is related to the process of socialization of the respective mathematicians. Publications in German alone are definitely not the decisive criterion for calling a mathematician "German-speaking," as German was still the leading language in mathematics at that time.[6] There are borderline cases of mathematicians such as Zygmunt Wilhelm Birnbaum (1903–2000), whom I decided not to include, since Polish seems to have been the main language during his mathematical training although his written German was excellent.[7]

Even though similar conditions of training made for certain shared mathematical traditions among "German-speaking" emigrants, one has to account for differences as well, particularly between Germany and Austria.[8]

Although systematic historical investigations are still lacking, it seems indisputable to me that the political and philosophical environment in Vienna supported a specific kind of mathematical research already in the 1920s differing markedly from the dominating mathematical trends in the Weimar Republic. Here shall be mentioned but *two directions* in which such research, yet to be conducted, would have to proceed:

Firstly, there is no doubt that the systematic claim of Hilbert's program of research in the foundations of mathematics, eventually refuted by Kurt Gödel's first "incompleteness theorem" of 1931, can only be understood

[4]Typical examples are mathematicians such as John von Neumann and Gabor Szegö, who originally came from Hungary.

[5]Among the "German-speaking scientific centers" one has to name also the "Deutsche Universität" in Prague, which had been left largely intact as a German-speaking institution by the Czech Republic and fell under Nazi rule in 1939. The Swiss system (in particular the ETH Zurich), which with respect to the educational principles can be considered to belong to a more general "German" system, is less central here, because it was not under Nazi rule, although we include G. Pólya among the refugees. For Austria, in particular Vienna, the two-volume Einhorn dissertation (1985) is the most important biographical source. See also Pinl and Dick (1974/76).

[6]Typical for a "non-German-speaking" mathematician in this sense is the Polish logician Alfred Tarski, who was mainly educated in Warsaw in the Polish and Russian languages. Nevertheless he had a good command of German, and communicated freely in German with Kurt Gödel and other Austrian mathematicians. Tarski's most important work on semantics and the notion of truth became visible internationally only after the German translation (1935) of the Polish original of 1933. Cf. Feferman and Feferman (2004).

[7]Birnbaum spent some time in Göttingen as assistant to Felix Bernstein. See Birnbaum (1982), Woyczynski (2001), and the Birnbaum Papers at the University of Washington in Seattle (USA), at http://www.lib.washington.edu/SpecialColl/findaids/docs/uarchives/UA19_14_5266BirnbaumZygmunt.xml.

[8]For the Austrian case and particularly emigration from Vienna see Sigmund (2001), the catalogue to an exhibition on the same topic in September 2001.

against a philosophical background much more neo-Kantian (retaining certain absolutes or a priori in its epistemology) than the philosophy of the Vienna Circle.[9]

Secondly, the deficiencies in Germany in several newer mathematical subdisciplines, such as topology, functional analysis, and some parts of mathematical logic, seem to have been conditioned by a certain self-sufficiency and by social hierarchies[10] in Germany and, in particular, by a politically motivated sealing off from Polish mathematics, which was much less typical of mathematicians in Vienna (D).[11] The close contacts that Wilhelm Blaschke (who was in Hamburg and had come from Austrian Graz) and his geometric school kept with the topologists in Vienna could apparently not make up for the partial international isolation of mathematics in Germany. Also, the Austrian emigrant Olga Taussky-Todd (1988a) reports on partially differing German and Austrian traditions even in core subjects of research such as algebra. For the impact of emigration one has also to consider the longer-lasting contacts of the Austrian and Prague mathematicians with mathematicians abroad, contacts that were restricted for German mathematicians after 1933.[12] For this reason it is necessary to differentiate between the various streams of German-speaking emigration. The existence of differences between two geographically and linguistically close mathematical cultures such as the German and Austrian ones may also explain the differing in which the emigrants adjusted to the American mathematical culture. In this latter respect one could imagine a triangle of different German, Austrian, and American epistemic traditions or "working units of scientific knowledge production" as recently investigated for topological research in Austria and the United States in the 1920s.[13]

Although, as indicated above, the cognitive dissimilarities among the German-speaking regions were partly related to differing political conditions, there were also "political" experiences the German-speaking émigrés had in common, and their political socialization was undoubtedly at variance with that of mathematicians in other countries such as Poland and

---

[9]This difference is still valid, if one compares Reichenbach's group in Berlin with the Vienna Circle. In Göttingen, the neo-Kantian L. Nelson was supported by Hilbert, who was opposed to most of the doctrines of the other schools of German idealistic philosophy. But Nelson's Kantianism was—from the perspective of the Vienna Circle—still affected by metaphysical beliefs. See Peckhaus (1990). Incidentally, Gödel, with his Platonist views, was himself increasingly distant with the Vienna Circle.

[10]See the short remarks in 3.D.4.

[11]See Menger (1994) and Szaniawski, ed. (1989).

[12]On the restriction of international contacts of German mathematicians after 1933 consult Behnke (1978) as an eyewitness report, and Siegmund-Schultze (2002).

[13]See Epple (2004).

France. The latter fell under German rule between 1939 and 1940, and French mathematicians suffered various forms of expulsion. The chances of emigration[14] worsened considerably at that time, mainly due to the current prevailing conditions of war. Although in Germany and Austria the expulsions had not been restricted to anti-Semitic purges either, in occupied countries such as Poland the Nazi policies of racial cleansing extended in many cases to whole social groups, in particular intellectuals. In fact, in occupied Poland the expulsions had the most deadly consequences for the victims.[15] For reasons mentioned these mathematicians are not primary subjects of this book. The task of describing their fates will be left to their compatriots who are better qualified to study the purges in detail. One might say that the fates of these mathematicians were in total even more tragic than those of German-speaking refugees. They shall therefore always be kept in mind in the following discussion as a comparative example and a background for this investigation.

Further restrictions and focus of this investigation have to be mentioned: Since the United States became the final host country for more than half of the mathematician-emigrants—which was a natural consequence of the course of the war but had additional historical reasons—this book will be focusing on immigration to the United States.[16]

Some authors distinguish between "emigration" and "exile." Historical research on "exile" concerns refugees "who went into exile in order to work politically, culturally or scientifically for a democratic future of

---

[14]Well-known mathematicians from the non-German area who survived and were able to emigrate are André Weil (France), Alfred Tarski (Poland), and Guido Fubini (Italy), Fubini a victim of the Fascist regime in his country after the introduction of the racist law of 1938.

[15]In a letter to the French Académie des Sciences on September 27, 1945, the Polish mathematician W. Sierpinski names the following thirteen Polish mathematicians murdered by the Nazis: H. Auerbach, C. Bartel, A. Hoborski, J. [or M.] Jacob, A. Lindenbaum, A. Łomnicki, S. Kempisti, A. Rajchman, S. Ruziewicz, S. Saks, J. P. Schauder, W. Stozek, and A. Wilk [Archives AS, Dossier Sierpinski]. As victims of the war, Sierpinski mentioned in addition S. Dickstein, A. Kozniewski, S. Kwietniewski, A. Przeborski, and W.Wilkosz. The list of victims published in *Fundamenta Mathematicae* 33 (1945): p. v., also names, the following four murdered: S. Kaczmarz, A. Koźniewski, J. Pepis, and J. Zalcwasser. According to later investigations one has to add J. Marcinkiewicz, S. Lubelski (*Acta Arithmetica* 4 [1958]: 1–2), and M. Presburger (Zygmund [1991]). Feferman and Feferman (2005), p. 129, remind of the fate of the female logician J. Hosiasson-Lindenbaum (1899–1942), wife of A. Lindenbaum. In 2003 R. Wójcicki also mentions logicians J. Salamucha and M. Wajsberg as murdered by the Nazis. See http://www.ifispan.waw.pl/StudiaLogica/PL.Logic.html. According to Kuratowski (1973), pp. 80–90, the following Polish mathematicians have to be added as well: Miss S. Braun, M. Eidelheit, S. Kolodziejczyk, J. Schreier, L. Sternbach, and M. Wojdysławski.

[16]But there will be side views on emigration to other countries as well, particularly in chapters 2 and 5. When there is no danger of misunderstanding, the United States will occasionally be called "America."

Germany."[17] Unlike many artists, the great majority of German academics forced to flee after 1933 did *not* belong to the *exile* in this sense but rather to the more general *emigration*, which is also attested by the fact that only a few of them returned to Europe after the war.

Furthermore, a distinction has to be made between *forced emigration* and *voluntary emigration*, depending on whether the lure of the host country or the pressure from the home country ("pull" or "push") were predominant. Both in pre-1933 emigration and in the employing of German and other European specialists in the United States and the Soviet Union after World War II, *voluntary* emigration was certainly dominant, although political pressures and economic hardships influenced the decisions as well. This kind of academic migration[18] or *brain drain*, has continued until today, with a peak in the 1960s.

Research on "academic emigration" includes the movement of persons and ideas and is not at all restricted to the investigation of individual biographies of academics. It has developed in Germany since the second part of the 1980s and has been particularly supported by a program of selected measures issued by the Deutsche Forschungsgemeinschaft (DFG).[19] Stimuli for that program not only came from research on the history of science during the "Third Reich," it was also stimulated by more general, partly epistemologically inspired, investigations into the acculturation of scientific styles, into the *gains* (for the host countries) and *losses* (for the countries of origin) due to academic emigration. This discussion developed in a context of controversial debates on the cultural and political consequences of emigration. Papcke (1988) referred to political tendencies in the United States that stressed the ambivalence of the impact of immigration and the possible loss of "original" American values.[20] Yet in Europe then and today one finds the articulation of a certain resentment against an exaggerated *Americanization* of the various national European cultures. Although Papcke does not share either kind of resentment (which in his opinion expresses either isolationist or nationalist thinking), he also stresses that "culture cannot be internationalized in a simple way" (p. 24 [T]). This

---

[17]Papcke (1988), p. 18. A similar distinction is also done in Pross (1955), p. 18.

[18]While *emigration* is reserved for movements between different countries, the more general notion, *migration,* is also being used for academic mobility within the same country. See Hoch (1987).

[19]See results in Strauss, Fischer, Hoffmann, and Söllner (1991), and a parallel program by the Volkswagen Stiftung, which led, e.g., to Kröner (1989). The DFG program continued an earlier one that went by the name of "Exilforschung." See Briegel and Frühwald, eds. (1988).

[20]A. Bloom, *The Closing of the American Mind* (New York: Simon and Schuster 1987). Papcke (1988), p. 22, mentions exaggerated self-criticism, relativity of values, and lack of orientations as those alleged consequences of immigration.

statement may sound irrelevant to mathematics at first sight. The investigation will, however, show that, even in mathematics, traditional judgments on success or failure of academic emigration have to be carefully evaluated, and the broader cultural and political context has to be considered.

As to academic emigration in the sciences, Papcke finds the following distinction: "Everywhere in the sciences there was a considerable transfer of knowledge. But a noticeable cultural impact can only be found in the USA" (p. 19 [T]). Coser, in the introduction to his book dealing with the impact and the experiences of émigrés in the United States, emphasizes that the transfer of knowledge requires direct and personal contacts: "The experience of being taught by a great scientist or a great humanist scholar cannot be duplicated by even the most diligent perusal of published works or by listening to even a major paper at an occasional international meeting."[21] In fact, the importance of this "oral communication" in the sciences was already apparent in the 1920s, and foremost U.S.-American foundations took account of that by granting stipends on an international basis. The foundation policies of the 1920s had a strong pro-American bias. However, the foundations also tried to promote American science indirectly, not just by supporting immigration but also through the support of European science on its home ground. This attracted American students in large numbers.[22] Contemporary witnesses before and after 1933, in particular some representatives of the Rockefeller Foundation, saw the drawbacks—due to emigration—of a loss of cultural diversity in world science, something that hitherto had stimulated science at large.[23] This policy, of course, had to be changed after Hitler came to power, but slowly, as argued by some concerned politicians and scientists. Some of them insisted that the United States should only temporarily host European scholars who later on intended reviving science in their countries of origin. The Rockefeller Foundation, for example, supported for a long time the sojourn of European mathematicians in their first host countries,[24] before global political developments made this less and less possible.

Evaluating *gains* and *losses* during emigration one has to be careful not to fall into the *post hoc, ergo propter hoc* trap, that is, to claim that developments in the host countries (the gain) would not have taken place without immigration.[25] The opposite assumption—that these de-

---

[21]Coser (1984), p. xi.

[22]See for details Siegmund-Schultze (2001) and the discussion in chapter 3.

[23]In retrospect, American mathematician Garrett Birkhoff saw the dangers of an "overkill" of mathematics due to emigration. See below.

[24]Gumbel at Lyon, Neugebauer at Copenhagen, Feller at Stockholm, etc.

[25]Particularly Fischer (1991), pp. 35–36, warns against making that mistake in historical methodology.

velopments would have taken place in the country of origin as well (the loss)—is equally illegitimate. This also shows that research on emigration cannot evade the dilemmas of "counterfactual" historical claims,[26] which can only be handled with extreme care in a historical investigation.

In this investigation I will mostly discuss *forced* emigration after 1933, when the great majority of mathematicians emigrated for strictly political reasons,[27] due to either racist policies (the dominating reason) or political dissent with the resulting pressure on them. However, in many cases the dividing lines between forced and voluntary emigration are blurred, and for historical reasons emigration has to be put into a broader perspective.[28] It is necessary to include some mathematicians who had emigrated before 1933 but who could also be considered forced emigrants, as they continued work in and for German mathematics after emigration, which was finally interrupted by the Nazi seizure of power.

A clear differentiation between forced and voluntary emigration is for instance not possible for Theodor von Kármán (1881–1963) and John von Neumann. The important *International Biographical Dictionary of Central European Émigrés, 1933–1945* (henceforth *IBD*), edited by W. Röder and H. A. Strauss in 1983, does not mention von Neumann and von Kármán. The latter had gone to the California Institute of Technology in Pasadena by 1929, mainly because he felt that anti-Semitism was impeding his career in Germany. Both men maintained contact with Germany until it was broken off in 1933; the much younger von Neumann, who at the time had a partial appointment in Princeton, even canceled his preannounced lectures in Berlin. In the case of von Kármán there is the additional problem of whether he can be justifiedly included among "mathematicians" (see chapter 2). It appears to me, therefore, that for a sensible definition of the (forced) "emigrant" to be used in this book, the dividing line should be drawn exactly between von Neumann and von Kármán, including the former and excluding the latter from the focus of the discussion.[29] There is, however, agreement between the *IBD* and the present book in treating the statistician and pacifist Emil Julius Gumbel as a (forced) emigrant, since he was a German-speaking mathematician who

[26]Thiel (1984), p. 228. "Counterfactual" is meant to signify the hypothesis that history could have developed otherwise, "contrary to the facts" that really occurred.

[27]Economic reasons, which in a certain sense are certainly also political, were becoming less an issue with the partial recovery of the economy in Nazi Germany in the late 1930s, when the chances for mathematicians, who were "Aryan" by Nazi standard, gradually improved.

[28]See particularly chapter 3 on early emigration.

[29]But von Kármán's relations with emigrants as documented in his rich archives at the California Institute of Technology are a crucial source also for the present book.

emigrated from Nazi-occupied territory (or Nazi-threatened in the case of southern France where Gumbel was in 1940).[30]

There is no way of considering refugees such as Richard von Mises as "voluntary" emigrants, even if, to the outsider, they were the ones who abandoned their appointments in 1933 or later. They were clearly under threat; they left in awareness of the impending developments and would have been dismissed later on anyway. As in the case of von Mises, they often had to leave their work and projects in shambles and unfinished.

There were, though, early emigrants in mathematics such as Theodor Estermann (1902–1991), Hans Freudenthal (1905–1990), Eberhard Hopf (1902–1983), Heinz Hopf (1894–1971), Chaim (Hermann) Müntz (1884–1956), Wilhelm Maier (1896–1990), and Abraham Plessner (1900–1961), who left for predominantly economic reasons and out of concern for their scientific careers. Some of them are—partly without their approval—treated as refugees from the Nazi regime in other historical accounts. This happened, for instance, with Estermann, Freudenthal, and Müntz (Pinl/Furtmüller 1973), although Estermann had left for London in 1926, Müntz for Leningrad in 1929, and Freudenthal for Amsterdam in 1930. Of course arguments pointing to academic anti-Semitism in pre-1933 Germany, which without any doubt hampered the careers of Müntz and Plessner,[31] and diminished their chances of return after 1933, could also be cited. The argument to count early Jewish emigrants as refugees from the Nazis is supported by the fact that non-Jewish early emigrants, such as Eberhard Hopf and Wilhelm Maier, returned to Hitler's Germany after 1933 and profited partly from the dismissals of their Jewish colleagues. Nevertheless, in accordance with this book's main restriction and for reasons of historical systematics, Estermann, Freudenthal, and Müntz do not appear in the list of emigrants (Appendix 1 [1.1]).[32] The Nazi seizure of power did not deprive them of an existing, immediate chance of returning to Germany or of a very important professional position, as it did for von Neumann. Freudenthal, who supported many a refugee from Germany before 1940,[33] shared the fate of other non-German emigrants in

---

[30]In a broader sense Gumbel could already have been included as a forced emigrant without that fact of renewed expulsion, because he was dismissed from the University of Heidelberg before 1933 for exactly the same political (Gumbel's antimilitarism) and racist "reasons," which after 1933 were used as a pretext by the Nazi regime.

[31]Gaier (1992). In Plessner's case as in others, like S. Bochner, the anti-Semitic prejudice was mixed with and partly hidden by concern for their lack of a German citizenship.

[32]Müntz was included in that list in the German edition of this book in 1998 due to erroneous information from Pinl and Furtmüller (1973), which has meanwhile been corrected by Ortiz and Pinkus (2005) and by recent findings in the Oswald Veblen Papers and the Bodleian Library (SPSL).

[33]Among them were Blumenthal and Rosenthal, and also Pinl, who was persecuted in Prague.

other occupied countries. After the German occupation of the Netherlands in 1940 he had to go into hiding. Müntz, however, was expelled
without the right to a pension from his professorship in Leningrad in
1937 (a professorship once occupied by P. L. Chebyshev), because he
had retained his German citizenship and because tension between Nazi
Germany and the Soviet Union was growing.[34] As both Freudenthal and
Müntz were German-speaking[35] and because Müntz was potentially
threatened in Sweden and therefore tried to get to the United States, both
of them are included as borderline cases in the list of persecuted German-
speaking mathematicians (Appendix 1 [1.3]). Other borderline cases are
Robert Frucht and Karl Menger. Frucht left Berlin in 1930 for economic
reasons and became an actuary in Italian Trieste. He can be considered
an "early emigrant," but also a part of the forced German-speaking emigration after 1933, since he had to leave Italy in 1938 when the racial
laws were passed. Also Menger can be categorized both as an early and a
forced emigrant, as the discussion in chapter 3 will show. I decided, however, not to include Henri A. Jordan (1902–?) among the forced emigrants, because he went from Germany to Italy in 1930, where he was
dismissed in Rome for reasons of restriction of staff at the International
Institute for Educational Cinematography (League of Nations) in December 1933.[36]

A very interesting and important borderline case between early and
forced emigration is the well-known set theorist Adolf Fraenkel, who immigrated to Jerusalem twice (in 1929 and 1933) and who later in Palestine called himself Abraham A. Fraenkel.[37]

A further historical problem lies in how far Switzerland and, in particular, the Eidgenössische Technische Hochschule (ETH) in Zurich, traditionally with strong ties to German mathematics,[38] can be regarded as a
host to refugees or—in contrast—as origin for forced German-speaking
emigration during the Nazi years. On the one hand, Switzerland offered
refuge to early emigrants such as topologist Heinz Hopf, and to forced

[34]Müntz to H. Weyl, Stockholm, August 8, 1938, OVP, cont. 32, f. Muentz, Hermann
1938–41. According to the same letter, Müntz fled through Estonia (Tallin), where he was
guest-professor for one term (but had to leave because he taught in German), to Sweden,
where he arrived in February 1938.

[35]The first-mentioned mathematician was later fluent in Dutch, even as a novelist; Müntz
probably acquired the Russian language during his eight years in the country.

[36]Jordan was born in Brussels, had acquired German nationality, and went to school and
studied at Frankfurt in the 1920s. He took his doctoral degree on Bessel Functions there in
1930. After 1933 he went through the United Kingdom to the United States in 1936. See his
file in SPSL, box 281, f. 1, Tobies (2006), p. 173, and a short note in OVP, cont. 31.

[37]See Fraenkel (1967) and in chapters 3 and 8.

[38]This connection, which goes back to the nineteenth century, is exemplified by the work
of Hermann Weyl at the ETH in the 1920s. See chapter 3.

Figure 3 *Hans Freudenthal (1905–1990). The noted topologist was an early emigrant from Berlin (1930); he helped émigrés to the Netherlands after 1933 and survived the Nazi occupation in hiding.*

emigrants such as logician Paul Bernays. Thus Switzerland, which—unlike Austria and Prague—never fell under Nazi rule, can primarily be considered a (rather exceptional and marginal) host country. Then, on the other hand, borderline cases such as that of Georg(e) Pólya (1887–1985), who emigrated from Zurich to the United States in 1940 because he saw the possible occupation of Switzerland as a real danger, point, once again, to the difficult problem of the definition of "forced emigration."[39]

---

[39]I nevertheless include Pólya among the German-speaking refugees from Hitler's domain because I do not consider it my business to decide in hindsight how strongly he felt threatened and whether there was maybe less danger for him than for Müntz in Sweden.

Another matter, quite apart from my decision to "classify" emigrants for reasons of historical systematics, is whether emigrants—either early ones such as Estermann,[40] or the ones who "asked for dismissal" after 1933 like von Mises and Hermann Weyl—would have liked to be represented as "refugees" or "emigrants," names that for some bore the stigma of the unsuccessful. Hermann Weyl, for one, occasionally described himself as a "voluntary emigrant," although the threats against his Jewish wife and children in Germany really left him with no choice.[41]

Even among the "forced emigrants" one has to differentiate in the historical investigation between the concrete "reasons" for their dismissal.[42] These "reasons," arbitrarily presented by the Nazis as relevant for academic careers, were important for the concrete fates and the self-image of the emigrants. They influenced their chances for acculturation in the host countries[43] and even had ramifications for post–World War II compensation claims.

In spite of the problems of definition just discussed, the present book will attempt to separate *early emigration* (chapter 3) from *forced emigration*, the latter being the main focus of the book. Chapter 2 will analyze how the extent of forced emigration within mathematics can be quantitatively measured. For this purpose the entire population (as far as known from the sources up to this moment) of mathematicians dismissed after 1933 is compared to the (large) subset of emigrants. The Appendix 1 (1.1) listing the emigrants contains only "successful" emigrants who made it to a host country outside the Nazi domain of power before 1945. Temporary refugees to Holland (Blumenthal, Remak, etc.), Belgium (Grelling), and Prague (Pinl), later caught by the Nazi Reich, appear as victims rather than as emigrants (appendices 1.2. and 1.3). Among the forced emigrants a further distinction will be made between those finally ending up in the United States and the (rather few) ending up in other countries. Special quantitative methods,

[40]"Professor Estermann died in December 1991 and he always made it very clear to people that he was never a refugee of any description and in fact would become very upset if anyone assumed he was. He apparently arrived in Britain firstly in 1926 and then settled here permanently from 1929" (R. Whiting to R. Siegmund-Schultze, July 6, 1993).

[41]See Weyl's letter of resignation written to the Nazi Prussian Ministry of Culture October 9, 1933, printed in Schappacher (1993), pp. 81–83, where Weyl describes these threats, not without reflecting sentimentally on his ties to Germany. For more on Weyl's self-image as "voluntary emigrant," see below in chapter 7.

[42]See also below chapter 11, where objections by emigrants against Pinl's meritorious report (Pinl 1969–72) are documented, criticizing that Pinl thought it advisable not to mention those reasons he found were no real "reasons" at all.

[43]On this see for instance the example of the differential geometer and son of an industrialist, Herbert Busemann, in chapters 5 and 7.

such as "co-citation analyses" of publications,[44] will not be used because of the relatively small size of the investigated population of mathematicians and because of the priority of presenting the unpublished material first. Further discussion in the second chapter will show that principal problems of historical methodology allowing certain types of emigration to fall into oblivion add to rather circumstantial problems of historical sources, something that will, hopefully, be partially repaired by the present publication. Both reasons, however, continue to make a complete representation of the German-speaking emigration in mathematics impossible.

[44]This has been partly used in Fischer (1991). This paper points to a possible extension of the present investigation, although the author Fischer acknowledges (pp. 52f.) that co-citation analyses are controversial as to their historical expressiveness.

The Notion of "Mathematician" Plus Quantitative Figures on Persecution

SEVERAL HISTORIANS and witnesses of the events have maintained that *mathematical* immigration has been of outstanding and singular importance to American science.[1] Geiger, for example, points out that in 1965, fourteen out of fifty-one members of the mathematical class of the National Academy of Sciences were from Europe.[2] Chapter 10 will give an overview of the effects of the German mathematical immigration to America and will also discuss some of the consequences to the development of mathematics in Germany in the 1930s as well. In some respects, given the then still important role of the German language and the strong part of symbolism in mathematics, the conditions of acculturation for the German mathematical emigration were relatively better than for scientists from other scientific fields or from other cultural and ethnic backgrounds. This does not mean, however, that language did not matter as a problem of acculturation among German emigrant-mathematicians. As a matter of fact, in the 1930s it was precisely in mathematics that teaching continued to be a major factor of legitimation for the discipline, and especially so in the United States. In addition, and as a natural consequence of emigration itself, English as the lingua franca of international science grew in importance during that time. The discussion of the specificity of the acculturation among emigrant-mathematicians will also show that there were relatively few problems integrating German mathematics into the host countries at the research level. However, other problems followed on the social level, such as increased pressure in the academic job market, something that worked against immigration. This points back to the problematic discussion of the *gains* that in particular American science could draw from immigration. Given the by-and-large *public invisibility* of mathematics[3] and the relatively well-developed infrastructure of American mathematical research around 1930, that question of the gains depends very much on the observer's point of view anyway. It would be interesting

---

[1]Fosdick (1952), p. 277.

[2]Geiger (1986), p. 244.

[3]Dawson Jr. (1997) describes this invisibility when pointing to the fact that one of the greatest logicians of the century, the emigrant Kurt Gödel, was all but unknown by the American public.

to ask whether American mathematics, which due to its historical origins was so closely connected to the development of German mathematics, could principally gain as much from immigration as other disciplines like musicology, which had been so far underdeveloped.[4] This question, which is closely related to the "problem of the two cultures" (i.e., the split between the humanistic and the scientific cultures as discussed for instance by C. P. Snow), can only marginally be touched in the present study.

In accordance with the widely held belief that the emigration of mathematicians and of mathematics was exceptionally successful, this discipline claims an important part in the standard encyclopedia of scientific and cultural emigration *IBD* (Röder and Strauss 1983) and is based on questionnaires distributed to the emigrants. Among a total of 980 scientists and engineers included as emigrants in *IBD*, 130, equaling about 14 percent, name "mathematics," "applied mathematics," "mathematical statistics," or "mathematical logic" as their field of competence. The present book will accordingly count representatives of these fields as "mathematicians."

I do *not*, however—with few exceptions[5]—include workers in "mathematical physics,"[6] even if they once obtained their doctorate in mathematics.[7] The same applies to scholars in "Information and Control" and, with few exceptions, "aerodynamics." This requires some explanation.

On the one hand there exists a marked esprit de corps in many disciplines defining who belongs or does not belong to it. Therefore self-assessments by scientists as expressed in *IBD* and in other documents cannot be ignored by the historian, not least due to the social consequences of that feeling of togetherness. Some emigrants such as Arthur Korn[8] and Hans Reichenbach,[9] who were at least as close to physics as to mathematics, considered themselves at the same time mathematicians.

---

[4]Compare in this respect a contemporary Gallup investigation that will be commented on in chapter 10.

[5]Exceptions are C. Lánczos and W. Romberg (see below) due to their contributions to numerical analysis.

[6]Even less so, of course, are experimental or theoretical physicists included. For the difference between "theoretical" and "mathematical" physicists see Schweber (1986). There the name "mathematical physicist" is described, following some distinction introduced by E. U. Condon, as "a name that characterizes him by his tools rather than by his function" (p. 66).

[7]The latter applies for instance to theoretical physicist Fritz London (1900–1954).

[8]See Korn (1945).

[9]The empiricist philosopher Reichenbach also published on the axiomatics of probability in volume 34 of *Mathematische Annalen*. On December 22, 1932 he asked Courant in a letter for a publication of a book on the foundations of probability ("Grundlagen der Wahrscheinlichkeitsrechnung") in the prominent Springer *Grundlehren* series. With a letter of February 1, 1933 Courant declined and expressed his hope that Reichenbach would publish his "investigations with sufficient completeness in more philosophically oriented journals." The discussion was obviously about a manuscript that appeared later on in English as Reichenbach (1949).

On the other hand it was natural that mathematicians in emigration had frequent and close contact with physicists who may have been emigrants as well. A clear separation between the two movements of emigration is therefore not possible, starting at the personal level. In the present work this is illustrated by the relationships between Courant and James Franck (1882–1964) and between Felix Bernstein and Albert Einstein. It is also clearly visible in the support physicist Paul S. Epstein (1883–1966), who had emigrated before 1933, gave to several mathematicians.

If one looks at the factual level of the mathematics produced, distinguishing clearly between mathematicians and, for instance, physicists seems even less possible. In fact, for the effects of *emigration for mathematics as a discipline* it does not matter whether the emigrants regarded themselves as mathematicians or physicists, as long as they contributed to mathematics or the application of such with work closely connected to mathematics.

In the period under investigation between the 1920s and 1940s many highly theoretical parts of mathematics such as functional analysis, differential geometry, and the theory of group representations proved to be of immediate importance to the application in physics. Nevertheless it is important not to automatically subsume practitioners in these fields of mathematical physics under the category of "mathematician," first of all for pragmatic reasons. In historical literature the emigration in physics has attracted much more attention than it has in mathematics. Thus those physicists working on the borderlines between the two disciplines (such as Wigner), and especially those working in the most spectacular fields of modern physics, such as quantum theory and relativity, have received a decent share of historical investigation. They will only be included if they also published independently in purely mathematical contexts such as von Neumann and Weyl.

It seems to me that some historically *younger* and more recent fields of applications of mathematics are even more closely connected to the process of emigration and its effects. One should mention in this context stochastics (as a collective term for probability and mathematical statistics), game theory, analytical philosophy, operations research, numerical analysis, and some parts of engineering, such as computer design and aerodynamics. In many of these fields, specifically mathematical methods of forming notions play an important role. One could even argue that during the process of emigration a change in the notion of "applied mathematics" occurred, not least influenced by the exigencies of the war.[10] At a human level this applies, for instance, to the theoretical physicist from

---

[10]Siegmund-Schultze (2003a and b).

Frankfurt, Cornelius Lánczos (1893–1973), who became one of the most influential workers in numerical analysis in the United States.[11] While Lánczos has to be excluded from the list of emigrants having left for the USA in 1931, others, physicists by education with mathematical achievements, such as Werner Romberg, have been included as an exception to the rule. Several of the more mathematically oriented aerodynamicists have also been listed as emigrant-mathematicians. This latter decision is— as in the case of "mathematical physics"—not unproblematic and again a pragmatic one, not independent of the self-images of the emigrants and evaluations of them by their colleagues. Again the borderlines are fuzzy if one considers more theoretical efforts in hydrodynamics and differential equations, as done in the work by Stefan Bergmann, Kurt Friedrichs, Alexander Weinstein, and others. Also, aerodynamicists and theoretical mechanics such as Ludwig Hopf, Gustav Kürti, Paul Nemenyi, Hans Reissner, and (a man who finally did not emigrate) Kurt Hohenemser, appeared as "mathematicians" in contemporary lists of emigrants. They were obviously considered "applied" mathematicians by persons who undoubtedly belonged to "core mathematics," although their training was often engineering. Support mathematicians gave to engineers like Eric(h) Reissner and Arthur Korn during emigration allows them to be included in the considered circle of mathematician-emigrants as well.

Another controversial decision concerns the inclusion of logicians, researchers on the foundations of mathematics, and epistemologists. Along with logicians like Kurt Gödel (who is no unproblematic case to include among the forced emigrants)[12] and Paul Bernays, mathematically oriented and educated philosophers such as Rudolf Carnap, Kurt Grelling, Paul Hertz, and Hans Reichenbach have been included. That inclusion and the exclusion of other emigrants in the field like Edgar Zilsel (1891–1944),[13] Friedrich Waismann (1895–1959), and Felix Kaufmann (1895–1949), is probably the most problematic choice made in this book and is in no small measure a subjective decision by me. The book wants, on the one hand, to represent mathematics in as broad as possible a thematic variety, and, on

---

[11]See Scaife (1974) and Stachel (1994). Lánczos's first publication on numerical analysis appeared in 1938 during emigration and contains the known Lánczos-Tau method.

[12]Gödel was not Jewish and hesitated to emigrate for quite a while, a fact that emigrants like Menger disliked. His decision to go to the IAS in Princeton as late as 1939 was influenced by economic rather than by political considerations. See Dawson Jr. (1997).

[13]In a letter to the director of the Institute for Advanced Study in Princeton, Hermann Weyl discussed on May 21, 1941 the matter of a possible grant for Zilsel, who had been in the country as a refugee from Austria since 1939. Weyl came to the conclusion: "Dr. Zilsel has never done any creative work in mathematics or physics. He is above all a philosopher, with a genuine interest in the philosophical problems of science. . . . We can give him no stipend" (OVP, cont. 33, f. Zilsel, Edgar, 1939–41.)

the other hand, tries to avoid an inappropriate inflation of "mathematician" that could disperse the historical focus.

As to "Information and Control," as a category within the dictionary *IBD*, it must be noted that these fields, with partial connections to logics, stochastics, and computer engineering, were among the youngest of the new fields in applied mathematics. For this reason the emigrants named in *IBD* belong mostly to the "second generation" of emigrants, meaning they are children of emigrants. This "second generation," which is included in *IBD*, shall for systematic reasons, and because of the strikingly different conditions of socialization compared to their parents, not be discussed in the present book.

Perhaps the most important criterion in this book for including an emigrant as a "mathematician" is his/her socialization, in particular the time and place where the mathematical education was received. More specifically the present work views those emigrants of the "first generation" (born in 1914 and before) who received the main part of their mathematical education (in general a degree for completed university studies) *before* emigration as "mathematician-emigrants."[14] Nevertheless, also the careers of scientists/mathematicians of the "second generation"—due to their European schooling and the fact that mathematical talent frequently runs in the family[15]—should not be underestimated in judging the effects of emigration as a whole. Perhaps the most prominent mathematician of the second generation of emigrants was Wolfgang Doeblin (1915–1940), who took his own life in war service for France in 1940, in a situation he deemed hopeless given the danger of being deported to Germany. Wolfgang was the son of the German-Jewish writer Alfred Döblin and left Germany for France shortly after his father in 1933. He studied mathematics mainly in Paris and went on to become, thanks in part to his work on Markov chains, "one of the greatest probabilists of [the twentieth] century."[16]

More for pragmatic than for principal reasons, the present book focuses on mathematics as a research subject, at that time largely carried out

[14]This factor of scientific socialization in Europe is also the starting point of the analysis in Coser (1984), p. xiii. See also Kröner (1989), p. 16. The 1914 date of birth to be used as point of reference for inclusion could shift somewhat in the direction of 1918 for emigrants from Austria and Prague, although I did not find any refugees of this young age. The youngest émigré listed in Appendix 1 (1.1) now deceased (2005) is Hans Samelson (1916–2005). Albrecht Fröhlich (1916–2001), an immigrant to England who later became prominent in number theory, is excluded, since he did not receive his mathematical education in Germany.

[15]To take just one example: the son of Issai Schur, Georg, became an actuary and was instrumental in founding the national insurance of Israel. See Brauer (1973), p. vi.

[16]As described by the editor in Cohn, ed. (1991), p. ix. For more detail on Vincent Doblin or Doeblin (as he spelled his name after emigration) see Bru (1991) in the same volume.

at universities, in spite of the tentative start of "industrial mathematics."[17] Mathematics teachers, who do not usually contribute to mathematics as a research subject,[18] are not principally excluded from this volume. However, the emigration of teachers in mathematics has rarely been investigated so far and is consequently difficult to evaluate as to its effects upon the host countries. It seems evident, though, that teachers had much less chance of emigration than research mathematicians. Their collective destiny, especially that of the Jews among them, is still in need of investigation.[19] The teachers who did emigrate—of which several figure prominently in a review by American educator Arnold Dresden (1942)—are included in the main list of emigrants as far as they are known to me (Appendix 1 [1.1]). In several places in the present study there will be reflections, as a kind of (rather atypical) example, on the emigration of the Freiburg teacher of mathematics Wilhelm Hauser, mainly due to the reliability of the historical material that is available in this case.[20] Hauser took his doctor's degree with Max Noether in Erlangen in 1907 summa cum laude. However, as an émigré to England and as one who returned to East Germany after the war—obviously for political reasons since it was not the region of his origin—Hauser was not typical of the group of people the present book aims at focusing upon.

After these remarks addressing the rather *qualitative* problem about whom to include among the mathematician-emigrants, now to the more *quantitative* side:

If one complements the "mathematicians" among the first-generation emigrants as listed in *IBD* by those given in Pinl and Furtmüller (1973) and in several additional sources,[21] one arrives at a number well above a

---

[17]There were some starting points with Siemens (H. Baerwald, L. Lichtenstein), with Zeiss (M. Herzberger), in the aviation industry and in the insurance business. Industrial mathematicians who emigrated are included in the list of emigrants (Appendix 1 [1.1]). As to the degree and character of "professionalization" there were considerable differences between mathematics and, say, medicine, which contributed to differing conditions of emigration. Cf. Kröner (1989).

[18]There were notable exceptions, though, particularly in subjects on the fringes of mathematics such as mathematical logics and the history of mathematics, where teachers did research (Ackermann, Grelling, Löwenheim, Mahlo, J. E. Hofmann). These fringe areas offered few research jobs to their contemporaries in other countries, too, as the cases of K. Popper in Austria and Alfred Tarski in Poland illustrate.

[19]Some conjecture on that in Appendix 1 (1.3). A useful newer source, which can help to shed light on the fate of teachers as long as they held a doctor's degree, is Tobies (2006).

[20]See Wirth (1982) and Hauser's estate at the Archives of the Konrad-Adenauer-Stiftung in St. Augustin near Bonn. I thank Günther Wirth (Berlin) and Manfred Agethen (St. Augustin) for their kind support.

[21]Among printed sources especially: Dresden (1942), Thiel (1984), and *List of Displaced Scholars*, 1936–37, reprinted in Strauss, Buddensieg, and Düwel (1987). Furthermore several estates (left papers) of mathematicians, particularly CPP. See also the list of the Rockefeller Foundation of fifty-six dismissed mathematicians in RAC, R.F. 2 (General Correspondence), box 140, f. 1045.

Figure 4 *Wilhelm Hauser (1883–1983). The student of Max Noether and pacifist mathematics teacher from Freiburg was dismissed both for political and racist reasons. Through the help of a Quaker organization he finally reached a position as a teacher in Newcastle (England) in 1939. (The school was evacuated during the war to Penrith.) He was one of the few émigrés to return, and along with Ludwig Boll was the only known mathematician-emigrant who went to East Germany after the war.*

hundred (in the present count 145),[22] among them fifteen women: C. Fröhlich (Froehlich), H. Geiringer (from 1943 von Mises), E. Helly (born Bloch), G. Hermann, G. Leibowitz, H. Rothe (born Ille), I. Brauer (born Karger),[23] E. Noether, R. Peltesohn, H. Reschovsky, A. Riess, K. Kober (born Silberberg),[24] K. Fenchel (born Sperling), O. Taussky, and H. von Caemmerer (Neumann). The lack of exactitude is due both to the still

---

[22]Which differs somewhat from the original German edition (1998) of this book.

[23]Ilse Karger took her PhD in physics in 1924 but then turned to mathematics, which she also practiced as a teacher during emigration. See Green (1978), p. 318.

[24]Käthe Silberberg appears also in the Rademacher-Tree with her number-theoretic dissertation in 1934, advised by H. Rademacher. See S. Golomb, Th. Harris, and J. Seberry (1997), p. 218.

continuing incompleteness of sources and the problems of defining the terms "forced emigrant" and "mathematician" as discussed above. Among the women who emigrated, the three most successful mathematicians in their own right were probably Hilda Geiringer, Emmy Noether,[25] and Olga Taussky. The last-named, as the youngest among them, had the relatively "easiest" fate as an early refugee from Austria and as a forced refugee from Göttingen; she soon won fellowships in England and the United States as a promising young mathematician. The hardest hit was apparently Hilda Geiringer, who was dismissed 1933 at Berlin University from her position in applied mathematics as an assistant to Richard von Mises. Rightly recognizing that she was in a disadvantaged position as a woman in the male profession of mathematics,[26] Geiringer apparently felt the need to formally diminish her age in 1933 from thirty-nine (nearly forty) to thirty-seven in order to improve her chances of emigration. She gave 1895 as her year of birth in all her correspondence with the British SPSL and with other refugee organizations, only to return to the correct year of birth, 1893, in her American marriage- and naturalization certificates of the mid-1940s. In spite of this very reasonable move, Geiringer fell nearly victim to the Nazis in October 1939 when desperately waiting for an American visa in Lisbon.[27]

Seven of the fifteen female refugees were (or in Geiringer's case became later) partners of well-known mathematicians. While Hilda Geiringer (later in 1943 married to Richard von Mises) and Hanna von Caemmerer (later Mrs. Bernhard Neumann) contributed substantially to mathematics themselves and are therefore listed by their maiden names, the fate of the five others might have remained hidden if not for it having been uncovered during research on their spouses. Among these five, most information seems to be available on Elisabeth Helly,[28] who translated the fifth edition of H. Lamb's *Hydrodynamics* into German (1931), and on Käte Sperling-Fenchel. There were probably other spouses or partners of emigrated math-

[25]On Noether's situation during her emigration see below 8.S.1.

[26]Geiringer alludes to this fact and feeling several times in her correspondence with the British SPSL. See SPSL, box 279, f. 3 (Geiringer), in particular her letter dated June 4, 1947, to be quoted in chapter 8.

[27]See below in Appendix 4.3 the report on R. von Mises's energetic and successful efforts to secure the visa for Geiringer in the last minute.

[28]She was for instance highly recommended as a mathematics teacher in an undated letter by Philipp Frank to the Institute of International Education after the death of her husband in 1943. EC, box 8, f. Frank, Philipp, 1939–43. The Hellys were good friends of Richard von Mises and Hilda Geiringer. According to a letter by Hilda Geiringer, Elisabeth Helly held a PhD in mathematics from the University of Vienna (W. Wirtinger). EC, box 10, f. Geiringer, Hilda, Geiringer to B. Drury, December 18, 1943. See also Sigmund (2004), p. 27, with a picture of Elisabeth Helly.

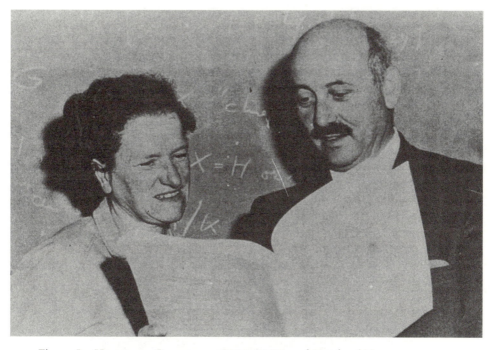

Figure 5   *Hanna von Caemmerer (1914–1971) and Bernhard Neumann (1909–2002). With the von Mises couple they stand here representative of companions who were both successful in mathematical research even after emigration. Hanna von Caemmerer was at the same time an example of a non-Jewish woman accompanying her Jewish spouse into emigration. The two went on from Great Britain to Australia after the war.*

ematicians themselves trained in mathematics of whose lives during emigration relatively little is known.

The great majority of emigrated and persecuted[29] mathematicians appearing in the lists in Appendix 1 belong to the "core discipline" mathematics and had completed the usual curricula at a German university with a diploma (Staatsexamen), doctoral dissertation, or had even obtained the venia legendi (habilitation). I have refrained from including emigrants and persecuted mathematicians no longer in mathematical positions at the time of persecution. This is justified by a historical perspective that looks for the impact on mathematics as a discipline. However, even here there are borderline cases. Emanuel Lasker, well known in mathematics for his

---

[29]On "persecution" see below.

Figure 6 *Käte Sperling-Fenchel (1905–1983). She married Werner Fenchel shortly after immigration to Denmark in 1933. In 1943 both had to flee to Sweden after the German occupation of Denmark. She published a few papers on algebra, but her career was apparently hampered by the additional burdens of emigration associated with being the spouse of a more prominent refugee-mathematician.*

proof of the Lasker-Noether decomposition theorem in ideal theory (1902) but even better known as a world champion in chess between 1894 and 1921, will be listed as an emigrant in Appendix 1 (1.1). After having been expelled from Berlin at the age of sixty-five he apparently returned to mathematical research in his first host country, the Soviet Union, before leaving for the United States in 1937. Albert Fleck, a trained mathematician who besides his main job as a physician volunteered in helping mathematicians in Berlin in the rejection of alleged "Fermat proofs,"[30] will be counted as "persecuted" in Appendix 1 (1.3). People like the novelist and pioneer of science fiction literature Curt Siodmak (1902–2000),[31] who had once studied mathematics in Zurich and Dresden and even taken a PhD, but who had abandoned this science long before his emigration in 1933, will, however, be excluded from the appended lists. Two men trained in mathematics who were connected to Vienna gained fame as novelists in the 1920s and 1930s and would emigrate in 1938; they were Robert Musil (1880–1942)[32] and Leo Perutz (1882–1957). In Vienna between 1905 and 1907 Perutz received an education in mathematics—one with a bias on insurance mathematics. This led to him working for some time as an actuary in an insurance company.[33] Like Siodmak, both Musil and Perutz will not be considered in this book as persecuted or emigrated mathematicians since they had left the field long ago. If one tried to include all Germans or Austrians who at some time or other had received training as mathematicians and who were later expelled by the Nazis, the lists would have to include even the former Reich chancellor Joseph Wirth (1879–1956), who had been the purged Wilhelm Hauser's predecessor as a mathematics teacher at the same Freiburg high school (see above).[34]

So far *emigration* has been the basic topic of discussion. However, it is important to look at the overall picture of *persecution* in order to gain a broader understanding of the emigration figures.

Besides the 145 emigrants of the first generation, found on the basis of sources such as *IBD*, I have counted eighty-nine mathematicians either expelled from German-speaking universities, and to a smaller degree from other areas of public service and industry, or otherwise persecuted during the Nazi years but who could not or did not want to emigrate.[35] Both

---

[30]See Biermann (1987) and Stürzbecher (1997). This was about analyzing erroneous "proofs" of the last number theoretic theorem by Pierre Fermat of the seventeenth century. A complete proof was found a decade ago (1994) by the British mathematician Andrew Wiles, who works in the United States.

[31]For Siodmak, the brother of the even more well known movie director Robert Siodmak, see G. S. Freymuth, "Er hat sein Gehirn nach Kalifornien verpflanzt," *Berliner Zeitung,* February 9, 1998, p. 3, and the Web site on émigrés in the film business at http://www .dartmouth.edu/~germ43/resources/biographies/index.html. The information on Siodmak's studies is based on a telephone interview I had with Siodmak on February 12, 1998. I thank Katrin Liebich (Berlin) for drawing my attention to Siodmak.

[32]Sigmund (2001), pp. 96–98. As a novelist, Musil also had relations to Richard von Mises.

[33]See Siebauer (2000), pp. 37–38, and Sigmund (2001), pp. 98–102. Thanks go to Leo Corry (Tel Aviv) who drew my attention to Perutz although the latter cannot be discussed here for systematical reasons.

[34]Wirth (1982), pp. 76 and 163.

[35]I include here as a borderline case Hans Freudenthal, as mentioned above.

groups of mathematicians are important for the evaluation of the conditions of emigration, although only the first group is included in the dictionary of emigrants *IBD*. However, the numbers of emigrants and nonemigrants in Appendix 1 are difficult to compare. The ability of many emigrants in research came to light only during their emigration, while the figures for mathematicians in Germany, Austria, and Prague barred or deterred from a career in the field remain unclear.[36] What is more: the notion of the *persecuted mathematician* is even more difficult to define than that of the emigrant, where at least the fact of emigration is not a point of dispute.

First, it is indubitably and historically proven that all those falling under the Nazi definition of "non-Aryan" and living in the Nazi domain of power were persecuted and have to be named as such. This applies no matter whether these persons still worked as mathematicians, already lived in retirement, finally emigrated, or stayed, or—in rare cases—survived the regime.

It is more difficult to evaluate the problem in the case of "non-Jewish" mathematicians. Among the persecuted and expelled mathematicians one has, of course, to count men such as O. Neugebauer, H. Schwerdtfeger, and P. Thullen. These were men who because they were unable to compromise accordingly went into emigration without them or their relatives actually being threatened by the Nazi regime.[37]

However, the situation of non-Jewish mathematicians remaining in Germany has to be judged in a different light. If one could count all of those barred from a career at a university for political reasons in spite of proven ability and interest as persecuted, one would have to list—given the highly politicized atmosphere in Nazi Germany—nearly all the mathematicians shying away from showing active support for the regime. This applies for instance to Erika Pannwitz (1904–1975) who had to take a job in mathematical reviewing (*Jahrbuch über die Fortschritte der Mathematik*) despite the very high standard of her mathematical dissertation.[38] If one inflates "persecution" even further, one could take Helmut Wielandt (1910–2001) and Kurt Schröder (1909–1978) into consideration as victims as well. Wielandt, like Pannwitz, worked for the *Jahrbuch*, while Schröder left the University of Berlin due to political pressure at the outbreak of war for an

---

[36]Among them there might have been many with a wish for emigration who were nevertheless not "eligible" for "non-quota-visa," because they had not been dismissed from permanent positions as scientists. See chapter 5.

[37]But very soon they were threatened due to their open opposition and had therefore no chance in Germany anyway. The refusal of some German authorities to acknowledge this kind of persecution and to give them compensation after the war was a very shameful point in the "coming to terms with the past" ("Vergangenheitsbewältigung"). See chapter 11.

[38]See Siegmund-Schultze (1993a).

institute in aerodynamics, not originally in the line of his research. Both Wielandt and Schröder had clearly demonstrated their ability in research and partially compromised their political beliefs by joining the Nazi Party NSDAP.[39] However, this proved insufficient for a career in academia. If one included all the mathematicians as "persecuted" who, at one time or another in their career within the twelve years of Nazi rule (or seven in Austria), had experienced political repression, including purges from the university, one would even have to take into account Nazi activists such as E. Tornier[40] as "victims." It is also problematic to regard—although we will do it here—Heinrich Grell as persecuted,[41] since he had apparently pinned his hopes on the Nazi movement before clashing with it.[42] In the case of Grell, another fact seems to have played a role—the strong discrimination of "deviant" sexual orientation by the Nazis.[43] Since this kind of social discrimination was not restricted to Nazi Germany one has to look at the details and the extent of persecution (for instance incarceration in a concentration camp) to be able to claim specific Nazi persecution.[44]

Admittedly just as much a problem is the question of who, among certain other mathematicians, *not* to consider as victims of the Nazis. Helmut Grunsky (1904–1986), as managing editor of the *Jahrbuch über die Fortschritte der Mathematik*, was one of the few among German mathematicians to openly resist the anti-Semitic policies of the Nazis and thereby not receive a lectureship (Dozentur) at the University of Berlin.[45] Obviously one has to simplify the criterion defining the "persecuted" (non-Jewish) mathematician. A viable and necessary criterion for persecution seems to be the exclusion or dismissal of the individual from a

[39]See more on Wielandt in Siegmund-Schultze (1993a) and Brüning et al. (1998), p. 45.

[40]Hochkirchen (1998).

[41]In a long letter to Louis Mordell, dated December 14, 1935 and written from the home of Heinz Hopf in Zurich, Grell reports on five months of incarceration in Nazi concentration camps. See Mordell Papers, 19.54.

[42]Particularly grotesque is a judgment of the mathematician H. L. Schmid on his colleague Heinrich Grell, dated July 22, 1948: "He is the one German mathematician who probably suffered the greatest persecutions and disadvantages from the past regime" (quoted from Grell's personal file in Siegmund-Schultze [1999], pp. 66–67). And this in view of the sufferings of the emigrants and of those murdered. See also the rather cold reaction on the part of emigrants vis-à-vis Grell's request for help, as documented in chapter 7. Nevertheless Grell will be included among the persecuted as mentioned.

[43]The sources are not quite unambiguous though with respect to the reasons for Grell's persecution.

[44]For instance, the well-known insurance mathematician and historian of mathematics Wilhelm Lorey (1873–1955) was forced into retirement in 1933 because of his sexual orientation. However, he never considered himself a victim of the Nazi regime and was not incarcerated like Grell. Kind communication by Gert Schubring (Bielefeld).

[45]But he did not suffer a long-lasting dismissal. See Siegmund-Schultze (2004a). Grunsky also joined the Nazi party after pressure in 1940.

position—usually in the public sector that was strongly politically affected—and the maintenance of that exclusion for a long period of time (official debarment = Berufsverbot) combined with the lack of previous clear political engagement for the Nazi regime.[46]

All in all the investigation of the pretexts for dismissal and of the political behavior of scientists under the Nazi regime requires much more detailed historical work. The 234 mathematicians named in Appendix 1 in the lists of émigrés and persecuted mathematicians shall stand symbolically for the political persecution of scientists under the Nazi regime.

The high percentage (about 62 percent) of emigrants among the total of 234 persecuted mathematicians points to a strong "pull-factor" by the host countries, although it was apparently no higher in mathematics than in the average of other disciplines.[47] A strong pull was to be expected due to the internationally recognized level of research mathematics in Germany. More striking, however, is the strong "push-factor" in mathematics causing the emigrations. Political activity before 1933 (often used by the Nazis as a pretext for dismissals) was not very pronounced among mathematicians. Thus it was mainly the high percentage of Jewish and foreign mathematicians in the German-speaking institutions as of 1933 that accounted for the dismissals. In fact, less than one-fifth of the persecuted mathematicians according to our lists, namely forty-five out of 234, and less than one-seventh of the emigrants, namely twenty out of 145, were undoubtedly of "Aryan" descent by Nazi definition. In addition, it must be noted that many mathematicians were forced to go because of their Jewish spouses (Artin, Weyl, Kamke, Friedrichs, etc.).

The generally much greater physical threat against Jewish scientists compared to most non-Jewish dissenters is indubitable. Among the seventeen mathematicians killed by the Nazis or driven to suicide (Appendix 1 [1.2]) only two, Eckhart and Haenzel, were not Jewish. The circumstances and possible causes, in a psychological respect, of the (partly untimely) deaths of Jewish mathematicians such as S. Jolles, L. Lichtenstein, E. Landau, and I. Schur remain so far unknown. The lists of the appendix also show that chances for emigration lessened the longer the Nazi regime remained in power. This was also due on the one hand to the immigration policies of the host countries, and on the other hand to the restrictive emigration policies in Hitler's Germany as practiced shortly before the out-

---

[46]In this sense Grunsky, Pannwitz, and Wielandt cannot be classified as "persecuted." There is still doubt as to the clarity of Grell's previous engagement in favor of the regime. Therefore he has been included as a victim.

[47]The average percentage of emigrants among the dismissed German university teachers was 60 percent, according to newer statistics. Personal communication by C.-D. Krohn from 1997.

break of the war. Thus about 90 percent of the mathematicians expelled from Berlin and Göttingen were able to emigrate, while only two-thirds of those from Vienna were able to flee after the occupation (Anschluss) of Austria, and only one-third of the dismissed were able to escape from Prague after the annexation of the Czech Republic.[48]

In spite of these terrifying numbers the historian has to be aware that scientists on average were "privileged" compared to other victims of the Nazi regime. On average, conditions for emigration were more favorable to them, even though there were hardly any actual privileges given to scientists within Nazi Germany.

At various points of time[49] during the Nazi regime, forty-five ordinary professors in mathematics were dismissed, from a total of about one hundred full professorships (Ordinariat) for mathematics existing at the German-speaking universities (including technical universities, excluding Switzerland) at the time of the greatest expansion of Hitler's empire.[50] This implies a much higher rate of dismissals among mathematics professors than among professors in other fields.[51] The expulsions in mathematics were very unevenly distributed over the German system of universities. The twenty-two full (ordinary) professors for mathematics (including applications) who emigrated from the German system of universities (excluding Austria) came from only thirteen out of thirty-eight universities, and the main brunt of these twenty-two emigrants had been teaching at the Prussian universities in Göttingen, Berlin, Breslau, and Königsberg. Since at the time of these dismissals there existed thirty-nine full professorships in mathematics at the thirteen universities, the rate of emigration and even more the rate of dismissals was over 50 percent.

If one wishes to evaluate the effect of mathematical emigration quantitatively one should look first at the distribution of the emigrants over the various host countries. Eighty-seven out of the 145 émigrés reached the United States or (in five cases) Canada before the war was over in 1945,

---

[48]In the case of Prague, the fact that some visas extended from countries such as England became invalid with the outbreak of the war in September 1939 seems to have played a particularly unfortunate role. Only a few Jewish mathematicians remaining there had a chance to survive. Heinrich Löwig, according to John von Neumann "one of the first to observe that the modern theory of Hilbert space carries over to non-separable spaces" somehow survived German occupation and fled to Australia only after the war, in 1947, when all of a sudden he counted as "German" and was again unwanted, this time in Czech eyes. OVP, cont. 31, f. Loewig, H. The opinion by John von Neumann on Löwig is also in this file. See another file on Löwig in SPSL.

[49]Particularly in 1933, 1935–36, and 1938–39.

[50]Scharlau (1990).

[51]Strauss (1991), p. 10, assumes an average *emigration* rate of 15 percent for all categories of professors. One must, however, take into account the difference between the rates of emigration and the (higher) rate of dismissal.

while only six went to South America. Among the eighty-seven going to North America were eleven mathematicians who had temporarily found refuge in Great Britain, which, with twenty more immigrants who stayed, became the second most important host country in mathematics. Palestine (10), Sweden (5),[52] the Soviet Union (3), Switzerland (3), Australia (2), the Netherlands (2), Belgium (1), France (1), India (1), and South Africa (1) were also final host countries until the end of the war.[53] Some of the countries mentioned (notably France) and others, too (Poland, Yugoslavia, the Czech Republic, Italy, Portugal), were stopping-off places on the way to further emigration, or they became holding areas for the emigrants before they were taken away to the extermination camps (O. Blumenthal, K. Grelling, R. Remak). The immigration to the Soviet Union, which in mathematical respects was an attractive country, was overshadowed by the Stalinist regime: at least one mathematician-émigré (F. Noether) was murdered there.

The *outward success story* of emigration correlated very much with the age of the emigrants. Several of the oldest mathematicians to be dismissed were unable to emigrate at all.[54] Some ended up in extermination camps, others committed suicide. The oldest among those who did manage to emigrate (F. Bernstein, M. Dehn, E. Hellinger, A. Rosenthal, H. Hamburger)[55] were the least successful and failed to get regular professorships in spite of their prominent history. Almost all of the others who had occupied permanent positions in Germany previously and who survived longer[56] obtained full professorships in their host countries, even if only after the war. Even before 1945 all but seven immigrants to the USA had gained positions in the university system.

In the main country of immigration, the USA, several of the immigrants helped in building up new mathematical centers.[57] Among those centers were the Institute for Advanced Study (IAS) in Princeton[58] and the mathematical departments of the universities Stanford, Berkeley, and New

---

[52]The case of Sweden is special because this country took in refugees during the last years of the war who had fled to the two other Scandinavian countries Denmark (two) and Norway (three) before.

[53]Some mathematicians did not live to see the end of the war, but the three immigrants to occupied France and the Netherlands, Pollaczek, Boll, and Freudenberg, who were not caught by the Nazis, did.

[54]The wish for emigration among them is documented for example in moving letters by Hausdorff, Blumenthal, and Remak to their colleagues. See below in particular chapter 5.

[55]See particularly chapter 9.

[56]Among those emigrants who died early on were L. Lichtenstein (in 1933), L. Hopf and E. Fanta (both in 1939).

[57]More on that in chapter 10.

[58]Porter (1988). The principal decision for funding the institute had been taken before 1933 already.

York. It is striking, however, that until the end of the war, with the exception of a few applied mathematicians (R. von Mises, W. Prager) and the historian of mathematics O. Neugebauer, none of the forced emigrants were called to the existing leading departments of mathematics at the universities of Harvard, Princeton,[59] Chicago, Brown, Yale, and the technical universities Caltech (Pasadena) and MIT (Boston/Cambridge). The exceptions mentioned were obviously related to special needs in American mathematics,[60] which shall be discussed later in chapter 10. A certain role seems to have been played by academic anti-Semitism as well. In any case there was institutional innovation caused by immigration, if partly outside the existing structures.

[59]Princeton is somewhat of an exception because the mathematical department cooperated closely, if not always harmoniously, with the IAS.

[60]In the case of immigrant logicians, such as R. Carnap, the appointment at bigger institutions was connected to the needs in American philosophy.

THREE

Early Emigration

> Exiles of German origin . . . as a class have contributed more to
> the upbuilding of science, agriculture and industry in the United
> States than the expatriates of any other nation.
> —C. A. Browne 1940[1]

> The prospects for an academic career, particularly for the first
> steps, are considerably better in America than in Switzerland.
> The educational system may be worse for the persons affected
> than with us. But for the academic teachers the gradual ladder of
> remunerated positions, assistent [sic], assistent professor, associ-
> ate professor, which all have their clearly defined tasks within
> the whole is undoubtedly very favorable, compared to our bread-
> less private docents [Privatdozenten] without obligations.
> —Hermann Weyl 1929[2]

> I would not have gone if it hadn't been for some people in Hei-
> delberg keen on saving my life. They brought disciplinary pro-
> ceedings against me thereby causing my dismissal, and so leav-
> ing me with no choice in the matter.
> —Emil Julius Gumbel 1964[3]

THE SINGULARITY AND SUDDENNESS of the Nazi occupation of Germany
in 1933 and of Europe since 1938, and the resulting collapse and reconstruction of the international communication network, caused *discontinuity*. Part of this discontinuity was forced emigrations. But there was
continuity as well, and academic emigration had a longer tradition. It is
therefore very important to stress the difference stated above between
forced emigration, and *voluntary emigration*. Emigration before 1933
had been more or less "voluntary," even though much of it had also been

---

[1]Browne (1940), 205.

[2]H. Weyl on July 20, 1929 in a letter (T) from Berkeley (California) to Michel Plancherel
in Zurich (Switzerland). Quoted from Frei and Stammbach (1992), pp. 127–28. The same
system of Privatdozenten was also in place in Germany at that time.

[3]Gumbel's ironic remark on the reasons for his early emigration from Germany in 1932
is from his German ms. "Memoirs of an Outsider" ("Erinnerungen eines Aussenseiters"),
23 pp., p. 9, which is contained in the Gumbel Papers, box 4, f. 6 at the Regenstein Library,
Chicago (T).

influenced by economic and political conditions. In order to fully evaluate the impact of forced emigration after 1933 on world mathematics and particularly on mathematics in the United States, one would also have to consider post–World War II emigration, which again can be called "voluntary," but was caused by new economic and political pressures as well.[4]

Before the outbreak of World War I in 1914, there had been a decline both in the figures of American freshmen studying in Europe, particularly in Germany, and in the number of immigrants to America.[5] This was partly due to political reasons, aggravated by the war, as well as being a sign of the growing strength and independence of the American university system.

The First World War created new, somewhat contradictory conditions in the relations between Germany and the United States. The resulting political estrangement, which for instance translated into a setback for German as a language in American schools,[6] was accompanied and partly neutralized by increased economic and cultural contacts, supported by globalizing technical developments in traffic and communication. Here we can only sweepingly refer to the very different structures of European and American science and educational systems that affected the process of emigration heavily and which are still partially in existence today.[7] For the general problems of academic employment and unemployment in Germany in the nineteenth and twentieth centuries and the combination of that phenomenon with periodic "crises of superabundance" ("Überfüllungskrisen") of graduates in particular fields, I recommend Titze's book (1990). The historian Weiner, pointing to the long-lasting structural weakness of the hierarchical, antidemocratic, and underfunded German science system, stresses the internal, social, and structural causes contributing to the rise of American physics and mathematics. Weiner and Schweber maintain that the rise would have occurred anyway, even without the landslide of forced emigration from Hitler's Germany.[8]

---

[4]This topic, which is even less covered by the available literature than early emigration, will only be occasionally discussed in the present book, for example in chapter 11.

[5]German-trained psychologist James McKeen Cattell (1860–1944), in the second edition of his *American Men of Science* (1910), regretted the decrease of scientific immigration to the United States as compared to the first edition of 1903.

[6]There are also certain isolationist policies in the United States to consider, which resulted in the nonmembership in the League of Nations, an organization that had been originally proposed by the United States itself. In this context conflicts of American scientists with their political authorities arose over the participation in international congresses.

[7]In particular the more centralized and state-regulated authorities in European educational systems. See also part D of this chapter for pertinent documents.

[8]Weiner (1969), p. 226; Schweber (1986), p. 81.

3.1. The Push-Factor

The early emigration of German-speaking mathematicians particularly to America began around 1900 at the latest and was often economically or politically motivated. There were the successful examples of the specialist in the calculus of variations, Oskar Bolza (1857–1942), and of the algebraist and geometer Heinrich Maschke (1863–1908), who found positions at the new University of Chicago in the 1890s after viewing their chances of finding similar employment in Germany as virtually nonexistent.[9] Years before, the mathematician and engineer Karl Steinmetz (1865–1923), who was driven out of Germany because of his social democratic beliefs, found an influential position in the American electroindustry.[10] Russian-born Solomon Lefschetz (1884–1972) went to the United States in 1905, partly owing to a lack of job prospects as a foreigner in France, and became one of the most important topologists in his new country.

While a combination of a lack of job prospects and political pressure had influenced academic emigration to the United States before the First World War, the pressure (push) to emigrate increased even more afterward, at least in Germany. The inflation of the German currency between 1919 and 1923 led to a corrosion of wealth in the traditional middle classes with intellectual interests (Bildungsbürgertum). Under the global economic crisis at the end of the 1920s the economic difficulties of German science became virulent again.

Throughout the 1920s, problems of remuneration and salaries did not just affect the young scientists taking their first unpaid career steps but also scholars of international fame, if they were unlucky enough not to have full positions. Thus the inventor of telephotography,[11] the applied mathematician Arthur Korn, and Ernst Zermelo—the latter having become famous for his work on the axiomatics of set theory early in the century[12]—encountered serious economic problems. Both scholars would be marginalized for different reasons after 1933 as well (D).

Salary cuts during the time of the "emergency decrees" ("Notverordnungen") since 1929 also affected full ("ordinary") professors.[13] Lack of prospects for an academic career in the overcrowded German university system strengthened the will to emigrate or caused scientists with less

---

[9]Parshall and Rowe (1994).

[10]Kline (1992).

[11]"Fernbildtelegraphie," which is a precursor to today's fax. See Litten (1993).

[12]On Zermelo see Ebbinghaus (2007).

[13]For Hermann Weyl, who was then in Göttingen, this was a reason to reorient his career toward the United States. See Sigurdsson (1996), p. 53.

self-confidence to seek alternative jobs such as in teaching, a career, however, usually reserved for German citizens.[14] Later on, under the conditions of forced emigration in 1933 (1938–39 for Austria and Czechoslovakia, in particular Prague), earlier orientation toward a school job sometimes backfired on the individual, since it was more difficult for teachers to be accepted abroad than for scientists.[15]

Economic problems, partly responsible for the emigration of the logician Kurt Gödel as late as 1940 from German-occupied Austria to the United States,[16] were, however, a general European problem, not just one restricted to the German-speaking domain. They affected the victorious powers of the First World War, such as France, as well.[17] Even in Sweden, a country less affected by the war, Einar Hille (1894–1980) had been forced to take a bread-winning profession as an actuary in public service foreseeing no future in scientific research. He received a stipend at Harvard University in 1920, which was partly financed by a Swedish American foundation.[18]

Early emigration was not exclusively directed toward the United States. In 1929, Adolf Fraenkel had been called from Kiel to a professorship at the new Hebrew University in Jerusalem.[19] Fraenkel shared—apparently motivated by Jewish religious beliefs—a feeling of responsibility for the development of a Jewish university in Palestine with Edmund Landau (1877–1938) from Göttingen who had accepted that position before. But both Landau and Fraenkel had to struggle with political and financial restrictions in Jerusalem. This caused Landau to return in 1928, while Fraenkel resumed his position in Kiel in 1931, after his proposal to establish an additional lectureship in applied mathematics in Jerusalem had been turned down. Thus Fraenkel had to suffer dismissal in Kiel after the Nazis took power in 1933. Fortunately, he could reclaim his position in Jerusalem, one that was subsidized in the following years by American money.[20]

Not only economic troubles peculiar to European and German science after World War I stimulated emigration. Also political problems such as

[14]See Frucht (1982), p. 101. Frucht, as a Czech citizen, was not accepted for the state examination (Staatsexamen) in Germany to become a teacher and therefore went to Italy.

[15]See Wasow (1986), p.108, and chapter 5 below.

[16]See Dawson Jr. (1997), p. 146, Einhorn (1985), p. 252, and Sigmund, Dawson, and Mühlberger (2006).

[17]Siegmund-Schultze (2001).

[18]Hille (1980), pp. 8–9. However, Hille had been born in New York and apparently had the advantage of an American citizenship.

[19]For the following see Fraenkel (1967), pp. 158ff.

[20]On the help by the American Friends of the Hebrew University mediated through the Emergency Committee in Aid of Displaced Foreign Scholars see chapter 8.

nationalism, xenophobia, academic anti-Semitism (D), and, in some instances, scientific drawbacks in certain fields in Europe led to an increased desire on the part of the individual to emigrate.[21]

The "pull-factor" (below to be discussed for the case of the United States as the main attractor), which drew young German mathematicians abroad, is of course in many respects not separable from the "push-factor." If one looks for instance on the ideological conditions, in particular anti-Semitism, so can both strong Jewish religious and Zionist ideology be considered as counterparts and remedies, which attracted some mathematicians, both students and established ones, to Palestine. This became relevant in the case of A. Fraenkel's early and temporary emigration. Zionist ideology seems to have played a major role well before 1933 for mathematics students such as Rafael Artzy (originally Deutschländer), Grete Leibowitz, and Dov Tamari (originally Bernhard Teitler).[22] Of course they would not have a chance to remain in Germany in 1933 anyway, but their convictions determined the direction of emigration.[23]

The differential geometer, Marxist, and future important historian of mathematics, Dirk Jan Struik (1894–2000), went to the Massachusetts Institute of Technology (MIT) in Boston in 1926. This decision was also influenced by his realization that chances for professional advancement in the Netherlands were bleak because of his political convictions.[24] Other emigrant-mathematicians who went to the United States from smaller countries included Hungarians such as Tibor Radó (1895–1965), who arrived in 1930, and Dutch topologist E. R. van Kampen, who arrived in 1931 (D).

Some problems encountered by the rising generation of mathematicians in the German-speaking area culminated in the biography of perhaps the most versatile mathematician of the first part of the twentieth century, Johann von Neumann. He was burdened by the double handicap of his Hungarian citizenship and his Jewish origin (D). Besides von Neumann, there were other capable young German-speaking scientists in purer mathematical domains, such as Eberhard Hopf and Aurel Wintner (1903–1958) who tried their chances in America. Physicists[25] and scholars from other disci-

---

[21]In mathematics this was true for instance for topology and statistics.

[22]The latter reports that "his Zionist activities were criticized by his mathematical mentors [i.e., M. Dehn and C. Siegel in Frankfurt; R. S.] as being incompatible with his professional development as a mathematician" (Tamari 2007, p. 333).

[23]There were other students active in the Zionist movement such as H. Kober who immigrated not to Palestine but to other places (England in Kober's case). See Fuchs (1975), p. 187.

[24]See Albers (1994) and a special issue of the *Notices of the American Mathematical Society* 48 (June/July 2001): pp. 584–92 (contributions by T. Banchoff, Ch. Davis, J. Tattersall, J. Richards, D. Rowe).

[25]One can think in this connection of P. S. Epstein (1921), Maria Göppert (Goeppert), E. P. Wigner (both 1930), and R. A. Ladenburg (1931) from the German-speaking area who emigrated in the years given in parentheses. See Holton (1983), p. 176.

plines[26] sought a future in the United States as well, again, economic and political problems being the motivating factors.

The appointments of Alexander Ostrowski (1893–1986) in Basel (1927) and Heinz Hopf in Zurich could still possibly be considered under the traditional aspect of interchange of scientific personnel between Germany and Switzerland while the immigration of T. Estermann to England, of Hans Freudenthal to Amsterdam, of A. Plessner to Moscow,[27] and of Karl Grandjot (1900–1979) to Chile can at least be partly interpreted as a sign of the decreasing ability of the German university system to retain several of its most talented graduates. As late as in 1933, the geometer of Hamburg, Wilhelm Blaschke (1885–1962), tried to find positions in America for his students Erich Kähler (1906–2000) and Emanuel Sperner (1905–1980), both very promising mathematicians.[28]

Given the widespread willingness to emigrate among the most talented young scientists such as von Neumann, the chances for less accomplished young German mathematicians to be able to emigrate or to come into close contact with American mathematicians were relatively poor. This was reflected in several rejections of German mathematicians applying for Rockefeller fellowships in the 1920s.[29] Some of those rejected seem to have reacted with German nationalistic emotion later on beneficial to a career under Nazi conditions.[30] In general terms, the feeling of being demoted or outclassed in and due to the process of internationalization was an important stimulus for Nazi ideology in science, not just in mathematics.[31]

---

[26]For instance the psychologist Kurt Koffka (1886–1941), and the philosopher and logician Herbert Feigl (1902–1988). See Ash (1985) and Feigl (1969).

[27]See the remarks on these three mathematicians above in chapter 1.

[28]W. Blaschke to O. Veblen, March 22, 1933. Veblen Papers, Library of Congress, cont. 2, general correspondence. Both mathematicians finally found positions in Germany, indirectly profiting from the purge of Jewish mathematicians. Sperner was guest professor in Beijing between 1932 and 1934.

[29]On the importance for international mathematical communication of the foundations connected to the Rockefeller family see Siegmund-Schultze (2001).

[30]In 1932, Max Steck (1907–1971) had unsuccessful contacts with the president of the Rockefeller Foundation, ex-mathematician Max Mason (RF, RG 2, box 77, f. 615). As late as June 1933 Steck offered O. Veblen the translation of Veblen's and J. W. Young's *Projective Geometry* of 1910/1916 (Veblen Papers, cont. 13, f. M. Steck), but did not find a publisher. Steck belonged to those mathematicians who tried to promote their careers in Nazi Germany with political means. In his book *Mathematik als Begriff und Gestalt* (Halle 1942), published during the war and dedicated to Nazi physicist Ph. Lenard, Steck called for the "liberation of the German spirit in science from Western influence" (p. vii).

[31]One is reminded of the campaigns against the theory of relativity launched by Ph. Lenard and J. Stark.

3.2. The Pull-Factor

As early as in the nineteenth century various efforts were made from the American side to secure the competence of European scientists for the development of a strong indigenous science system.[32] Subliminal and open religious prejudices and academic anti-Semitism—widespread in Europe at that time—were seen and used by some Americans as a chance to enrich their own culture.[33] Although academic anti-Semitism and restrictions in admission policies were noticeable at American universities in the 1920s, these policies never reached the open and terrorist level of post-1933 Germany.

The initiative to win European scholars was taken mainly by individual scientists or their corporations, such as the American Mathematical Society (AMS). Systematic efforts barely existed. Those that did were organized to some extent by the New York Institute of International Education (IIE) under the direction of Stephen Duggan (1870–1950). The limitations imposed upon the Duggan institute to be able to provide German lecturers for American universities in the 1920s are revealed by the unsuccessful attempt of Hans Reichenbach to emigrate at that time (D). In the IIE, an "Emergency Committee" (EC) was founded after 1933, originally with the specific purpose of helping dismissed *German* scholars and, later, gradually extending its activities to cover other foreigners.[34]

In mathematics, visiting lectureships of the AMS encouraged the mutual acquaintance of American and European scholars.[35] The first four holders of that lectureship between 1927 and 1931 were European mathematicians (C. Carathéodory, H. Weyl, E. Bompiani, W. Blaschke), three of them from the German-speaking area. International research fellowships issued by the International Education Board (financed by the Rockefeller family) led in several cases to the appointment of European mathe-

---

[32]One has to mention again the appointment of Maschke and Bolza, but also Eduard Study's (1862–1930) temporary position in the United States (1893–94) and the failed effort to appoint Felix Klein in Baltimore as a successor to James Sylvester (1814–1897), who went back to England in 1884. See Parshall and Rowe (1994).

[33]Probably the most important early example of the beneficial influence of European academic anti-Semitism on American mathematics is the appointment of the Englishman James J. Sylvester to build up the graduate school for mathematics at Johns Hopkins University in 1876. See Parshall and Rowe (1994). In 1907, the American W. F. M. Gross, who had been sent to Göttingen, reported that it was particularly easy and cheap to hire Jewish mathematicians (University of Illinois [Urbana-Champaign] Archives, 15/1/3 box 9, f. "Physics." Gross to E. J. James, November 21, 1907. Kind communication by Susann Puchta, Hof-Germany.)

[34]See details in chapter 7.

[35]Archibald (1938), p. 21.

maticians such as E. Hopf, A. Wintner, W. Maier, and Isaak Schoenberg (1903–1990). In Princeton it was Oswald Veblen (1880–1960) who, by maintaining close contacts with the English mathematician Godfrey H. Hardy (1877–1947) and by temporarily exchanging his professorship with the one occupied by Hardy, was greatly instrumental in paving the way for the appointment of European teachers in mathematics and physics.[36]

American scientific institutions after the First World War, usually well equipped due to superior financial funding, tried to keep abreast of European science by inviting guest lecturers. American-European scientific contacts in mathematics were strongly shaped by an informal net of European and American mathematicians (G. D. Birkhoff, H. Bohr, R. Courant, G. Hardy, O. Veblen, H. Weyl) who gained influence especially in the selection process of grantees for the International Education Board (1924–28) of the Rockefeller family and who would finally play a crucial role in the forced emigration after 1933. Several early emigrants, such as physicist P. S. Epstein in Pasadena, tried to convince other Europeans of the benefits of emigration (D). Special "talent scouts" were sent to Europe (D).

The large proportion of applied mathematicians among the "early emigrants" leaving Europe for the United States in the 1920s,[37] is striking. A strong need in academic applied mathematics, a field that developed rather late in the United States and that could only be filled through the forced emigration of the 1930s and conditions under the Second World War, seems to have been the driving force for that part of the immigration.[38]

As to emigrants from areas other than German-speaking areas, one can mention the Russian mathematicians and mechanics Stephen Timoshenko (1878–1972),[39] Ivan Sokolnikov (1901–1976), and Jacob Tamarkin (1888–1945),[40] along with English-speaking mathematicians such as John L. Synge (1897–1995). The Swede Thomas Hakon Gronwall (Grönwall) (1877–1932) trained as an analyst in Sweden and who afterward worked as a civil engineer in Berlin, went to the United States in 1904 and became one of the few early applied mathematicians there.[41] The early immigration of the physicist of Frankfurt, Cornelius Lánczos, was important for the development of numerical analysis in the United States, as mentioned in the previous chapter. Michael Sadowsky (1902–1967), from the Technical

---

[36]Aspray (1988).

[37]Many names of early emigrants are mentioned in the book by L. C. Young (1981), pp. 327ff.

[38]See Siegmund-Schultze (2003a). More systematic discussion of the influence of German applied mathematics on American mathematics below in chapter 10.

[39]See Soderberg (1982) on Timoshenko for a most detailed description of the problems of cultural and mental adaptation to the American society felt by some foreigners.

[40]These are for instance mentioned in Birkhoff (1977), p. 64, and in Hille (1980).

[41]See Gluchoff (2005).

University of Berlin, went to the University of Minnesota in 1931 to occupy a temporary position as assistant professor.[42] The most important applied mathematicians among those emigrating at an early stage from a German-speaking area were probably the aerodynamicists Max Munk (1890–1986), who went from Göttingen to the Langley Laboratory in Virginia in 1920),[43] and Theodor von Kármán who went from Aachen to Pasadena in 1929,[44] von Kármán being more successful than Munk in establishing a school of research.

In other fields of "applied mathematics" in a broader sense, such as philosophy and economics, special relationships of Americans with Austrian science seem to have played a role in early emigration (H. Feigl, K. Menger) as well as after 1933/38 (F. Alt, A. Wald).

As to the attractiveness of the offers from America to talented European scientists in the 1920s, it must be taken into account that the step over the Atlantic was still considered a very big one and was often seen as containing unpredictable risks. These risks concerned the value of the currency; different living conditions, for instance the lack of domestic help; problems of language; and, last but not least, a fear of isolation and insufficient scientific communication. As to the latter, quite early on the United States was considered a relatively attractive place while immigration to South America was a kind of last resort, as shown by the failure of the career of Karl Grandjot who had immigrated to Chile in 1929, where his reputation on an international level due to a previous Rockefeller fellowship proved worthless (D).[45] Experience with foreign cultures and a longer stay abroad did not automatically improve chances of a position at home.

Finally, there was—already in the 1920s—resentment on the part of American scientists against unhampered scientific immigration and its possible consequences, among other things on the prospects for young American scholars. Resentment was especially strong in the "Division of

[42]Knobloch (1998), pp. 32–33. In 1933 Sadowsky was denied the maintenance of his status as Privatdozent in Berlin while his appointment in Minnesota came to its end January 1, 1934. See Sadowsky's undated vita of (latest) June 24, 1937 (= covering letter to Courant), when he was unemployed in Berlin, in Courant Papers, Bobst Library, New York City, box 3. According to Dresden (1942), p. 422, he came back to the United States in 1938 and obtained a position at the Illinois Institute of Technology. This is confirmed by the Web site of IIT. As is clear from SPSL, box 284, f. 5, Sadowsky had appointments in Brussels in 1934 (as also mentioned in his correspondence with Courant) and in Russia between 1934 and 1937. Expelled from there, he stayed in Palestine in 1937 from where he reported to the British SPSL about discrimination against him because he was non-Jewish and of Greek Orthodox religion.

[43]Eckert (2006).

[44]Hanle (1982).

[45]Other mathematicians immigrating to South America, such as Peter Thullen, suffered a similar fate in their mathematical careers. On Thullen see below chapter 4 and Appendix 6.

Figure 7 *Theodor von Kármán (1881–1963). The distinguished aerodynamicist moved in 1929 from Aix-la-Chapelle (Aachen) to the California Institute of Technology. He is considered an "early emigrant" in this book although his future connections with Aachen from 1933 were doubtless curtailed due to the Nazi regime. Von Kármán became a key figure in research in mathematical engineering and in the training of aviation engineers in the USA.*

Mathematics" at Harvard University where W. F. Osgood talked about the "dangers in importing a foreigner" as early as 1920.[46] A jealous repulsion of European feelings of superiority and a conviction of the priority of building a specifically American mathematical culture combined in making some people less keen to hire foreigners.[47]

In any case, the still existing dominance of European countries in most scientific fields—at least in the more theoretical areas, if less in the material aspects of equipment—restricted the chances of hiring foreigners in the 1920s. The greater economic possibilities and political freedom in the United States had to be complemented by scientific attractiveness, which already existed in some mathematical subdisciplines at the time. In the years of the economic crisis after 1929, economic difficulties experienced by American universities, lasting well into the 1930s, contributed to restrictions on appointing foreign scientists (D). The influence of the Depression on the American colleges, with the worst year being 1933–34, is reported in Rider (1984), 123–29.

3.D. Documents

3.D.1. The Economic Troubles in German Science as a Stimulus to Emigration

Economic problems caused the head of the Berlin student organization Mathematisch-Physikalische Arbeitsgemeinschaft (Mapha), Hans Rohrbach (1903–1993), to undertake a "propaganda trip" ("Propagandareise," as he himself called it) to the United States in the early 1920s, in order to collect money for Berlin students.[48]

The Berlin mathematician Eberhard Hopf tried to avoid returning to Germany after his Rockefeller stipend in Cambridge (Massachusetts) had elapsed, as no financial support was in sight. He reported on that to Tamarkin, the latter an immigrant to the United States from Russia: "I got some letters from Berlin, containing of course not agreeable news. The budgets for scientific purposes are cut down by the ministry, and I do not know whether I will have a salary in Berlin."[49] John von Neumann—unlike most

---

[46]See Siegmund-Schultze (1998), Appendix 2.1, p. 306. This appendix is not reproduced in the present book.

[47]This feeling is particularly well expressed in a letter of 1928 by G. D. Birkhoff to his colleagues at Harvard, commenting on the search for talent in Europe. See Appendix 2.

[48]Rohrbach (1929/1998). "Working students" (Werkstudenten) who financed their study by parallel physical work (as for instance future topologist Hans Freudenthal) were an increasing phenomenon in the German student body.

[49]Hopf to Tamarkin, August 18, 1931. Tamarkin Papers, Providence, box correspondence (A–H), f. E. Hopf.

private lecturers (Privatdozenten) in Germany—was financially well situated. With this in mind, on January 31, 1931, Courant discouraged a request by the Muslim University Aligarh (British India), apparently triggered by the French mathematician André Weil (1906–1998): "V. Neumann has a half position in Princeton and is a rich man anyway, so he will hardly take over responsibilities of this kind" (CPP). However, von Neumann did not delude himself about his prospects in economically troubled Germany, where he was at the time a private lecturer in Berlin and Hamburg. His friend Stanisław Ulam (1909–1984) would later write in his autobiography of 1976:

> In the German universities the number of existing and prospective vacancies for professorships was extremely small—something like two or three in the entire country for the next two years. With his typical rational approach, von Neumann computed that the expected number of professorial appointments within three years was three, whereas the number of docents was forty. This is what had made him decide to emigrate, not to mention the worsening political situation, which made him feel that unhampered intellectual pursuits would become difficult. In 1930, he accepted an offer of a visiting professorship at Princeton University, and, in 1933, shortly after the creation of the Institute, he was invited to become the youngest member of the permanent faculty of the Institute for Advanced Studies.[50]

Economic hardships were not restricted to young scholars. An example was honorary professor Arthur Korn at the Technical University Berlin-Charlottenburg. The specialist in potential theory and student of Hilbert, Oliver D. Kellogg (1878–1932), from Harvard University, inquired about Korn in a letter to Courant, dated May 31, 1932, wanting to know "why, if he is as good a man as he seems to be, he has not arrived at a better position at this time" (CPP). In a later letter to Courant (August 2, 1932) Kellogg found it, however, given the crisis in the USA, "quite out of the question to find anything for him as long as the depression continues" (CPP).

The same applied to Ernst Zermelo, who had been an honorary (i.e., unpaid) professor at Freiburg since 1921. He thanked Courant on February 4, 1932 for having been elected a corresponding member of the Göttingen Academy of Sciences. Having just completed his edition of Georg Cantor's works for the Springer publishing house (to which Courant was closely connected), the sixty-one-year-old Zermelo offered his services for other remunerated work:

> In Freiburg and in Karlsruhe I have no prospects at all for a teaching job, after even salaries for assistants have been dramatically cut due to the present economy

---

[50]Ulam (1976), pp. 68–69.

measures. If you see any opportunity for translations or similar I would be grateful indeed for information. In a time of wide-spread unemployment one should not, however, have illusions.[51]

3.D.2. National Isolation, Xenophobia, and Anti-Semitism as European Phenomena

Examples abound confirming the international isolation of the French scientific system in the 1920s and even later.[52] As to the situation in Britain, it was G. H. Hardy who wrote in a letter to Hermann Weyl in the 1920s:

> The just reproach which all continental mathematicians have addressed to English mathematics for many years, has been its extreme insularity.[53]

National isolation and ideological prejudices usually went hand in hand.

On xenophobia directed against foreigners, Hermann Weyl wrote in 1925, in an expert opinion for the International Education Board in favor of Alexander Weinstein:

> The difficulties which every foreigner has to face today in European countries prevented him—as a Russian—from finding a scientific position adequate to his abilities in Switzerland.[54]

Judging by publications in German journals and employment figures at universities, Germany had perhaps the most internationalized culture of all European countries prior to 1933, at least in fields like mathematics.[55] Nevertheless, nationalistic and anti-Semitic prejudices existed in Germany as well and obstructed careers within academia or elsewhere in the civil service. The Kiel mathematician stemming from Croatia, Willy Feller, alluded to the preference given to ethnic Germans in obtaining German citizenship, as exemplified in the case of the Austrian-born Hitler, who had a police record from his putsch in Munich in 1923. Asking Courant for a positive opinion in favor of his application for German citizenship, Feller wrote ironically in May 1931:

---

[51]CPP (T). For Zermelo and in particular the loss of his honorary professorship in 1934 due to his refusal to extend the Hitler salute in a "proper" way see the recent book by Ebbinghaus (2007).

[52]Several such examples with respect to mathematics are given in Siegmund-Schultze (2001).

[53]Fragmentary letter from Hardy to Weyl, dated October 18, 192?, exact year unclear. Weyl Papers, ETH, Hs 91:589.

[54]Weyl to IEB, February 21, 1925 (T). Weyl Papers, ETH, Hs 91:725.

[55]The Munich mathematician Walther von Dyck (1856–1934) wrote to Ludwig Bieberbach in Berlin on January 19, 1925: "By the way one believes oneself to be rather in Hungary, Poland, Russia, and Bulgaria judging by the names of those who now publish in the Annalen" (Bieberbach Papers, Oberaudorf [T]).

My naturalization is converging well but the guys need some confirmation that I spent my time in Göttingen as a good Christian. I would be very grateful for the confirmation that . . . my naturalization is almost as desirable for the Reich as Hitler's.[56]

Indeed, being stateless was a serious obstacle to an academic career in Germany, as Emil J. Gumbel in 1933 retrospectively confirmed in a letter to the Russian-born topologist Boris Kaufmann, whose habilitation in Heidelberg had been prevented prior to his immigration to England:

Even in the time before Hitler it was practically impossible for a stateless former Russian to take the habilitation degree. I was always of the opinion that such extra-scientific arguments have no place at a university and that injustice has been done to you.[57]

In Kaufmann's case anti-Semitism probably was another point. In his letter to Courant, Feller had also alluded to anti-Semitism in Germany, which was often coupled with xenophobia. This also applies to the case of Alexander Weinstein, who—even after having obtained Swiss citizenship— wrote to Courant from Breslau (December 8, 1932) on the "problems which I am facing as a foreigner in Germany" (CPP [T]).

The problem of academic anti-Semitism in Germany is expressed in a letter written by Courant to Gabor Szegö in 1925 requesting Szegö confidentially for a curriculum vitae because

the problem of successors for the mathematical chairs is becoming very acute now, and I am feeling that our Universities ought not—in the long run—give in to unobjective motivations in appointments as much as is unfortunately still the case today.[58]

Maybe in anticipation of a position in Germany, Szegö—who was finally appointed as full professor in Königsberg in 1926—declined an offer from American Dartmouth College in 1925.[59]

Another example of academic anti-Semitism was the blocking of the Göttingen mathematician's Kurt Mahler's habilitation in Greifswald, an experience that was shared by at least three more Jewish mathematicians at other

---

[56]Feller to Courant, May 12, 1931 (T). Courant Papers, Bobst Library, New York City, box 4. Hitler had been stateless since 1925 and, due to the efforts of his National-Socialist friends in Brunswick, gained German citizenship in the nick of time for the elections for Reich president (which he lost to Hindenburg) in March 1932.

[57]SPSL, box 281, f. 2 (Kaufmann, B.), folio 81. E. J. Gumbel to B. Kaufmann, Paris, October 3, 1933 (T).

[58]Courant to Szegö, November 30, 1925 (T). Szegö Papers, Stanford University, box 5, f. 15.

[59]Szegö recommended instead Tamarkin (Askey and Nevai [1996], p. 15).

places.[60] Hellmuth Kneser (1898–1973), who had been asked for support by Courant, wrote to Courant on December 1, 1932, that anti-Semitism in Greifswald (particularly the attitude of the mathematician K. Reinhardt) prevented Mahler's habilitation. Mahler, who became known for his classification of transcendent numbers, immigrated to England in 1933.

Emmy Noether, in a 1929 letter to G. D. Birkhoff, recommended her student Jakob Levitzki (1904–1956) who at that time was unemployed in Palestine. She used defensive vocabulary, obviously due to her own experiences with anti-Semitism:

> In Germany it is also very difficult for foreigners to find a position. The available assistantships are barely sufficient even for Germans. I hope that the conditions in this respect are better in America and that the Jew from Palestine will not cause trouble. Levitzki is a very fine, agreeable person.[61]

Due to anti-Semitic prejudices, even Theodor von Kármán, well known internationally, did not deem his chances of being appointed to a position in Göttingen favorably, particularly in respect to the temporarily vacant leading chair in applied mathematics, once occupied by Carl Runge (1856–1927). Moreover, he did not share the nationalistic views of some of his colleagues (Ludwig Prandtl [1875–1953], von Mises); these views making the preparation of International Congresses for Mechanics difficult. Kármán's decision (1930) to stay in Pasadena for good, was, however, predominantly based on two facts: considerably better economic conditions and a greater influence afforded by his new position at GAL-CIT (Guggenheim Aerodynamical Laboratory at the California Institute of Technology), thus making him the undisputed leader in the field of aerodynamics in the United States.[62]

Compared to Germany a maybe even stronger and more open academic anti-Semitism existed in other European countries such as Poland[63] and Austria. The logician Herbert Feigl (1902–1988), who as a Jew had no chance of advancement in Austria, accepted an appointment in the United States in 1931.[64]

---

[60]These are the cases of S. Bochner in Munich, M. Herzberger in Jena, and B. Kaufmann in Heidelberg. On the latter two see their files, kept at the Bodleian Library in Oxford: SPSL, boxes 280, f. 3, and 281, f. 2. On Bochner see 3.S.1 below.

[61]Noether to Birkhoff, November 19, 1929 (T). Birkhoff Papers, Harvard University 4213.2, box 9, file: M-Q (1928–29). Noether wrote another letter for Levitzki to Oswald Veblen, but received an immediate negative response due to the lack of academic jobs in the United States. See Veblen to Noether, December 4, 1929. OVP, cont. 9, f. Noether, Emmy. As a temporary assistant in Kiel, Levitzki prepared A. Fraenkel in Hebrew conversation for the latter's position in Jerusalem. See Fraenkel (1967), p. 156.

[62]Hanle (1982), pp. 104ff. and p. 131.

[63]See, for example, Feferman and Feferman (2004), pp. 41–54.

[64]Feigl (1969) and Thiel (1984), p. 229.

In 1993, the immigrant to the United States Franz Alt wrote retrospectively on the situation in Austria that had forced him into a job in an insurance company after earning his PhD in 1932:

> You are right in your conjecture that it was with unpleasant feelings when I went into insurance mathematics. It would certainly have been more natural for me to enter an academic career. But this was impossible given the dominating, quite open anti-Semitism. In America anti-Semitism was also occasionally existent, but almost never officially admitted. I myself have never suffered from it.[65]

3.D.3. Personal Risks with Early Emigration

Scientific work abroad—even if only temporary—did not necessarily promote future career chances at home at that time. This is underlined in a letter from 1925, written by the Berlin full professor for mathematics, Ludwig Bieberbach (1886–1982), as advice to Heinz Hopf, one of the few active topologists in Germany at the time:

> It is of course a double-edged thing to go abroad. To advance there is always somewhat difficult, and the career at home is not facilitated by leaving the country either.[66]

These problems led to a reluctance on the part of several scientists who were, in principle, willing to emigrate. With respect to this theme, physicist Paul S. Epstein wrote from California to the Hungarian-Swiss mathematician Georg(e) Pólya in 1922:

> It is strange that many persons want to have positions in America, but are reluctant to give me information which is needed to provide the positions. For instance, I don't know, whether Weinstein speaks English.[67]

Karl Grandjot went to Chile from Göttingen in 1929, a move that apparently led to the end of his career as a research mathematician.[68] Head of the Göttingen Mathematical Institute, Courant, wrote on February 20, 1929 to the officer in the Prussian ministry of education, Windelband, that Grandjot had a gift for languages and organization, but:

[65]F. Alt to me, July 12, 1993 (T). On xenophobia and anti-Semitism in the United States cf. below chapter 7.

[66]L. Bieberbach to H. Hopf, September 23, 1925 (T). Hopf Papers, ETH, Hs 621:250.

[67]P. Epstein to G. Pólya August 23, 1922 (T). Pólya Papers, ETH, Hs 89:140.

[68]Grandjot was listed officially as Privatdozent in Göttingen between 1928 and 1936–37 (Schappacher 1987, p. 370). As becomes clear from Grandjot's curriculum vitae dated December 30, 1964, Grandjot did not give classes after 1929. His daughter Sigrid does not list any mathematical publication by Grandjot after 1928. This is based on a letter by Grandjot's daughter, dated August 8, 1988, to H. Wefelscheid (Essen), whom I thank for this information.

The only thing that worries Mr. Grandjot is whether a stay of several years in Chile would possibly be damaging to his future chances in Germany. Maybe one can reassure Grandjot on this matter for once and for all. Anyway I have the impression that Grandjot's services in Chile will greatly benefit German science. (CPPW [T])

In 1931, however, one reads in the diaries of a Rockefeller official about the erstwhile IEB fellow Grandjot:

Grandjot is in Göttingen for two months on vacation. He is a professor in Chili, but is too much isolated to prosper. He lacks contacts and has too much routine teaching there.[69]

3.D.4. The Ambiguous Interconnection between Social Hierarchies, Traditions at Home, and Internationalization in Mathematics

By international comparison, several areas even of "pure" mathematics were underdeveloped in Germany around 1930. This did not necessarily mean that there was no interest for these fields in Germany.[70] But there were deep-going traditions and social hierarchies in German mathematics that, under the conditions of academic unemployment, impeded a rapid familiarization of the new generation of mathematicians with foreign work. On the connection between the drawbacks in topology and functional analysis and the prospects for young mathematicians, the American mathematician Heinrich Guggenheimer wrote to me on September 9, 1993:

Not only algebraic topology did not fare well in Germany, but also e.g. functional analysis. While all these branches go back to some aspects of Hilbert's work, nobody was willing to risk displeasing the established professorial network and risk not having a job.

In some scientific fields, due to shortcomings in Germany, scientists felt compelled to seek positions on a more international level. Aurel Wintner and Eberhard Hopf left Germany for Baltimore (1930) and Cambridge, Massachusetts (1930), (a) because their areas of research, in particular celestial mechanics, had strong traditions in the United States,[71] and (b) also

---

[69]RAC, R.F., IEB 13, box 50, f. 757, July 4, 1931.

[70]One may think in this connection of the enthusiasm with which the topological work of the Russians P. S. Aleksandrov (1896–1982) and P. S. Urysohn (1898–1924) was received in Göttingen, which led to the publication of five articles by them in volume 92 (1924) of the *Mathematische Annalen.*

[71]Rockefeller functionary Tisdale reported on March 16, 1932 to W. Weaver in New York: "Hopf is extremely enthusiastic about his contact with Shapley and Birkhoff. As you may know, Birkhoff has recently made substantial contributions to a classical and fundamental problem in advanced dynamics, namely in the precise formulation of an ergodic

due to the lack of job prospects in their own country. After his Rockefeller fellowship, Hopf went to the Massachusetts Institute of Technology in 1932, and later, in 1936, he could not resist the temptation of taking up an appointment as a professor in Leipzig.[72]

Insisting on nonconformism in social or scientific respects could not only hamper careers at home but—due to the lack of formal credentials resulting therefrom—could prove to be a liability during emigration, as the tragic fate of the outstanding algebraist Robert Remak shows.[73]

3.D.5. The American Interest in Immigration (Pull-Factor)

Some early immigrants to the United States tried to pave the way for other Europeans. In a letter from 1921 by physicist Paul S. Epstein[74] to G. Pólya, the desire to enrich the American culture through immigration is jokingly mentioned:

> After having gained some perspective over the American conditions I am convinced that the future for you is here. . . . Since my desire to impoverish Europe has not abated, you will soon hear from me.[75]

Duggan's New York Institute of International Education (IIE) frequently sent out lists with the heading "Foreign Professors Available for Teaching Engagements" to interested institutions in the United States. The list of February 1, 1923, for instance, contains under mathematics Paul Dienes (1882–1952, Aberystwyth, Wales), G. Y. Rainich (Odessa), and Otto Szász (Frankfurt am Main), saying that all three of them spoke English fluently.[76] The mathematicians mentioned all originally came from Eastern Europe; Rainich soon succeeded in his immigration to the United States. The relative ineffectiveness of the IIE and of one of its counterparts in Europe, the Berlin Amerika-Institut, is documented by the unsuccessful attempt at emigration by Hans Reichenbach, who experienced political

---

principle, and in the properties of such ergodic systems. Dr. Hopf reports that this type of research is inadequately represented, at present, in Germany" (RAC, R.F. 2, box 77, f. 615).

[72]See Norbert Wiener's opinion on that below, in chapter 6.

[73]See Merzbach (1992), and chapter 5 below.

[74]Not to be confused with the mathematician from Frankfurt, Paul Epstein, who fell victim to the Nazis.

[75]P. S. Epstein to G. Pólya, November 4, 1921 (T), Pólya Papers, ETH, Hs 89:137. However, Epstein reported in a later letter that it was difficult to get information about the available positions because American universities did not know much about one another (E. to P., February 22, 1922, Hs 89:139).

[76]Richardson Papers, Providence, Correspondence, f. 6 I, 14 pp. This assertion was, at least with respect to Szász, an exaggeration, as later sources for the time of Szász's emigration reveal.

problems in Berlin that, as physicist Albert Einstein (1879–1955) commented on rather harshly, were "too foolish to communicate to a non-Prussian ear."[77] Einstein inquired with K. O. Bertling from the Amerika-Institut in Berlin, but found out, as seen from a letter to Reichenbach, that "the connections of the gentleman to American Universities are meagre."[78] Bertling had written to Einstein in 1925, pointing to information from the IIE:

> It remains . . . a fact that, firstly, the demand for European teachers in the U.S.A. is rather small—one is very concerned about the American young scholars—and, secondly, Americans constantly try to procure for their respective needs in almost all fields of academia [Wissenschaft] with the help of emissaries who are in Europe. It would be advisable to draw the attention of one of these Americans to Prof. Reichenbach and create a connection. In this, we from the Amerika-Institut might be able to help.[79]

The "emissaries" mentioned by Bertling did not seek out German speaking scientists specifically. The young Norwegian Øystein Ore (1899–1968), for instance, was contacted in 1926 by James Pierpont (1866–1938) in Oslo, while Pierpont was on a trip to Europe to engage mathematicians for research at Yale University.[80] The Dutch topologist E. R. van Kampen (1908–1942) had been hired in Delft in 1929, when he was just nineteen years old, by the chairman of the mathematics department of Johns Hopkins University, Francis D. Murnaghan (1893–1976). Due to his youth, van Kampen could not take up the professorship until 1931.[81]

In Princeton, Veblen and others wished to appoint a prominent European for the newly created Thomas D. Jones Professorship of Mathematical Physics. After Einstein had declined in 1927,[82] Veblen offered the position first to Werner Heisenberg (1901–1976) and then, in 1928, to Hermann Weyl. Both held the professorship consecutively for one year each, before returning to Europe.[83] In 1927, an effort by Columbia University in New York to hire Hermann Weyl failed.[84] Efforts to appoint a European to the Jones chair in Princeton remained unsuccessful until 1937.[85]

---

[77]Einstein to Epstein, June 10, 1926 (T). Einstein Papers, Jerusalem, 20 089-1. Reichenbach had once been a leftist student activist. On his problems see Hecht and Hoffmann (1982).

[78]A. Einstein to H. Reichenbach, March 31, 1926 (T). Einstein Papers, Jerusalem, 20 116.

[79]K. O. Bertling to Einstein, May 10, 1925 (T). Einstein Papers, Jerusalem, 20 088-1.

[80]Anon. (1970), p. i.

[81]Kac (1987), pp. 84–85.

[82]A. Einstein to O. Veblen, September 17, 1927: "One should not uproot and move an old flowering plant because it is going to die untimely" (T). Einstein Papers, Jerusalem, 23150.

[83]Frei and Stammbach (1992), p. 101.

[84]Ibid., p. 78.

[85]Butler (1992), pp. 159–61. E. Wigner accepted the chair in 1938.

1924

Figure 8 *Oswald Veblen (1880–1960). The important geometer and topologist from Princeton was also a proficient organizer of his science. He appointed among others Kurt Gödel, John von Neumann, and Hermann Weyl at the Institute for Advanced Study (IAS), which had been founded in 1932–33. He was also an unprejudiced supporter of immigrants to the United States.*

Figure 9 *John von Neumann (1903–1957). The universal mathematician from Hungary gave up his lectureship at the University of Berlin in 1933 and went to Princeton, where he had already been teaching during winter terms. Later he became one of the most influential American mathematicians, for example as a member of the U.S. Atomic Energy Commission.*

Winning John von Neumann over to American science was also prepared by emissaries. Von Neumann wrote in July 1929 from Hamburg to Weyl about a proposal from Murnaghan:

Eight weeks ago I was negotiating here with a professor Murnaghan from Baltimore, who wanted to import Europeans to Johns Hopkins. He made me an offer of an "associate" position with 3000 $ per year, for the time being for a year (with an option on my part to stay). The offer didn't seem too lucrative to me, and I declined it without further negotiations later, because I didn't want to leave Europe after my father's death in the winter. Now Prof. Murnaghan is offering me the same position for 1 February until 1 June at 1500 $ (for the 4 months). Actually, I would like to undertake such a short trip to America, but the offer still seems financially unattractive. I would be grateful if you could advise me on that, based on your experiences with America.[86]

Apparently Weyl advised von Neumann against the offer. Von Neumann reported in another letter to Weyl about an offer from Veblen to teach quantum mechanics Princeton as a successor to Weyl.[87] This offer, which von Neumann accepted, came with $4,000 for the period February 5, 1930 until July 1, 1930 and was thus much better than Murnaghan's $1,500.

3.D.6. The Start of Economic Problems in America around 1930 Foreshadowing Later Problems Incurred during Forced Emigration

Emil Julius Gumbel, dismissed in Heidelberg (Germany) for political reasons even before 1933, desired to go to America in 1932, preferably to the new IAS in Princeton. However, he was not certain about the demand for mathematical statistics in the United States, particularly because "due to the economic crisis which they call 'depression,' there are general tendencies towards cutbacks at the universities."[88]

Richard Courant, too, was skeptical due to the economic crisis, and in 1932, writing while journeying through the United States, he advised his colleague von Kármán, who had maintained contacts with his former university at Aix-la-Chapelle (Aachen), against a permanent position in America:

> People here believe in a rapid recovery in Germany and predict for America a number of difficult years with adaptation to a low level.[89]

Also the failure of appointments for C. Carathéodory and S. Bochner to Harvard in the late 1920s—described in the following case study—might partly reflect the start of economic problems within American universities.

---

[86]Von Neumann to Weyl, July 22, 1929 (T). Weyl Papers, ETH, Hs 91:683. In order to get an impression of the purchasing power of the U.S. dollar at that time, one needs to look at the stipends for Rockefeller fellows that were between $1,000 and $2,000 for a whole year and were usually considered sufficient.

[87]Von Neumann to Weyl, November 24, 1929. Weyl Papers, ETH, Hs 91:684.

[88]Gumbel to Einstein, November 18, 1932 (T). Einstein Papers, Jerusalem, 50 123-1.

[89]Courant from Berkeley to Kármán, July 30, 1932 (T). Kármán Papers, Caltech, Pasadena, 6.15.

3.S. Case Studies

3.S.1. The Failed Appointments of C. Carathéodory and S. Bochner at Harvard

In 1928 the head of the department for mathematics at Madison (Wisconsin) and future Rockefeller functionary, Warren Weaver (1894–1978), tried to win the prominent Greek German analyst Constantin Carathéodory (1873–1950)[90] for his department. Carathéodory—as a former visiting lecturer of the AMS—was well known to the Americans. G. D. Birkhoff of Harvard, who had been approached for advice by Weaver, proposed a young American mathematician instead. Weaver answered with the following letter:

> It seems unfortunate that it is so, but it is easier to get a large salary for a foreign appointment. The administration is likely to feel that there is more advertising value in a foreign appointment.[91]

However, the advertisement point of view was—at least for Birkhoff—not a decisive factor, and he apparently agreed on that point with his colleagues at Harvard University. In a meeting of the Division of Mathematics there it was discussed on May 7, 1928, whether Carathéodory, at the time filling in for Birkhoff at Harvard, should be offered a permanent position. Birkhoff, who had been asked for his opinion, telegraphed from abroad:

> Excellent man, excellent appointment, if not preferring Germany and not delaying young men.[92]

Nothing came of the plans for an appointment for Carathéodory; perhaps the evolving economic crisis played a role. One year later the Harvard mathematician Julian L. Coolidge (1873–1954) reviewed Carathéodory's lectures. In a letter to the Board of Overseers of Harvard College, he wrote on May 9, 1929:

> Not only was his scientific standing of the highest but he showed a remarkable aptitude to inspire every student who came near to him. We never made him an offer for we never saw the money available. . . . Recently an American University of high standing has offered him $10,000.[93]

---

[90]Georgiadou (2004), p. 224.

[91]Weaver to Birkhoff, October 15, 1928. Birkhoff Papers, 4213.2.2, box 10, file: V–Z.

[92]HUA, Math. Dept. UAV 561, Minutes, volume 3 (1924–28).

[93]HUA, Math. Dept. UAV 561, Correspondence and Papers 1911–62, UAV 561.8, box 1930–1939, f. visiting committee. By the "American University of high standing," Coolidge meant probably Stanford. According to van Dalen (2005), p. 624, Carathéodory considered temporarily an appointment at Stanford in order to evade the struggle over the *Mathematische*

At about the same time, the Division of Mathematics at Harvard discussed the possible recruitment of young European mathematicians. In June 1928 they talked about an appointment for Salomon Bochner, whose career had been hampered in Munich due to xenophobic resentments. Birkhoff and Kellogg used the International Congress of Mathematicians in Bologna that same year to inquire about younger European mathematicians. They reported to the Division of Mathematics in September 1928. Birkhoff had already written from Paris to his colleagues at Harvard in July 1928 stating in this letter how very reluctant he was as to the possible appointment of Europeans and referred to the still existing arrogance in Europe, a "feeling (not rare here) that any second rate European youngster is good enough for us."[94] The file at the Harvard University Archives documenting the negotiations about Bochner, which finally failed, contain a letter by the Munich mathematician Oskar Perron (1880–1975) to Carathéodory from July 6, 1928. Perron reported that a higher political authority, the Bavarian Council of Ministers (Ministerrat), had thwarted the intention of the Bavarian Ministry of Culture to appoint Bochner at least as a private lecturer (Privatdozent). Perron then said:

> The only thing which could still help would be an invitation to Bochner from America. Thus we would have new means at hand to show to the ministry even more clearly that Bochner really is somebody, and that it is embarrassing in the scientific world to make life difficult for him here.[95]

3.S.2. Early Emigration from Austria as Exemplified by Karl Menger

The noted topologist Karl Menger (1902–1985)[96] from Vienna had been guest lecturer at Harvard University in 1930. In 1932, both Menger and his former teacher Hans Hahn (1879–1934) sounded rather upbeat in their letters with regard to the political situation, looking forward to Veblen's visit to Vienna that same year. However, the situation within Austria in

---

Annalen, which was in full swing in 1928. There is a file in the Archives of the Ludwig Maximilians University Munich documenting negotiations that Carathéodory had with the Munich authorities to honor the offer from Stanford with improvements at home (Abwendungsverhandlungen). In 1936–37 Carathéodory assumed for one year the Carl Schurz Professorship in Madison (Wisconsin). See for the latter Georgiadou (2004), pp. 323–27.

[94] A longer passage from the letter is reproduced below as Appendix 2.

[95] HUA, Dept. of Mathematics, UAV 561.8, box 1920s, f. Miscellaneous. In a recent talk on Bochner's case (2006), Ulf Hashagen (Munich) revealed shocking details about state-supported anti-Semitism and xenophobia in Munich at that time.

[96] The letters quoted in this case study in original English are from the Oswald Veblen Papers (OVP) of the Library of Congress, Washington, DC. They are in OVP, cont. 8, f. Menger, Karl, 1930–39. See also Menger (1994).

1933 changed radically, closely linked to the developments in Germany at the same time. Two months after the Nazis had seized power in Germany, Menger wrote to Veblen on March 28, 1933:

> What do you say to the world situation? I am very pessimistic and the European situation seems to me also politically extremely serious and at least as threatening as it was in spring 1914 (not to speak of the economical situation). I wonder whether you are informed in U.S. how serious the preparations for a war are especially in the South-East of Austria, first of all in Serbia, whose interior situation is indescribable! And we in Austria, though completely neutral in most of the conflicts are much troubled by the idea that a war could invade our territory.

Despite this gloomy tone, Menger was, in the same letter, still optimistic about his own work and hoped, too, for a Rockefeller fellowship for his gifted student Georg Nöbeling (1907–2008), who planned to take his habilitation in Vienna in 1934. However, on October 27, 1933, Menger was worried about the political censorship of letters and therefore wrote to his colleague Veblen from Geneva rather than from Vienna:

> What I could not write you from Vienna is a description of the situation there. You know how fond I am of Vienna and how many things I started there in the intention of staying there still a good many years. But the moment has come when I am forced to say: I hardly can stand it longer. First of all the situation at the university is as unpleasant as possible. Whereas I still don't believe that Austria has more than 4% Nazis, the percentage at the universities is certainly 75% and among the mathematicians I have to do with, except of course some pupils of mine, not far from 100%.

Apparently alluding to blatant manifestations of anti-Semitism in the Vienna newspapers, Menger added:

> Not that I have particularly weak nerves. But you simply <u>cannot</u> possibly find the concentration necessary for research if you read twice a day things in the newspapers (whose reading is indispensable for man in a public position) which touch the basis of civilization of our country as well as your personal existence.

At the same time Menger acknowledged in this letter of October 27, 1933, that—for all his "fondness of Vienna"—he had found "a good many years" a move to the United States as being in the interests of his scientific career:

> I never did anything to move to America (v. Neumann, for instance, told me in September that he sometimes was wondering if I would like to go), though it has been my desire since a good many years, . . . since I realized that in some of the most important matters of culture particularly in sciences, America is the presence [*sic*] and Europe the past.

And Menger asked Veblen for help:

> So I feel, I have to speak at last, so difficult it be: If you think it of some advantage for an American university that, from now on, I devote all my forces to the culture of mathematics there, I shall be glad to do so.

In April 1934 a political system, sometimes called "Austrofascism," established itself in Austria. Here the authority of parliament was severely curtailed.[97] One year after his last letter to Veblen, on November 3, 1934, Menger had to report that even among the "pupils of mine," for whom he had vouched before, there were losses to be lamented. Obviously alluding to Nöbeling, who was a citizen of the German Reich and whose Rockefeller fellowship had not materialized, Menger wrote to Veblen:

> You certainly remember my assistant N., whom I considered as one of my best friends. In fall 1933 he had to go home due to the political tension. . . . There he had to join different organizations and had to do a lot of things till, finally, he came under the complete mental influence of this current and, since some months, does, quite spontaneously, more things than he even had to. Considering that such changes are possible <u>after</u> June 30th in <u>young</u> men,[98] who certainly may be ranked among the best young scientists of their nation (just recently N. succeeded in proving the triangulation of all manifolds!) my deep personal sorrow is still surpassed by fear as to the future.

Menger had to wait three more years before finding a position at Notre Dame near Chicago; but he came in January 1937,[99] one year before the Germans occupied Austria, making him technically speaking an "early" emigrant according to my definition. Due to the political pressure he was exposed to prior to emigration and the imminent threat of the occupation of Austria by the Nazis I decided to include him among the "forced" emigrants as well. It is from Menger's time in America that his next letter (now in the Veblen Papers, dated "Berkeley, summer 1938") expresses concern about other Austrians such as Franz Alt and Hans Thirring and their fates under German occupation.

---

[97]The assassination of chancellor Engelbert Dollfuß in July 1934 by Austrian National Socialists, and the erection of an authoritarian regime by Kurt Schuschnigg afterward were important dates in this context. See also in chapter 4 below A. Duschek's letter from November 1934 on academic anti-Semitism in Austria.

[98]Menger is here referring to the so-called Night of Long Knives, i.e., brutal slaughter of Hitler's previous "storm division" of the SA, some of them acting with socialist demagoguery, by Hitler and his now favored SS.

[99]Menger (1994), p. 215.

3.S.3. The Problems of Early Emigration as Exemplified by Hermann Weyl

The destiny of Hermann Weyl as an emigrant and his opinions on German American mathematical relations are of preeminent importance to the current publication. Weyl had, even during the time of his position at the ETH in Zurich, a variety of contacts within the German scientific culture, not just restricted to mathematics but also including physics and philosophy.[100] Weyl's return to Göttingen in 1930 as successor to David Hilbert had been the long-planned aim also of Hilbert himself. In 1918 Hilbert had even advised Weyl in nationalistic words, which otherwise were quite untypical of him, to stay in Zurich—which, at the time, offered better working conditions—in the interest of the "ideology of the German Reich" [im Interesse des Reichsdeutschtums].[101] Technically speaking, Weyl was no "early emigrant" to the United States in the definition of the present book, since he could not make up his mind before 1933 in spite of the tempting offer from Princeton. He only fled Germany in 1933 in view of the threat to his Jewish wife and his children. His hesitation in emigrating before 1933 is exactly what makes him a crucial case in gauging what constitutes the problems of early emigration. Weyl finally became a key figure at Princeton in organizing the flood of emigration in the 1930s. His uncontested mathematical brilliance made him an ideal person to handle the many delicate cases among the emigrants, not a few of whom had an inflated ego that could only be calmed down in view of Weyl's superior personality.

Back in the 1920s, in a 1929 letter to his colleague in Zurich, Michel Plancherel (1885–1967), Weyl gave his motives for his decision in declining, after long deliberation, the financially attractive offer by Princeton University:

> The deepest reasons, home country [Heimat], language, relatives, friends and something akin to loyalty to Europe weighed heavily in favor of my return, so be it got our own good.[102]

However, in the same letter to Plancherel, Weyl made it clear that for younger mathematicians not being in the favorable position of being wooed from all sides, the decision could be quite different. What Weyl was saying about Switzerland (in the epigraph above) was—as this present

---

[100]On Weyl see especially Sigurdsson (1991) and (1996).

[101]Frei and Stammbach (1992), p. 27 (T).

[102]Weyl to Plancherel, July 20, 1929 (T). See Frei and Stammbach (1992), p. 127. The epigraph in the present chapter is also taken from that letter. Also, in his letter of resignation to the Prussian Ministry of Culture in 1933, which was mentioned in chapter 1, Weyl emphasizes his closeness to the German language and culture. See Schappacher (1993).

Figure 10 *Hermann Weyl (1885–1955). A student of Hilbert's and universal mathematician (Lie groups, Riemann surfaces, mathematical physics), Weyl left Göttingen in 1933. He went to the Institute for Advanced Study in Princeton because he feared discrimination against his wife and children in Nazi Germany. In the United States he became one of the most active and influential immigrants, helping, with O. Veblen, H. Shapley (Harvard), and other Americans, to ease the problems of adaptation for the refugees.*

chapter illustrates—just as valid with regard to Germany, where, toward the end of the 1920s and in view of unemployment outside of the universities as well, a host of talented scientists lacked appointment. Apparently a similar situation also existed in Austria at that time.[103]

Nevertheless—and this makes Weyl's example typical—as late as four days before the Nazis came to power in Germany, the future spring tide

---

[103]Forcing many young Austrian mathematicians to take unwanted jobs in insurance companies or as schoolteachers.

of emigration and its consequences were not at all foreseeable. Richard Courant congratulated Weyl on January 26, 1933 on his decision *not* to go to the new Institute for Advanced Study:

> I firmly believe the two of you will not regret that decision. . . . For Göttingen this turn of events means a lot . . . particularly appropriate at a time when our loss would have implied an even more visible gain on the other side of the Atlantic. (CPP [T])

FOUR

Pretexts, Forms, and the Extent
of Emigration and Persecution

> These laws are being changed with more ease than for example
> a mathematician replaces one system of axioms by another.
> —G. Szegö 1934[1]

THIS BOOK gives priority to the topic of emigration of mathematicians, including the conditions of immigration in the host countries, especially the United States. The detailed picture of the circumstances in Germany and in German-occupied countries that led to emigration should rather be presented in a book on mathematics under National Socialism, since the behavior of the "unconcerned" colleagues would be of particular importance in such an investigation.[2] However, the two processes, expulsion from Germany and immigration in the host countries, cannot be neatly separated from each other, and therefore the fundamental conditions underlying those processes have to be discussed at least briefly. This necessity follows also from the fact that not all mathematicians expelled and persecuted by the Nazis succeeded with emigration, and that emigration itself can only be understood against the background of political oppression in its entirety.

There are detailed reports on the situation in those places most affected by the expulsions, particularly Berlin, Göttingen, Prague, Vienna, Hamburg, and Frankfurt.[3] Further accounts from the universities in Heidelberg and Munich, which were also strongly affected by the persecutions, give selected information but do not focus on the period of the Third Reich.[4] Publications on less affected places such as Aachen, Bonn, Freiburg,

---

[1]Szegö (Königsberg) on May 23, 1934, to the American Tamarkin (Szegö 1982, p. 4 [T]).

[2]For details see Segal (2003).

[3]See Brüning/Ferus/Siegmund-Schultze (1998), Knobloch (1998), Schappacher (1987), Siegel (1965), Maas (1991), and Einhorn (1985). On the expulsion of physicists and mathematicians from Göttingen see also Beyerchen (1977) and Sigurdsson (1996). Still unpublished is the preprint by Schwarz and Wolfart (1988). The most complete report on the conditions at the University of Prague is so far Pinl and Dick (1974).

[4]The fate of A. Rosenthal is well described in Mußgnug (1988). For Munich see Toepell (1996).

and Köln report on the fate of individual mathematicians.[5] Scattered biographical work on expelled and persecuted mathematicians complements the picture.[6] Until recently access to some of the archives of originally German universities (Wrocław-Breslau, Kaliningrad-Königsberg) has been difficult to obtain.[7] First reports on the total losses in personnel due to emigration are given by Schappacher and Kneser (1990) and Schaper (1992). The process of the political coordination (Gleichschaltung) of mathematics is well described by Mehrtens (1989) using the example of mathematical societies in Germany after 1933. A vivid description of the general atmosphere of terror and denunciation from the perspective of a contemporary mathematician in Münster can be found in the book by H. Behnke (1978), himself not persecuted. That atmosphere is also described in a report to the Rockefeller Foundation, compiled by Harald Bohr (1887–1951) and published as Appendix 3.1. The fate of one mathematician and Catholic political dissenter (P. Thullen) is revealed in his diaries for the period, reproduced in Appendix 6.

4.1. The Nazi Policy of Expulsion

The immediate and visible cause of the expulsions was of course the terror regime of National Socialism, with its firm ideological components, namely anti-Semitism, anti-Marxism, and anti-Communism. Contemporary observers and historians alike have often wondered why the Hitler regime could be so "shortsighted" as to deprive itself of its most capable scientists. The question cannot be answered other than in the total context of the regime's exertion of power. Preparing a nation for war requires "irrational" means and a strategy to bind individuals irrevocably to the system by entangling them in guilt and giving them small or big advantages over other persecuted persons. Even if pragmatic political motivations led sometimes to a curtailment of too blatant and self-destructive measures,[8]

---

[5]Brieskorn, ed. (1996) on F. Hausdorff, Remmert (1995) on A. Loewy, Butzer and Volkmann (2006) on Blumenthal. In Golczewsky (1988) the expulsion of H. Hamburger is described in detail.

[6]See the reference to biographical sources in the various lists of mathematicians given in Appendix 1.

[7]I am informed by Renate Tobies (Berlin) that access to the archives in Polish Wrocław is easy now. As to Russian Kaliningrad the situation seems unclear. I have chosen to make up for missing information by some documents in this chapter below (D), in particular an important letter by Szegö to Tamarkin.

[8]The ideological "Deutsche Mathematik" movement, spearheaded by L. Bieberbach, and the "Deutsche Physik" (Ph. Lenard) found less support by the regime in its later years. See Mehrtens (1989).

there was no way of casting doubt on the primacy of the dogma of anti-Semitism also in the international relations of the regime.[9] The coordination (Gleichschaltung) in the scientific-cultural domain also fulfilled a kind of ideological compensatory function in view of the fact that the Nazis did not attack the real foundations of German society, in particular the economic power structures—contrary to the socialist demagoguery of their program.[10] Moreover, one has to consider the relatively lesser—compared to today—importance of science, in particular mathematics and physics, to economics and warfare at that point in history. Finally, it has to be taken into account that the Nazis demagogically took advantage of certain structural problems in the German science system (overcrowding of universities, academic unemployment) that existed around 1930.

The form and concrete shape of the expulsions from the universities often corresponded with the seemingly chaotic and aimless methods of the Hitler regime. However, the latter methods aimed rather deliberately at destroying solidarity with the scapegoats, Jews and foreigners. They created subliminal guilt feelings among those who were privileged and not persecuted, while at the same time leaving nobody in perfect security. The pseudo-legalism of the new Nazi university laws, relying on the traditional submissiveness of the civil servants to the state, was complemented by an atmosphere of gratuitous accusation and denunciation, and by student boycotts,[11] thereby continuing the Nazification of the student body begun in the last years of the Republic of Weimar. The central pseudo-legal instrument used for the expulsions was the infamous Law for the Restoration of the Professional Civil Service of April 7, 1933 (henceforth called BBG, following the German short title, Berufsbeamtengesetz), which, together with the ordinance for its implementation from April 11, 1933, arbitrarily formed the notion of "non-Aryan descent."[12] The "law" did not show consideration for the religious confessions or political positions of the "non-Aryan" scientists, who had often converted to Christianity and even sometimes held German-nationalist positions in the First

[9]See Siegmund-Schultze (2002) for the effects of the Nazi rule on the international participation of German mathematicians.

[10]This is also astutely observed by the emigrant Hans Reichenbach, who in a letter to psychologist and early immigrant to the United States, Kurt Lewin (1890–1947), on February 23, 1933, wrote: "While the Nazis are impeded in the realization of their economic aims due to their alliance with Hugenberg [Alfred Hugenberg (1865–1951), leading German industrialist and right-wing politician], they will have their fling in the cultural domain" (translation of the German original as quoted in Hoffmann [1993], p. 396).

[11]See the Documents part of this chapter and Appendix 3.4 with the report by A. Rosenthal in Heidelberg.

[12]An English translation of the "Gesetz zur Wiederherstellung des Berufsbeamtentums" (BBG) and of its ordinance for implementation, which will be partly used here, can be found in Hentschel, ed. (1996), pp. 21–26.

World War as well as afterward in the Weimar Republic. The law stipulated in its central paragraphs 3, 4, and 6:

§ 3 (1) Civil servants who are not of Aryan descent are to be placed in retirement. . . .

(2) No. 1 does not apply to officials who had already been in the service since the 1st of August, 1914, or who had fought in the World War at the front for the German Reich or for its allies, or whose fathers or sons had been casualties in the World War. . . .

§ 4 Civil servants who, based on their previous political activities cannot guarantee that they will always unreservedly support the national state can be dismissed from service. Their previous salary will be maintained for the duration of three months following their dismissal. From this time on they shall receive three-fourths of the pension. . . .

§ 6 To simplify administration, civil servants may be placed in retirement, even when they are not yet unfit for service. If civil servants are retired for this reason, their posts may not be refilled.

In the *First Ordinance on the Implementation* of the BBG one reads in the *Reichs Law Gazette (Reichsgesetzblatt)* of April 11, 1933:

Re § 3 (1) Anyone descended from non-Aryan, and in particular Jewish, parents or grandparents, is considered non-Aryan. It is sufficient that one parent or one grandparent be non-Aryan. This is to be assumed especially when one parent or one grandparent had practiced the Jewish faith. . . .

On May 6, 1933, the BBG was extended in its range of application to Privatdozenten, who were not civil servants and usually received no salary except for student fees. The arbitrary "pseudo-legalism" of the Nazi laws is further underlined by such extensions and also by the fact that the authorities in many cases did not adhere to their own laws, but dismissed people in spite of the exemption clause in § 3, then reinstated them temporarily, and so forth. In the same vein, § 6 was often arbitrarily used to expel politically unwanted persons and/or to remove undesirable areas of research (D). By a cynical play with the different stipulations for pensions to which the dismissed were entitled according to §§ 3 and 4 of the BBG,[13] by arbitrariness in granting or refusing pensions for scientists who went abroad,[14]

---

[13]Since § 4 stipulated a reduction of the pension by one-fourth, scholars were sometimes dismissed according to this more severe paragraph by suitable construction of political "reasons," thus making dismissal according to the "Aryan" § 3 almost desirable for its victims. Otto Blumenthal in Aachen, for one, was dismissed in accordance with § 4 and was thus punished for his liberal views in the Republic of Weimar and for his contacts to Soviet science. See Butzer/Volkmann (2006).

[14]There was harassment in this respect against H. Hamburger in Köln (Golczewski 1988) and against Richard von Mises in Berlin (D). That the battle of the emigrants for their pensions

the Nazis reinforced anxious maneuvering and political lip service on the part of the scholars threatened with dismissal, which also led to weakened solidarity between them (D). The Nazi laws remained even partly effective after the war ended in 1945 and frequently served as a basis for decisions on compensation for the persecuted and dismissed.[15]

Students' boycotts of lectures complemented Nazi laws and were often, at least passively, tolerated by the Nazi authorities, and only seldom rebuked.[16] The student boycotts were often just as much motivated by racist as by political resentment, and—a fact that seems to have been typical for mathematics—were frequently directed against docents who were seen to be scholarly "too demanding" in the eyes of some students but who could not easily be dismissed according to Nazi laws. Such more politically and scientifically motivated boycotts are known at least in the cases of the mathematicians O. Blumenthal, H. Grötzsch, E. Landau, H. Liebmann, W. Prager, H. Reichenbach, K. Reidemeister, A. Rosenthal, and F. Willers.[17] The most widely known and infamous boycott was the one against number theorist Edmund Landau in Göttingen, who continued to practice his Jewish religion, but could not, as a prewar civil servant, be dismissed according to the BBG. By the concerted effort of the Nazi ministry and Nazi students, Landau was finally forced into "voluntary resignation." In the case of Landau there are also the most shocking documents attesting to the anti-Semitic blindness and fanaticism of students (O. Teichmüller) and the collaboration of colleagues (L. Bieberbach) with the Nazis (D).

After the promulgation of the BBG in April 1933, anti-Semitic legislation continuously tightened in the following years—something that only a few scientists realized at an early stage.[18] "Non-Aryan" students were exposed to ever more restrictive conditions for admissions and exams—beginning with the "Law against the over-crowding of German schools and universities" of April 25, 1933.[19] As a consequence of the Law on

---

was not principally hopeless is evidenced by, for example, the fact that Hellinger's mother in Germany received money from his pension as late as 1940, one year after his emigration. In any case the pensions had to be spent in Germany and could not be transferred to foreign accounts.

[15]See chapter 11.

[16]Typical was the university rector's reaction to a student boycott against Rosenthal and Liebmann in Heidelberg, as documented in Appendix 3.4.

[17]See in these cases the bibliographical information given in the lists of Appendix 1, and D.

[18]Among the more prescient ones was R. von Mises, who left Berlin in 1933, although temporarily protected by the exemption clause of the BBG.

[19]This Nazi law restricted the portion of "non-Aryan" students to 1.5 percent of the student body. The percentage of women was set to be 10 percent at most. See Jarausch (1984), p. 177.

Figure 11 *Oswald Veblen, Edmund Landau, and Harald Bohr. In happier times, probably in Göttingen around 1930, the American Oswald Veblen and the half-Jewish Dane Harald Bohr take into their midst the famous German-Jewish number theorist Edmund Landau. Later on, Veblen and Bohr did much to help refugees from Germany, Landau's lectures were boycotted by Nazi students in 1933, and he died in his place of birth Berlin.*

German Citizenship (Reichsbürgergesetz) in September 1935—part of the infamous Nuremberg laws—the exemption clauses of the BBG were canceled. This implied that prewar civil servants and participants in the war were no longer protected against dismissal. The somewhat more restricted Nazi definition of "non-Aryan," according to the Nuremberg laws, was of no help to scholars who had been dismissed in 1933 but who may not have been affected in 1935. From 1937 on, the few remaining Jewish students who were German nationals lost their right to obtain a doctor's degree.[20] From that same year on, the "racial descent" of scientists' spouses was being increasingly scrutinized by the Nazi authorities, and loyalty to their relatives cost many scholars their jobs and drove them to emigration.[21] At about the same time, mathematicians were also being increasingly dismissed from nongovernmental or semigovernmental organizations and institutions such as the Deutsche Mathematiker-Vereinigung and the Preussische Akademie der Wissenschaften, often through the active participation of nonthreatened members (D). The annexations of Austria in 1938 and Czechoslovakia in 1938–39 caused new waves of persecution and emigration of German-speaking mathematicians, mainly from Vienna and Prague.[22]

The extent of emigrations and persecutions is documented in the three lists of Appendix 1, which are, however, presumably not complete. The geographic distribution was very uneven. For the special population of persecuted mathematicians, as defined in the previous chapters, the emigrations/persecutions were spread over various places as seen in the following table.[23]

---

[20]See Jarausch (1984), p. 180. The Nuremberg laws of 1935 had introduced a distinction between "citizens" ("Reichsbürger"), which Jews were no longer considered to be, and the lesser "members" of the Reich ("Reichsangehörige").

[21]In the lists of Appendix 1, mathematicians who were persecuted due to the racial descent of their spouses will be marked with "PR" (Partner of racially persecuted). They were officially called to be in "Jewish clan" ("jüdisch versippt"). At that time at least the word "versippt" probably sounded less disparaging than it appears to us today against the background of the historical experience of National Socialism.

[22]The situation in Czechoslovakia immediately after the Munich Dictate (sometimes euphemistically called the Munich agreement) of September 29, 1938, is described by Max Pinl in a letter to H. Weyl, reproduced and translated as Appendix 3.5.

[23]The first figure denotes the number of emigrants, the number after the slash denotes the total number of those expelled and persecuted, including emigrants. Some differences compared to the total number of those persecuted result from double counting of certain persons or from uncertainty as to the place of expulsion/persecution. As is generally the case in this book, only German-speaking mathematicians are included, which is important to note for Amsterdam (persecution of Freudenthal), Trieste (Frucht), and Stockholm (threat to Müntz). Places outside Germany, where Germans were persecuted, are set in parentheses.

TABLE 1
Places of expulsion/persecution

Aachen	1/2	Göttingen	24/28	Munich	4/5
(Amsterdam)	0/1	Graz	0/1	Münster	1/1
Berlin	41/62	Greifswald	0/1	Prague	5/13
Bonn	1/3	Halle	1/2	Rostock	0/1
Brunswick	1/1	Hamburg	4/4	Saarbrücken	0/1
Breslau	8/11	Heidelberg	4/5	Schweidnitz	0/1
Cologne	1/2	Karlsruhe	2/4	(Stockholm)	0/1
Dresden	0/1	Kassel	0/2	(Trieste)	1/1
Elsterwerda	0/1	Kiel	2/4	Tübingen	0/1
Essen	0/1	Königsberg	7/8	Vacha	0/1
Frankfurt	9/14	Landsberg	0/1	Vienna	20/27
Freiberg	0/1	Leipzig	2/2	(Warsaw)	0/1
Freiburg	4/6	Mansfeld	0/1	Würzburg	0/2
Gießen	0/2	Marburg	1/4	(Zurich)	0/1

It becomes clear that ninety out of a total of 145 emigrants, and 130 out of a total of 234 persecuted (including nonemigrants and killed) came from four cities: Berlin, Göttingen, Prague, and Vienna.

4.2. The Political Position of Mathematicians, Affected and Unaffected by Persecution

Because about 90 percent of those persecuted fell under the arbitrary Nazi definition of "non-Aryan,"[24] one is sometimes tempted to ignore the fact that the persecutions were also directed against politically undesirable scholars, in particular those—a clear minority—who had actively supported the "system of Weimar," greatly loathed by the Nazis. In order to understand the reactions to the Nazi policies on the part of the scholars,

[24]There are specific sociological reasons (strong intellectual traditions and the fact that Jews were pushed away into "free professions" outside the civil service) that cannot be discussed here and accounted for the high percentage of Jews in the fundamental, nonideological sciences such as mathematics.

Figure 12 *Karl Löwner (later Charles Loewner, 1893–1968). The noted specialist in function theory was expelled from Prague in 1939, from where he had reported on the conditions of his colleagues in Germany (Appendix 3.2). After the war, he was a professor at Stanford University in California.*

both by those affected and by those nonaffected by persecutions, and also in order to understand the later behavior of the emigrants in the host countries, it is imperative to consider the political positions of mathematicians and their history. However, it should not be forgotten that even the persecution of the apolitical "non-Aryan" scientists was a political act and fulfilled political purposes.

The traditionally "apolitical" and state-loyal attitudes of German university professors as civil servants had been temporarily shattered by the First World War and by the ensuing hyperinflation that destroyed the fortunes of the middle classes. Many scientists laid the blame for defeat in the war and its consequences on the leading politicians of the Republic of Weimar, who were often represented as compliant with the policies of unconditional fulfillment of the reparations and disarmament clauses of the Treaty of Versailles. Leading scientists such as Max Planck had a stronger loyalty to the prewar monarchy and the idea of restoring Germany's greatness than to the Republic of Weimar, which has therefore been drastically described as a "republic without republicans." Scholars such as the mathematician Richard Courant, who had shown republican tendencies immediately after the war, were forced into political silence and adaptation to the predominant opinion among academics.[25] The latter defined themselves as "apolitical," which did not rule out their tolerance and indirect support of outspoken anti-Republican actions such as the ones uttered by the mathematician Theodor Vahlen (1869–1945), the future Nazi functionary and president of the politically "coordinated" Prussian Academy of Sciences.[26] The conservative majority never went as far as the early National Socialists Vahlen and Philipp Lenard in openly obstructing governmental measures. In this respect they were really "apolitical," but they were usually inept at resisting the following Nazi pressure as well. The anti-Republican feelings of the clear majority of the professors, which were also shared by many Jewish scholars (D), gave growth to an early Nazification of greater parts of the student body. All these—resentment, nationalism, apolitical aloofness—were reasons for the relatively "unproblematic" political coordination of the German universities by the Nazis in 1933 and for the incredulous horror of many of those dismissed over what was being done to them by a purportedly "national" government.

However, one has to qualify this general judgment in view of the behavior of individual persons and with respect to different scientific disciplines

---

[25]See Reid (1976). Conversely, persons who did not adapt to the majority opinion, such as H. Reichenbach and, above all, Emil Julius Gumbel, were persecuted by their own colleagues during the years of Weimar.

[26]On Vahlen see Siegmund-Schultze (1984).

and different places of action in Germany and Austria. Regarding the later persecution in Austria—in spite of the political turmoil and increasing anti-Semitism and anti-Republicanism there before 1938—one has to consider the very different political traditions in the two countries, for example the stronger influence of social-democratic positions among some academics in Austria, particularly in Vienna. Also, among the mathematicians and physicists in Göttingen during the 1920s there were hardly any decidedly German-nationalist and revanchist positions to be found. This was obviously above all due to the highly internationalized research atmosphere (unlike the situation among Göttingen humanists). Such an internationalized atmosphere did not exist to the same degree in the German capital. Berlin mathematicians such as Ludwig Bieberbach, Richard von Mises, and Erhard Schmidt (1876–1959) openly opposed the International Congress of Mathematicians in Bologna in 1928 because the organizers had not, in their opinion, given satisfactory guarantees for the participation of German mathematicians on an equal par.[27] In the two leading mathematical cities of Germany, Göttingen and Berlin, liberal and republican feelings among students—at least with students of mathematics and physics—were not untypical.[28] Forman, in his very well known paper, goes as far as claiming that the "Weimar culture" was dominated by the much more nationalist and revanchist humanists and social scientists and that the natural scientists felt forced to adapt to these feelings.[29] Even the more liberal Göttingen scientists had to adapt to certain norms of apolitical attitude and of abstention from open pacifist or antimilitarist action. Among the few prominent scientists and mathematicians in Germany who abstained from living by this norm were Albert Einstein and Emil Julius Gumbel, the statistician who published material about anti-Republican undercover organizations and their connections to the military. Other mathematicians who were open sympathizers of the Republic of Weimar included Hans Rademacher, Emmy and Fritz Noether, and Felix Bernstein. The Nazis dismissed all of them in 1933—except for Rademacher they did not meet the Nazi standard of acceptable "race" either. The Freiburg

[27]Dalen (2005), pp. 587ff.

[28]This applied for instance to the supporters of the scientifically oriented philosopher L. Nelson (1882–1927) in Göttingen, whose works were brought to a partial completion in England by emigrant Grete Hermann. See Schappacher (1987). The future emigrant Ludwig Boll belonged to a group of communist students in Göttingen. In Berlin many students were gathering around H. Reichenbach and his Berlin Society for Scientific Philosophy. See Danneberg et al. (1994). See also Erhard Schmidt in "Ansprachen 1951," pp. 19–21, in his response to H. Freudenthal. Schmidt is reporting on actions of the mathematics students, which prevented a Nazi student boycott against his lectures at the University of Berlin in 1929.

[29]Forman (1971). Forman even sees cognitive consequences of this adaptation in the theories of the physicists, leaning to an abandonment of strict causality.

mathematics teacher Wilhelm Hauser and the mathematics student Ludwig Boll, then in Frankfurt, were persecuted by the Nazis for both political and racial reasons.[30] Liberally oriented young mathematicians such as C.-G. Hempel, R. Lüneburg, M. Zorn, who were not affected by the racist laws,[31] found a further stay in Nazi Germany unbearable or were driven out for political reasons. There were further dismissals and "voluntary" resignations because of political nonconformity, not necessarily in the sense of leftist deviation (Baule, Heesch, Mohr, Mahlo, Naas, Neugebauer, Pinl, Rembs, Romberg, Thaer, Thullen, Zermelo).

The reaction of nonpersecuted colleagues to the dismissals was often influenced by the devilish anti-Semitism of the Nazi ideology that produced the reassuring and egotistic feeling of not being concerned, of belonging to a privileged "race." Additionally, the traditional animosities of the scholars that have been mentioned above often made them ignore the brutal methods of the Nazis' exertion of power; sometimes they even realized an identity of interests based on the so-called successes in foreign policy by the Hitler regime (D). Some career chances that arose through the dismissals influenced the behavior not only of eligible younger scholars but also of their teachers who were eager to help them. Although a clear majority of scientists objected to the interference of the scientific discussion caused by racist pseudo-theories such as "Deutsche Mathematik" and "Deutsche Physik," these theories played their political role in supplying "reasons" for the dismissals (D).

The main reaction to the persecution on the part of the unaffected scientists was the anxious concern to maintain—despite political turbulence—the scientific enterprise and its institutions at all costs and to come to a compromise with the regime if necessary. Against this backdrop it is not surprising that there was almost no openly articulated protest against the dismissals. Van der Waerden's obituary of his teacher Emmy Noether in 1935—in which he avoided any political commentary on the circumstances of her emigration—was the most one could expect in the way of public statements against the system from mathematicians.[32] A courageous stand, similar to the public resignation of the Göttingen experimental physicist

---

[30]Not quite accidentally—for political reasons—Boll and Hauser went to East Germany (later to be the DDR) after the war.

[31]Regarding Lüneburg, the information in the history of the Göttingen University is ambiguous (Becker et al. 1987). I assume, in accordance with Beyerchen (1977), p. 32, and based on Courant's correspondence CPP that Lüneburg was not affected by the racist laws.

[32]Van der Waerden (1935). It is well known that van der Waerden resisted the regime nonpublicly at the faculty in Leipzig on various occasions, but also, that he compromised with the Nazis on other occasions. See Soifer (2004/5).

James Franck (1882–1964), who published a letter of protest in a newspaper, is not known from mathematicians, who reacted rather ambiguously to Franck's decision (D). Mathematicians occasionally tried to save their cherished colleagues and teachers from the worst by writing letters to the Nazi ministries. A case that became rather well known—even if only after the war—was a petition by twenty-eight friends and students of Courant's of May 1933 to the ministry of cultural affairs, defending Courant against "rumors . . . about his political position."[33] Another petition by twelve students of Emmy Noether's, also from 1933, paradoxically stresses Noether's "notion of the essence of mathematics that is very much in accordance with Aryan thinking."[34] Both petitions obviously tried to appeal to existing political prejudices. Also the new director of the mathematical institute in Göttingen, Helmut Hasse (1898–1979), felt the need to defend abstract algebra politically, though with little success. The English mathematician, Harold Davenport (1907–1969), wrote to Mordell on January 14, 1934 after a Hasse lecture: "The Rektor and the Studentenführer [Nazi student-leader] attended Hasse's sample lecture in Göttingen the other week and were not convinced by his arguments that abstract algebra etc. is the perfect expression of the Third Reich."[35] The political taboo of anti-Semitism caused both petitions in favor of Courant and Noether to fail. Not surprisingly, a petition for Kurt Reidemeister, who had been dismissed from Königsberg, was more successful. After all, Reidemeister was "Aryan" according to the Nazis.[36] The letter in favor of Reidemeister, reproduced in facsimile below in the document part D, was initiated by W. Blaschke in Hamburg, who on other occasions came to compromises with the Nazis. Several of these compromises were outrageous, because they were unnecessary and even aggravated the situation (D).

In some instances, for example in Reidemeister's case (D), there were attempts to mobilize foreign mathematicians for the cause of persecuted colleagues. However, this was a two-edged sword that could lead to even more political suspicion against the threatened colleagues in view of the regime's increasing international isolation.[37] Some reports and many of the letters sent abroad contain information about the situation in Nazi Germany and about the dismissed mathematicians (Appendices 3.1 and 3.2). In view of threats from the terror regime, the reports and letters were partly written anonymously or by visitors to Germany such as Karl Löwner, or partly sent from outside Germany.

[33]Reprinted in *Exodus Professorum* (1989), pp. 22–24. The quote is from page 24 (T).
[34]Reprinted in *Exodus Professorum* (1989), pp. 26–27, p. 27 (undated 1933) (T).
[35]Mordell Papers, Cambridge, 4.38.
[36]Oswald Veblen, in his letter of support for Reidemeister, expressly alludes to the fact that this case did not touch any of the Nazis' political taboos (D).
[37]This ambiguity in its consequences for the DMV is stressed in Mehrtens (1989).

4.D. Documents

4.D.1. The Pseudo-Legalism of the Methods of Expulsion

The mathematician from Königsberg, Gabor Szegö, who was threatened with dismissal, discussed in a letter on May 23, 1934, to the American Russian mathematician, J. D. Tamarkin, the arbitrary Nazi practices:

> I point to the many who have been placed in retirement recently, often without previous proceedings (for example Rademacher, late February this year), further placements into retirement for simplifying administration, however with the hidden goal to remove unwanted personalities, who otherwise are protected by the civil servant law. . . . With mathematics in Königsberg things are as follows. Reidemeister has been moved, a successor for him is not yet there. That's why they need me temporarily, because I am alone in a responsible position. But as soon as a successor appears . . . I do not believe that I will stay for long, although as a participant in the war and front officer I am purportedly not affected by the law. Anyway it was assumed last summer that the law had a time limit, such as sooner or later legal certainty would be restored. Since then the law has been extended twice already, and there are no restraints to extend it ad inf. In addition, even if the law should be lifted, there are a thousand other possibilities to make life here impossible for one. These laws are being changed with more ease than for example a mathematician replaces one system of axioms by another.[38]

4.D.2. Student Boycotts as a Means of Expelling Unwanted Docents

The leader of the student boycott in Göttingen, the brilliant mathematician Oswald Teichmüller (1913–1943),[39] wrote in a "letter of explanation," dated November 3, 1933, to his victim and teacher Edmund Landau, who had asked for reasons of the boycott:

> You stated the opinion yesterday that it had been an anti-Semitic demonstration. I held and continue to hold the view that an individual anti-Jewish action should rather be directed against anyone else than you. It is not about making life difficult for you as a Jew, but only about preventing German students of the second term from being taught precisely in differential and integral calculus by a racially totally foreign teacher. I would not dare more than any other to question your ability to teach international mathematics to suitable students of arbitrary descent. . . . However, the chance of you being able to communicate the

---

[38]Szegö (1982), p. 4 (T). The introductory heading in this chapter is taken from this letter.
[39]See also the document in Appendix 3.4 on the expulsion of H. Liebmann and A. Rosenthal in Heidelberg. K. Hohenemser, in a letter to me from December 23, 1997, reported on a student boycott against the newly appointed W. Prager in Karlsruhe 1933.

essentials of mathematics to your listeners without your own national heritage being apparent is as unlikely as it is certain that a skeleton without flesh does not walk, but slumps rather and withers away.[40]

Rafael Artzy (born Deutschländer) on the reasons of the temporary dismissal of his teacher Kurt Reidemeister in Königsberg:

> In the years just before the Nazi takeover, the University students organized riots. . . . Reidemeister was furious, and in one of his classes he "proved," in his naive way, that the behavior of the students had not been logical. The result was his dismissal right after Hitler took over.[41]

4.D.3. The Racist "German Mathematics" (Deutsche Mathematik) of Ludwig Bieberbach as an Ideology Supportive of the Expulsions

Ludwig Bieberbach, full professor of mathematics at the University of Berlin, in April 1934 on the student boycott in Göttingen against Edmund Landau:

> A few months ago differences with the Göttingen student body put an end to the teaching activities of Herr Landau. . . . This should be seen as a prime example of the fact that representatives of overly different races do not mix as students and teachers. . . . The instinct of the Göttingen students was that Landau was a type who handled things in an un-German manner.[42]

4.D.4. Personal Denunciations as Instruments of Expulsion

The escape of Hanna von Caemmerer, who was not Jewish, from Berlin after her 1936 state exam, was commented thus in 1980:

> When she was (rightly) suspected of being friendly to Jews, one of her professors at the University of Berlin made it impossible for her to continue there.[43]

The denunciation of the differential geometer and high school (Gymnasium) teacher Eduard Rembs is contained in a letter from February 13, 1936, written by the dean of the Philosophical Faculty at Berlin, Ludwig Bieberbach:

> My faculty received an application by Dr. Eduard Rembs for a habilitation in mathematics. He . . . writes in the questionnaire that he was a member of the

---

[40]Schappacher and Scholz, eds. (1992), pp. 29–30 (T).

[41]Artzy (1994), p. 2.

[42]Bieberbach (1934), p. 236 (T).

[43]*IBD* microfilm. The information is probably from B. H. Neumann, whom Caemmerer married after their joint immigration to England in 1938. Between 1936 and 1938 Caemmerer continued her studies in Göttingen.

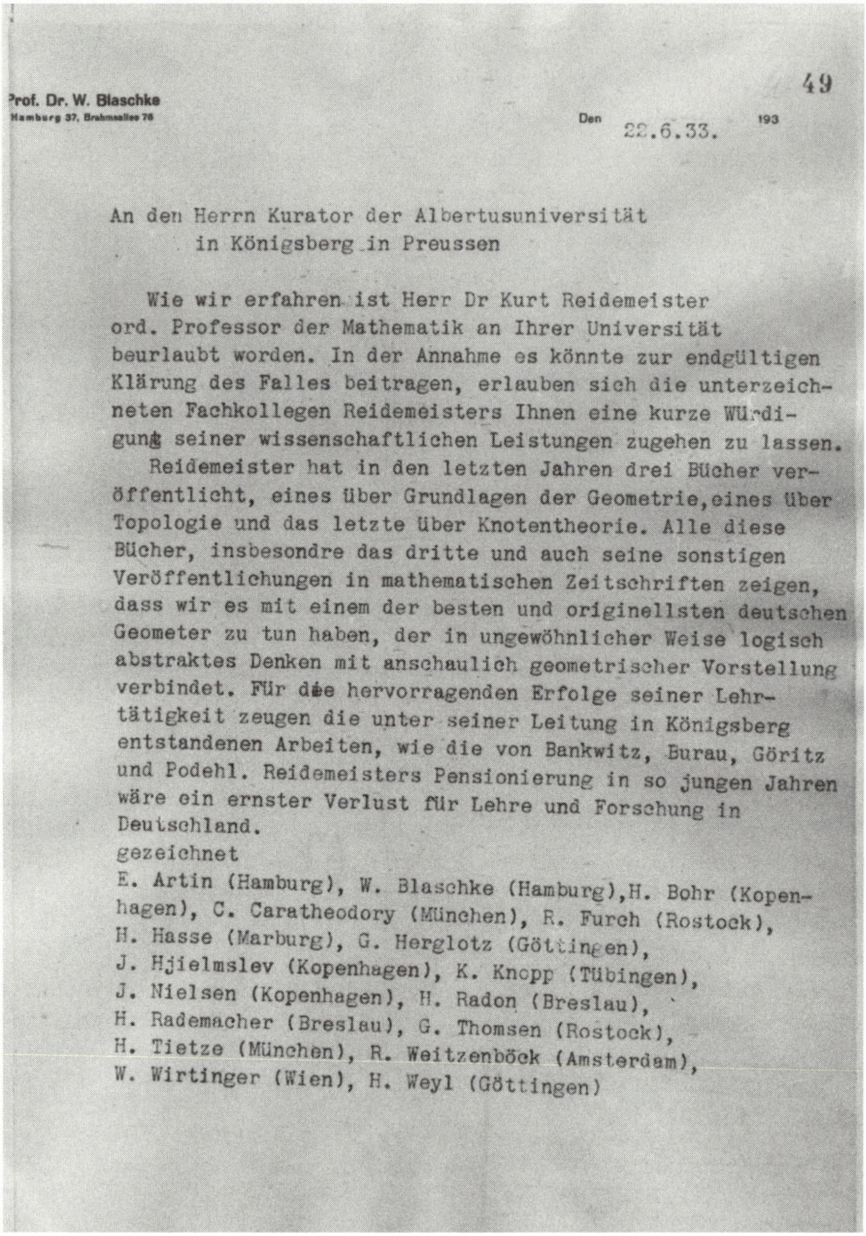

Prof. Dr. W. Blaschke
Hamburg 37, Brahmsallee 76

49

Den 22.6.33. 193

An den Herrn Kurator der Albertusuniversität
in Königsberg in Preussen

Wie wir erfahren ist Herr Dr Kurt Reidemeister
ord. Professor der Mathematik an Ihrer Universität
beurlaubt worden. In der Annahme es könnte zur endgültigen
Klärung des Falles beitragen, erlauben sich die unterzeich-
neten Fachkollegen Reidemeisters Ihnen eine kurze Würdi-
gung seiner wissenschaftlichen Leistungen zugehen zu lassen.
Reidemeister hat in den letzten Jahren drei Bücher ver-
öffentlicht, eines über Grundlagen der Geometrie,eines über
Topologie und das letzte über Knotentheorie. Alle diese
Bücher, insbesondre das dritte und auch seine sonstigen
Veröffentlichungen in mathematischen Zeitschriften zeigen,
dass wir es mit einem der besten und originellsten deutschen
Geometer zu tun haben, der in ungewöhnlicher Weise logisch
abstraktes Denken mit anschaulich geometrischer Vorstellung
verbindet. Für die hervorragenden Erfolge seiner Lehr-
tätigkeit zeugen die unter seiner Leitung in Königsberg
entstandenen Arbeiten, wie die von Bankwitz, Burau, Göritz
und Podehl. Reidemeisters Pensionierung in so jungen Jahren
wäre ein ernster Verlust für Lehre und Forschung in
Deutschland.
gezeichnet
E. Artin (Hamburg), W. Blaschke (Hamburg),H. Bohr (Kopen-
hagen), C. Caratheodory (München), R. Furch (Rostock),
H. Hasse (Marburg), G. Herglotz (Göttingen),
J. Hjielmslev (Kopenhagen), K. Knopp (Tübingen),
J. Nielsen (Kopenhagen), H. Radon (Breslau),
H. Rademacher (Breslau), G. Thomsen (Rostock),
H. Tietze (München), R. Weitzenböck (Amsterdam),
W. Wirtinger (Wien), H. Weyl (Göttingen)

Figure 13 *Petition by Wilhelm Blaschke. A petition organized by the Hamburg mathematician Wilhelm Blaschke (1885–1962) dated June 22, 1933, directed against the dismissal of Kurt Reidemeister in Königsberg.*

Social Democratic Party from 1919 until early 1933, furthermore of the study group of Social Democratic teachers from about 1926 until early 1933, of the German Peace Society [deutsche Friedensgesellschaft] for about one year (1930?), then of the German Peace Association [deutscher Friedensbund] until early 1933. In his curriculum vitae he reveals that he is still a senior teacher [Studienrat] at the Kantgymnasium in Spandau [part of Berlin]. I find this striking [auffällig] given his political history. I therefore ask to look into why Dr. Rembs is still in office as a senior teacher.[44]

Needless to say that Rembs's application for a habilitation came to nothing and that—quite the contrary—he was dismissed from his position as high school teacher.[45]

A similarly unbelievable and merciless act was the denunciation by Bieberbach that led to the dismissal of Issai Schur from the academic commissions of the Prussian Academy of Sciences.[46]

In March and early April 1938 mathematicians and physicists of the Academy who belonged to the academic commission for the edition of Karl Weierstrass's works signed a circular, beginning with the signatures by Erhard Schmidt and Issai Schur, who both wrote: "read" [gesehen]. The following signatures were [see facsimile below]:

29 March, Bieberbach: "I find it surprising that Jews are still members of academic committees"
30 March, Th.Vahlen: "I propose modification"
3 April, M. Planck, who was Secretary of the Academy: "I will take care of it."

In the respective file of the Academy, Schur's resignation from the academic commissions follows immediately. Half a year later Schur had to resign from the Academy altogether. In 1928 Bieberbach and Schur had, all the same, published a well-known joint article in the Proceedings of the Academy.[47]

---

[44]Partial estate L. Bieberbach, Oberaudorf (Germany) (T).

[45]The denunciation had consequences for the relationship between Berlin mathematicians after the war, especially within the Berlin Mathematical Society. Rembs left the BMS in 1953, after Bieberbach had been accepted as a member. See Knobloch (1998), p. 51.

[46]The incidence was first mentioned in Quaisser (1984), p. 38. See the facsimile below.

[47]"Über die Minkowskische Reduktionstheorie der positiven quadratischen Formen," *Sitzungsberichte der Preussischen Akademie der Wissenschaften* 1928, Physikalisch-mathematische Klasse, pp. 510–35.

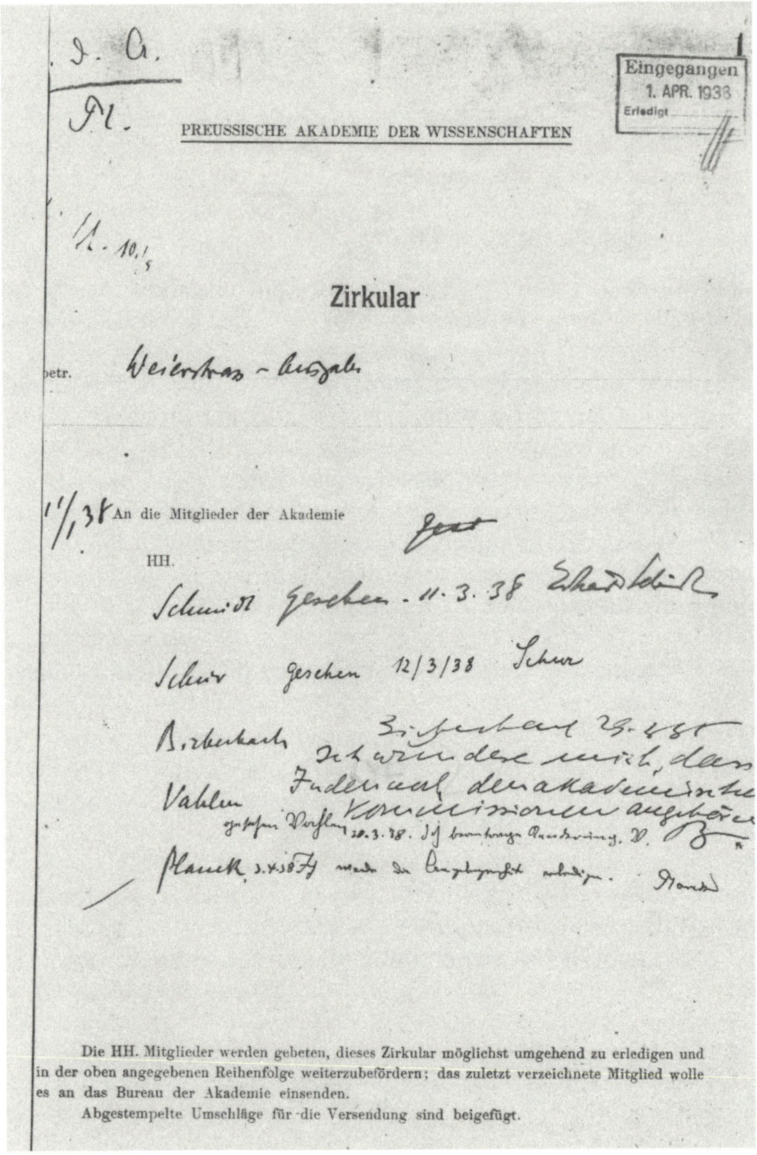

Eingegangen
1. APR. 1938
Erledigt

PREUSSISCHE AKADEMIE DER WISSENSCHAFTEN

Zirkular

betr. Weierstrass-Ausgabe

An die Mitglieder der Akademie

HH.

Schmidt gesehen 11. 3. 38

Schur gesehen 12/3/38 Schur

Bieberbach

Vahlen

Planck

Die HH. Mitglieder werden gebeten, dieses Zirkular möglichst umgehend zu erledigen und in der oben angegebenen Reihenfolge weiterzubefördern; das zuletzt verzeichnete Mitglied wolle es an das Bureau der Akademie einsenden.

Abgestempelte Umschläge für die Versendung sind beigefügt.

Figure 14 *Circular of the Weierstrass-Commission. Circular of the Weierstrass-Commission of the Prussian Academy of Sciences in Berlin March/April 1938, which shows the roles of Bieberbach, Vahlen, and Planck in sacking Issai Schur from the academic commissions.*

4.D.5. Political Reasons for Emigration beyond Anti-Semitism

Some emigrants, among them C.-G. Hempel, O. Neugebauer, W. Romberg, H. Schwerdtfeger, P. Thullen, and C.-L. Siegel, left Germany without immediate threat to themselves or to their relatives.

Werner Romberg, trained by the mathematical physicist Arnold Sommerfeld (1868–1951) in Munich and who, after his immigration to Norway, became known as a numerical analyst, wrote the following to me in 1998:

> I was close to the SAP [Socialist Workers Party] as it supported the joint fight of SPD and KPD against the Nazis. We were about 10–20 students and therefore known to the Nazis. In 1932 Sommerfeld formulated a prize competition for the University of Munich and suggested I should participate. I submitted the solution and received the following response: "The assignment was completely solved by the sender. However, the sender lacking the necessary maturity of mind [geistige Reife], the prize cannot be awarded." Sommerfeld suggested I submit it as a PhD and urged me to hurry. Accordingly I was able to pass the examination with magna cum laude in the summer of 1933.
>
> Sommerfeld had heard about requests for theoretical physicists from the USSR. By way of curing me of my leftist illusions, he recommended me.[48]

The specialist in function theory, Peter Thullen, did not want to return to Germany from Italy in 1934, even though he had offers of work. Immigration to Austria, which he considered for a moment, was out of the question for Thullen, although he was not Jewish. The differential geometer Adalbert Duschek then Privatdozent at the Technical University in Vienna, who—after the German annexation of Austria in March 1938—would himself be dismissed both for political reasons and due to his Jewish wife, wrote in November 1934:

> Quite confidentially, and in order to spare Mr. Thullen a disappointment, I want to remark that he has no prospects at all here, if he happens to be a Jew. To be sure they are not yet as rigid in this point as in the German Reich, but a certificate of baptism (not a recent one) and corresponding looks are also here a prerequisite.[49]

In New York City, Richard Courant wrote after the war:

> Thullen was a very active member of the German Catholic Youth Movement and from the outset a bitter foe of the Nazis. Although the German authorities

---

[48]W. Romberg to R. Siegmund-Schultze, October 1, 1998 (T).

[49]A. Duschek to "Herr Professor" (probably H. Behnke in Münster), November 11, 1934 (T), from Thullen's estate in the possession of his son Georg Thullen Genthod (near Geneva). On Duschek see Einhorn (1985), vol. 2, pp. 403–11, and OVP, cont. 30, f. Duschek, Adalbert, 1938–39.

Figure 15 (a, b) *Peter Thullen (1907–1996). The talented function theorist and Catholic dissenter did not return to Germany from a research year in Italy, but immigrated instead to Ecuador. The images show the front and back of a legitimation for Thullen from the Italian Ministry of Education in the year XI of Fascist rule (1935).*

built golden bridges for him he decided to leave Germany as soon as the Nazis took charge and went to Quito, Ecuador.[50]

The widow of Hans Schwerdtfeger mentions as the reason for his emigration "the clear conviction that Nazism would lead into disaster."[51] In 1937, Max Born, then in Edinburgh, wrote in a letter to Albert Einstein on Schwerdtfeger:

> Dr. Hans Schwerdtfeger: young mathematician from Göttingen. Lone wolf, earned his university education by doing factory and similar work. Pure "Aryan." Was not popular with Weyl and Courant, as he used to go his own way. I believe him to be talented, but lacking in self-criticism; his enthusiasm has up to now been greater than his achievement. Herglotz had a good opinion of him, but he does nothing for his people. . . . Schwerdtfeger was a violent opponent of the Nazis right from the beginning, and has therefore no chance of a position in Germany in spite of his "spotless" ancestry. It is people such as this we should help.[52]

The mathematics student Ludwig Boll, who had already received a topic for his dissertation from Hellinger in Frankfurt, was arrested due to his membership of the Communist Party on April 6, 1933 and interned for five weeks in the concentration camp Osthofen[53] near Worms. He succeeded in fleeing to the Netherlands in 1934, only to be interned again by the German occupiers in 1943 in the concentration camp Westerbork, from which he escaped deportation to Auschwitz again. He survived in Amsterdam, similar to the early emigrant Freudenthal.[54]

4.D.6. Cheating Emigrants out of Their Pensions

The Berlin mathematician Richard von Mises wrote to the ministry on December 21, 1933:

> In my application dated October 12, I requested acceptance of my resignation from the civil service according to the appropriate legal regulations. After twenty-four years of service I fail to see any reason for an explicit renouncement

[50]Courant's letter of recommendation to the Catholic University of America in Washington, DC, dated June 20, 1945. CIP New York, file: P. Thullen, 1944–48. Thullen's diary on his experiences in Germany immediately after Hitler's seizure of power is quoted below in Appendix 6. See also Thullen (2000) for the original German version of the diary, and Siegmund-Schultze (2000) for further commentary on it.

[51]Hanna Schwerdtfeger to the author, undated, received July 21, 1993.

[52]Born, ed. (2005), p. 124. Letter to Einstein, January 4, 1937. On Einstein's response to Born's letter see chapter 6.

[53]The camp is described by Anna Seghers as "Westhofen" in her famous novel *The Seventh Cross* (1942), made into a film with Spencer Tracy.

[54]Information based on an interview I had with Boll, August 29, 1983, and on Arnold (1986).

on my part of claims to which I am entitled according to the law. I request a decision as soon as possible enabling me finally to accept the position offered to me in Turkey.[55]

Von Mises's application for a pension was turned down, although Theodor Vahlen, the old Nazi and mathematician in the ministry, had temporarily given hope to von Mises.[56]

4.D.7. Increasing Restrictions Imposed upon "Non-Aryan" Students

Even if in individual cases "non-Aryans" could go on to take their PhDs until 1937, the restrictions in admissions and other harassment induced many students to emigrate immediately after 1933. Otto Blumenthal in Aachen wrote on November 18, 1933 to his former colleague Theodor von Kármán in California about the diminishing chances for his daughter:

> Of my children Margarete continues to study in Köln, where she has made notable progress with her [Anglicist] dissertation. . . . She is in a hurry to complete it as nobody knows if and when non-Aryans will be barred from a PhD.[57]

Rafael Artzy's untimely departure from Königsberg without completing his PhD was also influenced by the transfer for disciplinary reasons of his teacher Kurt Reidemeister, which was mentioned above:

> During my sixth semester [third year], namely in 1932, Reidemeister informed me it was a good idea to give me a topic for a doctoral dissertation because "nobody knows what could happen to the Jews." Thus I began to work on the topic (on Gewebe [topological notion]). Then, immediately after Hitler's seizure of power, Reidemeister was dismissed. Since I had been very active in the Zionist movement a while back, I had decided to go to Palestine as soon as possible anyway; I also had a good knowledge of Hebrew.[58]

4.D.8. Political Position of Emigrants before 1933: German Nationalism, Illusions, and General Lack of Prescience

Alfred Barneck, who was dismissed from the Technical University in Berlin-Charlottenburg in 1933 due to his Jewish descent, had written in an obituary of his teacher Jahnke in 1922:

[55]GSA, Rep. 76 Va, Sekt. 2, Tit. IV, Nr. 68c, fol. 349 (T).

[56]On the consequences of this decision for von Mises after the war see chapter 11. On the role of Vahlen cf. Siegmund-Schultze (1984).

[57]Kármán Papers, Caltech, Pasadena, 3.10 (T).

[58]Letter by Artzy to me, January 11, 1998 (T). A similar case was Dov Tamari, then Berhard Teitler, who had to leave Frankfurt in 1933 before finishing his PhD with C. L. Siegel and left for Palestine because he had been active for Zionism before (Tamari 2007).

Figure 16 *Rafael Artzy (1912–2006). Rafael Artzy (then Rafael Deutschländer) had been active in the Zionist movement before 1933 and could not finish
his PhD with Kurt Reidemeister in Königsberg due to the latter's dismissal in
1933. The geometer Artzy went to Palestine, was temporarily in the United
States, and then lived in Haifa (Israel).*

Eugen Jahnke, a marvelous, genuine through and through German patriot [kerndeutscher Mensch] has left us. . . . He felt the plight of our country deeply and tried to alleviate the problems where he could.[59]

The Jewish emigrant from Göttingen Kurt Mahler remembers in 1971:

Needless to say that I was at this time [1923] and long into the 1930s still a very patriotic German![60]

The director of the Institute for Applied Mathematics and future emigrant Richard von Mises said in 1930 in his address before the University of Berlin:

We remember with deep reverence the immeasurable procession of the dead [Zug von Toten], of those who fought in battle with us but did not return, who in braveness, in unshakeable discipline and loving enthusiasm helped to drive away the horrors of the war from the Rhineland, but who did not succeed in sparing it the heavy rigors of occupation by the enemy after the war. We remember in sorrow the lost and not yet liberated country, which even now we cannot enter.[61]

After a short stay in Germany, the early and temporary emigrant Eberhard Hopf reported in America about the political situation in Germany in 1932:

We are amazed how many Germans voted for Hitler. . . . Most of the people who voted for Hitler are dissatisfied with the general and their own situation. They follow anybody who promises them impossible things.[62]

Hermann Weyl, 1932, to Einstein on a possible appointment at the Institute for Advanced Study in Princeton:

The political conditions in Germany are becoming increasingly unpleasant (I should be in prison according to the National Socialists because of "defilement of the race" [Rassenschande]).[63]

On January 6, 1932, Weyl deplored in a letter to Oswald Veblen that he had gone from Zurich to Göttingen as Hilbert's successor and not straight to Princeton:

[59]Barneck (1922), p. 39 (T).
[60]Poorten (1991), p. 368.
[61]Mises (1930), p. 885 (T).
[62]Hopf to Tamarkin, May 1, 1932. Tamarkin Papers, BUA, box correspondence (A–H), f. E. Hopf.
[63]Einstein Papers, Jerusalem, 24098-1/2, June 22, 1932 (T). Weyl's wife Helene was "non-Aryan" according to the Nazis.

For entering into an Aryan-Jewish marriage the National Socialists plan 15 years severe prison.[64]

Edmund Landau in 1932 after a councilor in the Prussian Ministry of the Interior had intimated to him that the Nazis planned a concentration camp in nearby Lüneburg Heath:

> In that case I had better reserve a room with a balcony, south view, as fast as possible.[65]

The applied mathematician of Darmstadt, Alwin Walther (1898–1967), began a letter to Courant who was about to leave for a trip to America in March 1932 with words that were obviously ironically paraphrasing Nazi slogans:

> Heil and Victory for America! [Heil und Sieg für Amerika][66]

In retrospect, the emigrant Wolfgang Wasow saw in 1986 an amazing lack of prescience on the part of the Austrian Jews shortly before the Nazi occupation [Anschluss] in 1938:

> It was then—and still is now—a mystery to me that most Austrian Jews were just as unprepared for what happened as the Jews in Germany had been five years earlier. Looking at the events in Germany, they should have taken as many of their possessions as possible abroad, while there still was time. Very few had done that. To get out with at least some of your money and to find a country that would let you in was much harder in 1938 than in 1933.[67]

4.D.9. First Reactions by the Victims: Readiness to Compromise and to Justify, Adoption of the Martyr's Role

In a letter to H. Kneser after his dismissal in 1933, Richard Courant defended his short engagement for the Republic of Weimar in the early 1920s: according to the letter it was Felix Klein who had encouraged him to join the Social Democratic Party, his membership was in the interest of the University of Göttingen and was meant to serve as a bulwark against Communism.[68]

---

[64]OVP, cont. 15A, file: Weyl. As the letter is written in the beginning of January and its prescience seems surprising, it cannot be ruled out that it was actually written in January 1933.

[65]According to Kluge (1983), p. 94 (T).

[66]Walther to Courant, March 14, 1932, CPP (T).

[67]Wasow (1986), p. 192. Among those who were prescient enough and saved their money from Austria was Richard von Mises, but he had the firsthand experience of Berlin in 1933.

[68]Courant to H. Kneser, April 28, 1933, CPP. A similarly apologetic passage from this letter is quoted in Beyerchen (1977), p. 22.

Figure 17 *Hans Rademacher (1892–1966). Rademacher was one of the few German professors before 1933 holding liberal, partly left-leaning political views. He was dismissed by the Nazis for that reason and went to the United States, where he brought "dormant number theory" back to life (Weyl). Until his retirement he was at the University of Pennsylvania.*

Courant to his friend James Franck, the Göttingen physicist, on March 30, 1933, reacting to Einstein's criticism of the Hitler regime from abroad:

> Even if Einstein does not regard himself as German, he has experienced a lot of good in Germany. So he should feel obligated to make amends for the trouble he has caused as far as he can. (CPP [T])

Hans Rademacher (Breslau) and Kurt Reidemeister (Königsberg) felt political pressure to justify themselves in order to avoid dismissal. Rademacher to the ministry of culture and science in Berlin, on December 17, 1933:

> Concerning information supplied in the questionnaire, I take the liberty of adding as an explanation that my membership in the League for Human Rights [Liga für Menschenrechte] was restricted to paying the membership fees. I never took part in meetings of that organization and did not pay much attention to it anyway. . . . Ever since my habilitation as Privatdozent in mathematics at the University of Berlin, I have devoted my energies exclusively to scientific research and academic teaching. My international relations are of a purely scientific nature stemming from the fact that mathematics transgresses the borders of language. As a Prussian civil servant I take great pains in fulfilling my duties to the people and to the state as conscientiously as possible.[69]

Reidemeister wrote to the Nazi ministry, on May 13, 1933:

> Above all I declare that I was never a member of a <u>political</u> party and that I was never (according to my notion of political activism) politically active. . . . I disapproved of the propagandistic advocacy of logistic philosophy and demonstrated this by preventing the formal participation of the Verein Ernst Mach in the meeting for exact philosophy in Königsberg. . . . Due to the introduction of practical exercises the wheat got separated from the chaff even more visibly, and some students who were to my knowledge barely average in their mathematical talent constituted a dissatisfied group. . . . When the negative position of the student body was openly expressed even among the mathematics students I resolutely retired from my professional guild lead [berufsständische Führerrolle].[70]

On August 30, 1935, Otto Toeplitz wrote in a letter to Courant in New York, writing from Arosa (Switzerland):

> It is my opinion that we have to hold out in the positions that they are still granting us until the last moment. Not because there is any improvement in sight—quite impossible—but because otherwise we will become, in one way or another, a burden for the whole of Jewry and deprive at least somebody of a position. I consider it a sacrifice to Jewry to hold out in this position. . . . I wanted to explain to you the basic principle . . . I could not have done it from Bonn—at the moment every letter from Bonn is opened under the pretext of "valuta problems." (CPP [T])

---

[69]GSA, Rep. 76 Va, Sekt. 4, Tit. IV, No. 5 1, fol. 3 96 (T).
[70]Szegö Papers, Stanford, SC 323, box 9, f. 15 (copy, 9 pages [T]).

The reactions of colleagues to James Franck's public retirement from the University of Göttingen were very different:[71] Hans Lewy to Franck (without date):

There are still men!

The leading Göttingen aerodynamicist, Ludwig Prandtl, to Franck (April 19, 1933):

With greatest consternation I read today in the newspaper that you relinquish your professorship! This must not be your last word.

Publisher Ferdinand Springer to Franck (April 19, 1933):

Your letter will not fail to make an impression everywhere where there still is the capability to see things as they are.

4.D.10. The Partial Identity of Interests between the Regime and the "Unaffected" German Mathematicians

Emigrant Menahem Max Schiffer related at second hand the following discussion between dismissed Issai Schur and unaffected Erhard Schmidt in Berlin in the year 1938:

When he complained bitterly to Schmidt about the Nazi actions and Hitler, Schmidt defended the latter. He said, suppose we had to fight a war to rearm Germany, unite with Austria, liberate the Saar and the German part of Czechoslovakia. Such a war would have cost us half a million young men. . . . Now Hitler has sacrificed half a million Jews and has achieved great things for Germany. I hope some day you will be recompensed but I am still grateful to Hitler.[72]

There are reported similar, if less drastic nationalistic remarks by Schmidt before and after 1933, and above all a certain lack of courage to stand up for colleagues who were threatened.[73] Yet Schmidt was known among colleagues as a critic of Bieberbach's racist "Deutsche Mathematik" and as an opponent of anti-Semitism. One Nazi activist, Werner Weber, wrote in a secret police report on Schmidt in 1938:

I think that Schmidt shows little or no understanding of the Jewish question.[74]

---

[71]The following from the James Franck Papers, Chicago, Joseph Regenstein Library, box 7, f. 7 (T). Beyerchen (1977), p. 22, quotes from Courant's letter to H. Kneser, April 28, 1933, where Courant rejects the idea that he had supported Franck's action of "voluntary" retirement.

[72]Schiffer (1986/98), p. 180. Schiffer's quotation has to be judged very cautiously due to the great distance in time and the indirect report.

[73]For instance, experienced by H. Grunsky. See Siegmund-Schultze (2004a).

[74]University Archives Berlin, UAB, NS-Dozentenschaft, no. 222 (E. Schmidt), folio 9 (T).

There were even colleagues such as Karl Löwner who gained the impression, at least in 1933, that Schmidt tried to avert Schur's dismissal.[75]

In any case, the partial identity of interests between the regime and some "unaffected" German mathematicians was much more pronounced and publicly formulated by the geometer of Hamburg Wilhelm Blaschke (1885–1962). He had been born in Austrian Graz and welcomed the annexation of his country in 1938 (henceforth called "Eastern marches" [Ostmark]), as the fulfillment of a "dream from my younger years."[76] In a review of a volume of the American Mathematical Society devoted to the fiftieth anniversary of the Society, Blaschke criticized that the publication "is shamefully silent about the national [völkisch] origin of the representatives of American scholarship." On the same page Blaschke wrote, maybe not without feeling of envy and of anger about the flight of many emigrants from Germany:

> The most surprising thing is the mathematical large-scale enterprise [mathematische Großunternehmen] in the little Negro village of Princeton, where almost one hundred mathematical docents, with no students to speak of, are laying their golden eggs.[77]

It is against the backdrop of such utterances from Blaschke, which have been concealed or minimized by some mathematicians for a long time,[78] or compared with the even more extreme statements by Bieberbach (see above), that Erhard Schmidt's position appears relatively objective and exemplary. The early emigrant Hans Freudenthal, who survived the Nazi occupation of the Netherlands by remaining in hiding, said on the occasion of the celebration of Schmidt's seventy-fifth birthday in 1951 in Berlin:

> It is very easy to exert the honesty that mathematics demands <u>within mathematics</u>, because if one fails to do so it backfires very soon and bitterly. It is much more difficult to remain true, also among people and friends, to that characteristic to

---

[75]See Löwner's letter to Silverman in Appendix 3.2.

[76]Blaschke: *Geometrie der Gewebe*, Berlin 1938, p. vi (T).

[77]"Negro" was of course a code word for "Jew" here. *Jahresbericht Deutsche Mathematikervereinigung* 49 (1939): p. 81 (T). The AMS volume under review was volume 2 of *American Mathematical Society Semicentennial Publications in Two Volumes*, New York: AMS 1938.

[78]In a historical paper on the mathematical institute of the University of Hamburg from 1991, one finds, with direct reference to Blaschke's racist quotation, the following rather ambiguous and trivializing remark: "One can certainly not conclude from that quotation that he made propaganda for National Socialism. The relevant question here is not what Blaschke wanted to reach for National Socialism, but what he wanted to reach for himself (and for people around him and for his science) by his attitude to National Socialism" (Maas 1991, p. 1094 [T]).

Figure 18 *Kurt Reidemeister (1893–1971). The versatile researcher in the foundations of geometry, topology, and in number theory was transferred for disciplinary (political) reasons from Königsberg to Marburg, after conflicts with National Socialist students.*

which one was trained in numbers and figures. That we, on the outside, to whom Germany was closed and hostile, are aware of that and that we never had doubts about you, is demonstrated by the huge number of contributions which have reached the editor of the Festschrift from abroad.[79]

4.D.11. Reactions to the Expulsions from Abroad[80]

Oswald Veblen (IAS Princeton) wrote to the German ambassador in Washington on June 11, 1933, intervening for Kurt Reidemeister:

[79]*Ansprachen* 1951, p. 18 (T).

[80]Concerning this point there is a huge amount of documents: declarations of termination of membership in the DMV, etc. This discussion, however, lies beyond the scope of a book that is primarily oriented toward the process of expulsion itself. See also Appendix 4.2.

Dear Sir:

It has been suggested to me that it might be worth while to intercede with you on behalf of Professor Dr. K. Reidemeister who has recently been "beurlaubt" from his chair of mathematics at Koenigsberg although he is neither a Jew nor a member of any of the parties of the left. Under these circumstances it might be possible to secure a revision of his case without raising any question of general principle.

May I therefore say that Professor Reidemeister has written books and articles on pure mathematics which are well known to the mathematicians of America, and that he is regarded as one of the important mathematicians of Germany. We in Princeton are especially interested in him because of the close relationship between his work and that of our colleague, Professor Alexander.

Since you, of course, do not know who I am, may I say that I have many ties of friendship with Germany and that I have taken great pride in the marks of esteem I have received from German colleagues, the latest being an honorary doctorate conferred only a few months ago by Hamburg University? With this background I venture to suggest that German Science can ill afford to lose the services of men like Reidemeister after having been so severely injured as it has been by the expulsion of so many brilliant and valuable Jews.

Yours sincerely Oswald Veblen[81]

[81]GSA, Rep. 76 Va, I. HA, Sekt. 11, Tit. 4, Nr. 37, fol. 52.

FIVE

Obstacles to Emigration out of Germany after 1933, Failed Escape, and Death

> The Germans—as far as I have heard—no longer let out of the country any of the dismissed professors. These are obviously too inferior to be of any service for the Germans, but too good to let other countries have them. So I am afraid the case of our colleague sorts itself out.
>
> —James Franck 1938[1]

> Auch Endenich ist noch vielleicht das Ende nich! [Even Endenich is perhaps not yet the End!].
>
> —Felix Hausdorff 1942[2]

THE DISCUSSION of the "acculturation" of emigrant mathematicians in the foreign (in particular American) societies, which will be the focus of the following chapters 6 through 10, has to be clearly separated from the preceding process or emigration. First, the emigrants had to overcome considerable legal, bureaucratic, material, and mental obstacles both in the countries they left and in the host countries. These hurdles proved to be insurmountable in many cases, particularly for the older would-be émigrés. Therefore the present chapter also includes remarks on the fates of those mostly elder victims of the Nazi persecution who did not succeed with emigration. Several of them lost their lives in concentration camps or committed suicide, like millions of their mostly Jewish fellow sufferers. As they did not get the chance to influence and enrich mathematics in the United States or in other host countries, they are often ignored in traditional accounts of the history of science and are bound to be easily forgotten, unlike the "successful" emigrants.

[1] James Franck on October 16, 1938, about Ludwig Hopf (Aachen) in a letter to G. Szegö (T). Szegö Papers, Stanford, box 5, f. 17. See also below in chapter 6.D.

[2] Felix Hausdorff's sarcastic play with the German word "Endenich," a place near Bonn, to which he and his wife were to be deported as their first stop. The quote is from his last letter, dated Bonn, January 25, 1942, before the Hausdorffs committed suicide (Neuenschwander 1996, p. 263). To make the pun even clearer Hausdorff left the letter "t" out of the word "nicht." Thanks go to Sanford Segal for pointing me to this.

Insufficient social adaptation prior to 1933 was for several older mathematicians—very apparent in the case of Robert Remak (see below)—an additional burden in the attempt to emigrate. In particular, schoolteachers of mathematics had significantly smaller chances of finding positions abroad because they could not, usually, offer an internationally "in-demand" product—namely research. Therefore the PhD in mathematics re-gained some of its appeal under the conditions of emigration. Young mathematicians such as Wolfgang Wasow successfully completed their doctoral theses during emigration as the teaching exam (state exam), qualifying them for the teaching profession in Germany, proved to be worthless in their new home.[3]

Emigrants from Hitler's domain of power—if they held German citizenship—first had to overcome considerable problems on the German side, in particular emigration visa, payment of a "Reich Flight Tax" ("Reichsfluchtsteuer"), and the relinquishment of almost all claims on property and pensions. The conditions for emigrants worsened during the 1930s even in the economical respect, particularly after the November pogrom of 1938, euphemistically called "Crystal Night."[4] The economic conditions of the prospective emigrant were important both to his chances of leaving and of being accepted in the host country (D).

The psychological problems of emigration, which in mathematics often derived from an emotional attachment to the venerable German mathematical tradition, but generally had much to do with acquiring a new language and such, will be discussed in more detail in chapter 7. Let it suffice to mention here that several dismissed mathematicians tried to put off emigration as long as possible (Schur, Toeplitz, etc.), while others (Landau, Liebmann, Jolles,[5] etc.) died of natural causes before almost certainly

---

[3]Wasow, who went to the United States in 1939, could not meet—on the basis of his German "Staatsexamen"—the requirements for a teaching job in France, the first step of his emigration (Wasow 1986, p. 159). In Germany, Wasow, as many others, had deliberately taken the state exam instead of a PhD because of the widespread academic unemployment around 1930.

[4]In 1934 Courant (D) succeeded in evading the Reichsfluchtsteuer (usually 25 percent of property to be paid cash), which had been introduced prior to Nazi rule in 1931 in connection with the German Emergency Decrees [Notverordnungen] and was later, in 1938, complemented by an additional 20 percent of taxes for Jews. See Mußgnug (1988), p. 177. A. Brauer reports that his teacher Schur had to rely on a sponsor in 1938 in order to be able to pay the tax on his emigration. On pensions see some remarks in the previous chapter.

[5]On Stanislaus Jolles (1857–1942), who apparently tried to emigrate at one point and died under unknown circumstances, one finds the following remark from the year 1938 by H. Weyl in the refugee files of the Oswald Veblen Papers, OVP, cont. 31, f. Jolles, S.: "Last survivor of the tradition of 'synthetic geometry' . . . He asks whether there is a haven in America for old people like him to die quietly. Wife could give lessons in French and German" [undated 1938].

having been deported by the Nazis. However, there are moving documents that reveal failed attempts at emigration by mathematicians such as Ludwig Berwald (Prague), Otto Blumenthal (Aachen), Walter Fröhlich (Prague), Kurt Grelling (Berlin), Felix Hausdorff (Bonn), Robert Remak (Berlin), and Alfred Tauber (Vienna). These seven mathematicians were murdered by the Nazis or committed suicide under immediate threat, as was also the case with Ludwig Eckhart (Vienna), Paul Epstein (Frankfurt), Gerhard Haenzel (Karlsruhe), Fritz Hartogs (Munich), Charlotte Hurwitz (Berlin), Margarete Kahn (Berlin), Paul Lonnerstädter (Würzburg), Nelly Neumann (Essen), Georg Pick (Prague), and Reinhold Strassmann (Berlin). In one case (Fritz Noether, the brother of Emmy Noether, from Breslau) an emigrant was murdered in his host country, the Stalinist Soviet Union,[6] which does not, however, exonerate the Nazis from blame in his case.

5.D. Documents

5.D.1. Obstacles to Emigration from Germany

THE WIDESPREAD WISH TO EMIGRATE

Richard Brauer to Gabor Szegö in 1935:

> The wish to emigrate is common among the Jews in Germany. My wife's brother and several other relatives also want to emigrate to America. It's just that it is so damned difficult to get the chance.[7]

FINANCIAL AND AGE-RELATED OBSTACLES TO EMIGRATION FROM GERMANY

In 1965 Carl Ludwig Siegel wrote about the inhibitions of his Frankfurt colleagues to emigrate:

> Dehn, Epstein, and Hellinger stayed in Frankfurt until 1939. In spite of the increasing oppression of the Jews in Germany, many of the older ones among them could not decide for emigration because this would have meant leaving all savings at home and starting emigration with 10 Mark in the pocket. Moreover, many academically trained people had already gone to America in the first years after 1933, so that it became difficult for an older professor to found a new existence there. In Europe several states allowed permanent residence only when the foreigner was rich and brought his fortune with him.[8]

---

[6]See Appendix 1 (1.2). The list of non-German mathematicians who were murdered by the Nazis, in particular Polish mathematicians, is even longer. See the respective footnote in chapter 1. Moreover, there is clearly a lack of information about murdered schoolteachers in mathematics. See Appendix 1 (1.3). For F. Noether see Schlote (1991).

[7]R. Brauer to Szegö, October 19, 1935 (T). Szegö Papers, Stanford, box 5, f. 20. Brauer's sister Alice was murdered by the Nazis in a concentration camp. Cf. Rohrbach (1988), p. 147.

[8]Siegel (1965), p. 14 (T). To take the fortune abroad was expressly forbidden by the Nazis.

While Dehn and Hellinger finally made it to America (without obtaining adequate positions there), their colleague Paul Epstein, shortly after receiving a summons to the secret police (Gestapo) in Frankfurt, committed suicide in 1939.[9]

THE RELATIVE VALUE OF EARLIER MATHEMATICAL WORK DURING EMIGRATION

Otto Toeplitz writes 1936 to Courant:

> It is one way to go immediately abroad, and seek a position based on the reputation [Geltung] I have. I am very suspicious of this way. I feel . . . that abroad one is not judged by reputation but by direct appearance [Impetus]. Given my rudimentary linguistic skills the impression I would make right now would be an unfavorable one.[10]

FINANCIAL CONDITIONS FAVORABLE TO EMIGRATION

The differential geometer Herbert Busemann's chances of emigration were good as he was the son of an industrialist. Richard Courant wrote 1935 from New York to Busemann, who was temporarily in Copenhagen:

> In order to be accepted here it is very advantageous not to be forced—as a Jewish immigrant—to accept a position at any cost, but to act instead as an independent human being, to adapt and wait for a chance.[11]

Veblen from the IAS wrote in 1940 to C. B. Allendoerfer, Haverford College, Pennsylvania, to help Busemann get a permanent job.

> He was not obliged to leave Germany because his father is in a high industrial position and is in good standing with the present Government. But he left because he disapproved of the Nazi regime. I would not guarantee that he has no Jewish blood, but I should think that if the Nazis don't object to him on this ground, no one else would. He certainly does not look like a Jew.[12]

John von Neumann and Richard von Mises found themselves in financial situations similarly beneficial to their emigration.

---

[9]Siegel (1965), p. 17.

[10]Toeplitz to Courant, March 11, 1936, CPP (T). Toeplitz eventually went to Palestine in 1939 to escape the life-threatening pressure.

[11]Courant to H. Busemann, September 26, 1935, CPP (T). See also below in chapter 7 the case study on the conflict between the two immigrants Busemann and Lüneburg.

[12]Veblen to Allendoerfer, February 1, 1940. IAS Archives, School of Mathematics, Member Applications: Busemann, Herbert. In contrast to Reid (1976), p. 153, I assume Busemann was not affected by the racial laws. Both Courant's letter above and another one by the same author to W. Fenchel, dated July 17, 1935 (CPP) and alluding to "einem nordischen Menschen, wie Busemann" seem to rule out Jewish ancestors.

In 1935 the student of the philosopher E. Husserl and emigrant from Göttingen, Moritz T. Geiger, wrote to the Emergency Committee[13] in New York about the "Reich Flight Tax" and about Richard Courant's ability to avoid it:

> The capital flight tax can be cancelled if the government is willing to grant the emigrant that his emigration is in the German cultural or economic interest. This favor was given to quite a few of the German scholars who left Germany, for instance to Professor Courant of Göttingen, who emigrated from Germany last summer. . . . We learned that on principle this favor is no longer given to Non-Aryans.[14]

PERSONAL RELATIONSHIPS SUPPORTING EMIGRATION

Erich (Eric) Reissner reflected in 1994 on the way of his emigration:

> My American existence started with a one-year student visa, after a letter by Issai Schur (my father's friend) to Eberhard Hopf had led to an invitation and a fellowship at the MIT mathematics department. After several months they promised me an assistantship (1937–1939) that allowed me to obtain an immigration visa by going to Niagara Falls.[15]

Otto Toeplitz in Bonn in his letter to Courant on July 31, 1933 (CP, T):

> I have no relationships in foreign countries, thus my future is considerably more insecure than yours.

5.D.2. Unsuccessful Attempts at Emigration, Mathematicians Murdered

Chapter 6 will report on the failed emigration of Walter Fröhlich and Kurt Grelling who finally perished in German extermination camps. However, several of the other mathematicians murdered by the Nazis tried to emigrate as well. The information on them is for the most part scat-

---

[13]The "Duggan Committee" to be discussed in more detail in chapter 8.

[14]January 5, 1935 (T), EC, box 6, f. R. Courant, 1934–43. More details on Courant's negotiations with the Nazi authorities to avoid the Reich Flight Tax can be found in his private correspondence CPP. Courant apparently succeeded in persuading the ministerial functionaries that his work abroad was important for the German publishing system, particular Springer. Courant's good standing due to his having raised money from the Rockefeller philanthropy also seems to have made an impression, the more so since it was not yet clear at that time whether Rockefeller would stand by his promise to build a physics institute in Berlin. This was finally carried out by the foundation in spite of the regime (Macrakis 1993).

[15]Letter by Reissner to me, March 18, 1994 (T). Niagara Falls lies on the Canadian side of the border. Also Fritz Herzog's student visa was converted into an immigration visa after he had stayed in Montreal for one year (*IBD* microfilm, reel 26).

tered. Next to nothing is known about two victims, Charlotte Hurwitz and Paul Lonnerstädter, of whom we do not even know the year of their death.

LUDWIG BERWALD IN PRAGUE

The Oswald Veblen Papers at the Library of Congress keep the following internal note of the IAS, dated February 6, 1942, on Berwald, who had been dismissed in Prague in 1939 after the Germans had occupied the "rest" of Czechoslovakia:

> Re Ludwig Berwald and wife:
> New address: An den Ältestenrat der Juden, Prager Transport C., Nr. 616 und 817 Warthegau, Franziskanerstrasse 21, Litzmannstadt [Łodz.] Ghetto Poland. Our information from April 4, 1940, that Professor Berwald was in England, was a mistake. He stayed on in Prague, and he and his wife have now been deported to Poland.[16]

Berwald, who had corresponded with Veblen in 1935 and 1936 on problems of projective and differential geometry,[17] had also sent several letters to Veblen in 1939 and 1940 announcing his dismissal and asking for expert opinions for a stipend from the British Society for the Protection of Science and Learning (SPSL). Apparently, Berwald received such a stipend but could not use it due to the outbreak of the war in September 1939. In the very same year (1942) as the note of the IAS was issued, Berwald (born 1883) and his wife, both relatively young, perished in the Ghetto in Łodz.[18]

OTTO BLUMENTHAL IN AACHEN (AIX-LA-CHAPELLE)

In November 1933, the intimate friend of Hilbert's and managing editor of the *Mathematische Annalen* for many years, Otto Blumenthal,[19] wrote to his former colleague in Aachen, von Kármán, about his dismissal on September 22. In the same letter the fifty-seven-year-old Blumenthal mentioned his wish to emigrate and approached von Kármán in his characteristically modest way:

> Sooner or later I will need the opportunity to teach again. It is through teaching that I get the most vivid stimulus for research. Therefore I need to go abroad. I do not dare think about a permanent position: that is too sweet a dream. But perhaps there is a chance for lectures or semester courses? Can you [Du] help me obtain such?

[16]OVP, cont. 30, f. Berwald, Ludwig, 1939–42.
[17]OVP, cont. 2, f. Berwald, Ludwig, 1935–36.
[18]Pinl (1965).
[19]See also Butzer/Volkmann (2006). Volkmar Felsch (Aachen) is currently editing Blumenthal's diaries, which he kept until his deportation from the Netherlands.

SOS. What I can or what I can't do, you probably know even better than I do. Lectures at a big American university are too demanding for me, but I might be of some use at a smaller one.[20]

Blumenthal was dismissed from his position as managing editor of the *Mathematische Annalen* since volume 116 (1938–39). In his last letter to von Kármán, Blumenthal wrote from Delft on January 10, 1940:

On July 13 we crossed the border [to the Netherlands]. We were allowed to take furniture with us but no money or valuables. . . . At first Weyl reacted very enthusiastically but it was Weylian enthusiasm without any real promise behind it. The only thing is that we got affidavits[21] for the USA, with which you helped us. But the affidavits could not (and cannot) help us, as our quota number is for in 10 years' time on.[22]

There is an application in English that Blumenthal sent to Weyl in 1939 from the Netherlands, asking for help to emigrate.[23] On October 16, 1940, Weyl wrote to Blumenthal about his failure to find a position for him:

Your age is against you. . . . Veblen fully shares my opinion that the mathematical world owes you—the editor of the *Mathematische Annalen* for many years—assistance of some kind or another.[24]

Blumenthal was finally deported from the Netherlands—where he had received temporary support from the Dutch Academic Assistance Council (Steunfonds) of fl. 100 per month[25]—to the concentration camp Theresienstadt, where he died from his suffering in 1944.

FELIX HAUSDORFF IN BONN

Richard Courant informed Weyl in February 1939 that he had received a "very touching letter" from Hausdorff[26] in which Hausdorff asked for a research fellowship in the United States.[27] Three months later, in May

---

[20]O. Blumenthal to Th. v. Kármán, November 18, 1933. Kármán Papers, Caltech, Pasadena, 3.10 (T).

[21]A financial guarantee by an American to be given to somebody who wished to immigrate.

[22]O. Blumenthal to Th. v. Kármán, January 10, 1940. Kármán Papers, Caltech, Pasadena, 3.10 (T).

[23]In OVP, cont. 30, f. O. Blumenthal.

[24]Ibid. (T).

[25]Ibid. Blumenthal to Weyl, July 28, 1940.

[26]See Brieskorn, ed. (1996) and Neuenschwander (1996). See also the epigraph in the present chapter. Currently Hausdorff's Collected Works are being edited in nine volumes in German (the Hausdorff Edition in Bonn), a worthy monument to this remarkable and versatile mathematician, philosopher, poet, and astronomer.

[27]R. Courant to H. Weyl, February 10, 1939. Veblen Papers, Library of Congress, cont. 31, f. Hausdorff. Hausdorff's letter is not included in this file. Expert opinions by Weyl and

1939, Weyl received a letter from his former colleague in Zurich, Georg Pólya. He was concerned about the prospects of Bernays, but also about Hausdorff:

A case which is very near to me is <u>Hausdorff</u>. He had written a few lines first to Schwerdtfeger, then to me. From that anybody who knows him realizes that he is in a very bad situation. One hope that I had for him based on a communication by Toeplitz, and which I was incautious enough to relate to Hausdorff as well, has proved to be totally illusory. He is over 70—and he is one of the nicest and most pleasant human beings I know—his direct and indirect students (through his book) are everywhere densely distributed [überall dicht verteilt]. Isn't there a chance of doing anything for him?

My heartfelt congratulations on your U.S. citizenship, by which you have left the combination of murderers, gangsters, and slaves of which we have the pleasure of being surrounded from three different directions.[28]

The emigration of Hausdorff, who was already seventy-four years old, failed. In 1942, the mathematician who had been so influential in American mathematics through his topology book of 1914, *Grundzüge der Mengenlehre*, committed suicide together with his wife when facing the threat of deportation.

FRITZ NOETHER, FORMERLY BRESLAU, LATER AT TOMSK (SOVIET UNION)

Weyl's efforts to get Noether out of the Soviet Union, where he had been arrested by the Stalinist secret police in 1937, failed. Weyl's letter dated October 3, 1939, to the Georgian mathematician N. I. Muschelischwili (1891–1976), whom he asked to involve "his friend Berija" (as Weyl wrote), the chief of the secret police, in the matter, could not prevent Noether's execution in 1941. Weyl's efforts to help were somewhat impeded by his effort not to compromise Noether in Russia because of his Western contacts.[29] In an interview of the EC (B. Drury) with Stefan Bergmann, who had also been in Russia, it is stated that Einstein even wrote to Stalin on Noether's behalf. Bergmann is quoted with the following remark:

Bergmann said he knew F. Noether well—a very close friend of his in Tomsk. Unfortunately, despite warnings, N. stayed in Russia too long (Tomsk); he disappeared.[30]

---

von Neumann on Hausdorff are in the Harlow Shapley Refugee Files in the Harvard University Archives, Shapley Papers, box 6B, file: Ha.

[28]Pólya to Weyl, May 29, 1939 (T). OVP, cont. 30, f. Bernays, Paul, 1939.

[29]Veblen Papers, Library of Congress (OVP), cont. 32, f. F. Noether. See also Schappacher and Kneser (1990), pp. 37–38, and Schlote (1991).

[30]EC, box 84, f. Noether, Fritz. February 26, 1940.

Figure 19 *Felix Hausdorff (1868–1942). Hausdorff was known worldwide for his book* Grundzüge der Mengenlehre *(1914). His attempts at emigration failed and he committed suicide, together with his wife, in 1942, threatened by deportation to a Nazi camp.*

ROBERT REMAK IN BERLIN

Efforts to save Robert Remak were unsuccessful. He was finally deported by the Germans from the Netherlands to Auschwitz. In 1936 Issai Schur, himself threatened by the Nazis, had written an expert recommendation for Remak to be used for a possible position abroad. In it he called Remak "undoubtedly (the) leading capacity in the beautiful and important field of the geometry of numbers."[31] Remak's wife Hertha repeatedly sent telegrams and letters to Weyl with requests for affidavits, apparently without even Remak's knowledge. On December 11, 1938, she wrote to Weyl that her husband had been away for over four weeks without getting

[31]OVP, cont. 32, f. Remak, July 10, 1936 (T).

in touch and that she was very concerned, particularly given his peculiarities of character[32] that could cost him is life:

> Apart from his follies which I do not want to deny, you will agree that my husband is a deeply honest and decent character and an able mathematician. I am Aryan, so you cannot interpret my letter as a Jewish impertinence.[33]

On January 20, 1939, Hertha Remak wrote to Weyl that Remak had meanwhile returned to Berlin from the concentration camp Sachsenhausen and that he had received a temporary permit for the Netherlands.[34] Weyl was informed about Remak by other sources as well. To Heinz Hopf in Zurich, he wrote on November 29, 1938, about Remak's suffering in Sachsenhausen:

> About R. I heard that comrades of his, who have in the meantime been released, are saying that he is suffering more than others. One can imagine what this means, also because it is clear that due to his character, Remak is incapable of adapting in any way. It is generally known that prisoners are released if they have complete emigration papers, ship tickets etc. The poor (and very clumsy) Mrs. R. desperately tries to achieve something to this effect.[35]

Remak's wife was unable to withstand the persecution in the long run. Staying in the marriage would have meant for her to be sent into a Jewish ghetto at some point.[36] She sought divorce, thereby apparently depriving Remak of any last vestige of protection.[37] However, the fate of Grelling and his wife (see chapter 6) shows that life in a "mixed marriage" (the Nazi concept of "Mischehe") did not offer either partner a guarantee against the terror of the Nazis.[38] Thus one should be cautious with a hasty

---

[32]Described for instance in Biermann (1988), pp. 209–10. Van Dalen (2005), p. 731, writes that due to his unconventional behavior, Remak was in danger of being extradited to Germany from the Netherlands even before the Netherlands was occupied.

[33]H. Remak to Weyl, December 11, 1938 (T). OVP, cont. 32, f. Remak.

[34]Ibid.

[35]Weyl Papers, ETH, Hs 91:287 (T).

[36]The suicide in 1938 of Siegfried Samelson, father of the later famous topologist Hans Samelson, helped his family, in particular his "half-Jewish" children, to survive. See Tamari (2007).

[37]F. Hartogs in Munich suffered a similar fate. See H. Freudenthal to H. Hopf, July 28, 1945, in ETH Hs 621:537. Freudenthal himself was partly protected in the Netherlands due to his marriage to an "Aryan" Dutch woman, but he had to go into hiding anyway. See Dalen (2005), p. 752.

[38]In 1943 the insurance mathematician Reinhold Strassmann refused to use his "mixed marriage" to escape deportation to Theresienstadt. After deportation he was later sent to the death camp Auschwitz. According to Strassmann (2006), p. 293, his marriage to an "Aryan" woman had become a formal one many years before.

Figure 20 *Robert Remak (1888–1942). The gifted Berlin mathematician, student of G. Frobenius and specialist in the geometry of numbers, was deported by the Nazis from his place of refuge in the Netherlands and murdered in Auschwitz. The authenticity of the photograph, which could be confirmed only by one contemporary witness, is subject to a remaining doubt.*

condemnation of Remak's wife who was, after all, also a victim of the Nazis. There is no real obituary of the important mathematician Robert Remak apart from the late and deserved appreciation by Merzbach (1992) and more recently Vogt (1998). Merzbach indicates that Remak's fate having been forgotten is mainly due to his social and partly[39] scientific nonconformism:

> His refusal—in mathematics and everyday affairs—to compromise, or to be "realistic," swept him out of the mainstream of mathematics and cost him his life.[40]

ALFRED TAUBER IN VIENNA

The mathematician who became known for the "Tauberian Theorems" (1897) in the theory of function series, was deported to the concentration camp Theresienstadt where he died on July 26, 1942. As late as November 1941, the nearly seventy-five-year-old Tauber had desperately tried to immigrate to South America, corroboration of which is given in the following letter to a relative:

> I still want to try to get an assistant teaching post in Quito, where the university has advertised positions for European applicants. There I might have a chance as a retired university professor both of mathematics and actuarial science in spite of my advanced age.[41]

---

[39]Cf. Remak's work on mathematical economy, where he is considered a forerunner of "activity analysis." In an article from 1929 he wrote: "I emphasize that I make no economic claims, only formulate problems and schemes of calculation. . . . It remains totally open whether calculation decides in favor of socialism or capitalism" (Merzbach 1992, p. 496 [T]).

[40]Merzbach (1992), p. 514.

[41]Quoted from Binder (1984), p. 160 (T). An excerpt from that quote also in Sigmund (2004), pp. 31–32.

SIX

Alternative (Non-American) Host Countries

> The risk of being captured by the Third Reich is becoming
> too great.
>
> —Richard von Mises 1939[1]

> It seems that in addition to the German, Austrian, and Czecho-
> slovakian files we have to open the French file for displaced
> scholars.
>
> —Theodor von Kármán 1940[2]

ACADEMIC immigration to the different host countries after 1933 in general, not restricted to mathematics, has been very unevenly covered by the available research literature. There is ample discussion on immigration to the United States and Great Britain, but considerably less so on immigration to Turkey, Norway, and Denmark.[3] There are huge gaps concerning both the immigration to, and the situation in Palestine, France, the Netherlands, the Soviet Union, and South America. A special case is the immigration to Czechoslovakia, a country that gave temporary refuge not only to writers and politicians but also to many a dismissed mathematician. The Czechoslovak Republic was finally occupied by the Germans in March 1939 and thus could not qualify as a final host country for German mathematicians. As the situation of refugees in the United States is generally (apart from the conditions for mathematics) rather well documented, it is all the more important to have a look—as a complement and for comparison—at the situation in the various other host countries (D).

For obvious historical reasons the second most important host country and therefore the most important alternative host beside the United States

---

[1]Richard von Mises (Istanbul) to Theodor von Kármán (California) on March 28, 1939. Kármán Papers, Caltech, Pasadena, 79.25 (T).

[2]Theodor von Kármán to Oswald Veblen, July 24, 1940, after being informed about Felix Pollaczek's problems in German-occupied France. OVP, cont. 32, f. Pollaczek, Felix, 1938–40.

[3]For the USA: Coser (1984), Weiner (1969), Holton (1983), Rider (1984). For Great Britain: Rider (1984), Hoch (1983). For Norway: Lorenz (1992). For Denmark: Dähnhardt and Nielsen (1993). For Turkey: Widmann (1973), Neumark (1980). Reisman (2006) on Turkey is a popular account and not very reliable.

was the United Kingdom. Twenty out of the 145 refugees counted in this book stayed permanently in Great Britain until 1945, while an additional seventeen were hosted temporarily and went on to other places, mostly the United States. Literature on refugee mathematicians to Great Britain includes Rider (1984) and Fletcher (1986). The most important archival source are the refugee files of the British Society for the Protection of Science and Learning (SPSL) kept at the Bodleian Library in Oxford. It is far from being exhausted in its value by the published literature. A separate study on immigration to Britain is desirable, while the current book focuses on immigration of mathematicians to the United States. The SPSL, which was the successor to the Academic Assistance Council (AAC) formed in 1933, can be considered an organization parallel and similar to the American Emergency Committee (EC or Duggan Committee) to be discussed in detail below. The AAC, where British scientists such as E. Rutherford, J. B. S. Haldane, and A. V. Hill had the say, together with the Notgemeinschaft Deutscher Wissenschaftler im Ausland (Emergency Committee for German Scientists Abroad, where German refugees where in the lead), served as a clearing house for information on displaced scholars. The AAC/SPSL and the Notgemeinschaft published the "Lists of Displaced German Scholars" in 1936–37 (LDS). Both organizations had close connections to the League of Nations in Geneva, which occasionally provoked suspicion from the Americans.[4] But in many cases the European and American committees cooperated closely, often securing a temporary stay for the refugees in the UK before their departure for the United States (D).

Examples from various host countries show how widespread economic problems and political resentment, such as anti-Semitism, made acculturation difficult. Some countries, such as Austria[5] and Poland, had to be ruled out as host countries from the outset, since they offered similar, if not quite as extreme, political conditions as Germany. Others, such as Italy and the Soviet Union, also ruled by dictatorial regimes, served nevertheless and somewhat surprisingly as temporary host countries. Hopes harbored by Turkey to profit from the German immigration for its own science system failed due to Hitler's expansion policies and the death of Kemal Atatürk in 1938; both circumstances forced the refugees to go on to safer places. Australia was a rather less attractive option for emigrants because of the rudimentary state of mathematics there at that time. Although some authorities involved in emigration tried to use Australia to ease the situation in other host countries, only two mathematicians finally ended up there before the end of the war (D).

[4]Rider (1984), p. 141.

[5]On the situation in Austria in 1933, particularly judged from the topologist Karl Menger's point of view, see chapter 3, esp. 3.S.2.

It becomes clear from the documents that in spite of the immigration barriers still existing on the American side, all major economic and political developments (which included the tightening of the Stalinist regime in the Soviet Union) necessarily led to an increase of immigration in the direction of the United States. Here, the mathematicians followed the aggravating situation in Europe with disbelief, which was characterized by Richard von Mises's exclamation quoted above.

6.D. Documents and Problems Pertaining to the Various—Often
Temporary—Host Countries outside of the United States

The order of the host countries is alphabetic, except for Australia, which will be discussed in connection with Great Britain. On the general situation in Great Britain, Norway, Denmark, and Turkey one may also compare the printed sources mentioned above. Continuation of migration after 1945 is not considered in the following.

Belgium[6]

The applied mathematician Michael Sadowsky, who had tried in vain to resume his teaching position at the Technical University in Berlin in 1934, wrote to the British SPSL in April that same year:

> I am staying in Belgium because it is the cheapest place on the continent to live. The people here have been awfully nice to me, but I could earn nothing until now.[7]

Some members of the Berlin Society for Scientific Philosophy (an ally of the Vienna Circle) went to Brussels, among them Kurt Grelling and Carl G. Hempel. They could not, however, hope for financial support in an academic teaching position because they had not obtained a teaching permit in Germany.[8]

Both the much younger Hempel and Zermelo's erstwhile student in Göttingen, Grelling, were apparently abetted in their escape by the German industrialist, chemist, and philosopher, Paul Oppenheim (1885–1977), who himself had to flee in 1933 and who coauthored articles with both of them.[9]

---

[6]Belgium was temporary or permanent home to C. Froehlich, H. Geiringer, K. Grelling, C. G. Hempel, M. Sadowsky, and J. Weinberg.

[7]SPSL, box 284, f. 5 (Sadowsky, M.), folio 238. Sadowsky to SPSL, April 10, 1934.

[8]Dahms (1987), p. 93 (T).

[9]See in particular Grelling's file in the New York Public Library, EC box 60, where correspondence between several Americans and Hempel reveals the failed effort to get Grelling to the United States.

Grelling, who had become widely known for his discovery of the so-called heterological paradox in logic, worked in Belgian emigration with Oppenheim on the logical analysis of principles of psychology, in particular for Gestaltpsychologie. Hempel, who had obtained his PhD in Berlin with a mathematics-related philosophical topic in 1934, fled just in time to the United States in 1939. In September 1940, Oppenheim, now in America, secured a two-year salary for Grelling at the New School for Social Research in New York.[10] However, obtaining a visa for Grelling proved to be more difficult, since he had not held a position as professor in Germany but had rather taught at secondary schools.[11] Meanwhile, in 1940, Grelling had been interned after the occupation of Belgium by Hitler's troops:

> Grelling apparently returned to Berlin before the outbreak of the war and fled to Belgium after that. In 1940, shortly after the invasion by the German occupation forces, he was sent over the border by the Belgians as an "unwanted foreigner" and came to Paris, where he was interned in a camp for enemy aliens. On September 16, 1942, Grelling, who was "non-Aryan," was deported to Auschwitz and murdered, together with his wife who was "Aryan."[12]

Canada

Canada[13] had a special role to play during emigration, for example in obtaining re-entry visas to the United States. In addition, though to a much smaller extent than in the United States, mathematics in Canada profited from immigration, with the group-theoretic school of Richard Brauer in Toronto being the biggest success, although Brauer left for the United States in 1948.

Regarding reactions in Canada after the "Crystal Night" pogrom, Richard Brauer wrote to Heinz Hopf in late 1938:

> First the prospects were good, newspapers wrote that the government planned to let in several thousand German emigrants. Then there was resistance, particularly

---

[10]Ibid. C. G. Hempel to Alvin Johnson, September 20, 1940.

[11]Ibid. Hempel to Alvin Johnson, October 7, 1940.

[12]Dahms (1987), p. 94 (T). It is somewhat unclear from this report whether the French (as the term "enemy alien" would suggest) or the Germans (who occupied Paris) interned Grelling. In any case he was sent to the camp Gurs in southern (Vichy) France. For Grelling's biography see Peckhaus (1994) and more recently Luchins and Luchins (2000), also as an extended version at http://gestalttheory.net/archive/kgbio.html, accessed February 27, 2008.

[13]Canada became temporary or permanent residence to R. Brauer, I. Karger, F. Rothberger, P. Scherk, and A. Weinstein.

Figure 21 *Kurt Grelling (back middle, 1886–1942). The logician from Berlin could not follow up on the invitation from the New School for Social Research in New York, as he did not receive the necessary visa. This photo is of Grelling (back middle) at the internment camp Gurs in southern France around 1941. He was deported to Auschwitz in 1942 and perished.*

from nationalistic French Canadians. . . . For a mathematician it is impossible at the moment to find something in Canada.[14]

Another immigrant to Canada was the former assistant to Edmund Landau and student of Weyl's in Göttingen, the number theorist and geometer Peter Scherk. His letters to Weyl are rather self-critical and defensive, and he wrote at one point, in 1939:

> I will never go beyond assistant mathematics in number theory and average craftsmanship [solides Handwerk] in geometry. But even if mathematics does not need me, I for my part need mathematics.[15]

But Weyl seems to have had a higher opinion of his former student than Scherk himself, and he was supported in this by an expert opinion in favor of Scherk written by geometer J. Hjelmslev (Copenhagen). Scherk had arrived "penniless" in the United States in 1939, so Weyl provided him with money from the Relief Fund and secured a temporary position for him at Yale,[16] until he was offered a position as instructor at the Canadian University of Saskatchewan in 1943. Later, in 1959, Scherk moved to Toronto. The strong attachment of Canada to the British Commonwealth led in at least one case, Friedrich (Fritz) Rothberger, to the deportation of a refugee-mathematician from Britain to Canada and to his internment there in 1940.[17]

Denmark[18]

Copenhagen was a stopping place for Otto Neugebauer and for the editorial offices of the Zentralblatt für Mathematik und ihre Grenzgebiete, the important reviewing journal edited by Springer and led by Neugebauer.[19] Copenhagen also provided temporary refuge for Herbert Busemann, who later went to the USA. Werner Fenchel and his wife Käte, née Sperling, had to flee temporarily to unoccupied Sweden toward the end of the war, as did their host Harald Bohr.[20]

[14]Brauer to Hopf, December 12, 1938 (T). Heinz Hopf Papers, ETH Zurich, Hs 621:307.

[15]OVP, cont. 32, f. Scherk, P. Scherk to Weyl, April 6, 1939 (T).

[16]Hjelmslev's German review, dated Copenhagen, August 22, 1939, is in the same file. OVP, cont. 32, f. Scherk, P. Also at the same place is documentation for Weyl's help.

[17]See the file for Rothberger in SPSL, box 284, f. 2. He was apparently released for academic occupation in Canada in 1943 (ibid., folio 124).

[18]H. Busemann, W. Fenchel, G. Hermann, P. Nemenyi, O. Neugebauer, and K. Sperling temporarily immigrated to Denmark.

[19]See Siegmund-Schultze (1994).

[20]Jessen (1993). On the biography of Harald Bohr see the Danish PhD dissertation with limited circulation by Ramskov (1995).

Figure 22 *Harald Bohr (1887–1951). The brother of the physicist Niels Bohr was a specialist for almost-periodic functions and a good friend of E. Landau and R. Courant. He helped refugees to Denmark before he had to flee temporarily to Sweden since he was "half-Jewish," according to the German occupiers.*

France[21]

In 1935 Emil Julius Gumbel wrote from Lyon to Einstein in Princeton:

> In almost no case has a German succeeded in finding a permanent position. Most of the German scholars are on the brink of hunger. . . . French universi-

---

[21]France temporarily or permanently hosted S. Bergmann, E. J. Gumbel, W. Hauser, I. Heller, G. Hermann, F. Pollaczek, H. Schwerdtfeger, W. Wasow, and A. Weinstein. It has to

ties would like to keep the Germans but have no possibility of paying them . . . in view of renewed xenophobia and academic unemployment.[22]

At about the same time, the American Norbert Wiener (1894–1964) remarked in an article in the *Jewish Advocate* that France was unable to help "in view of the miserable salaries paid to French scholars themselves."[23] Wiener appended a recent undated letter to him from the leading French mathematician Jacques Hadamard (1865–1963), which reads as follows:

Dear Professor Wiener,

The task of affording help to German intellectual refugees—a heavy one, for as you know refugees of every kind have come to us in great number—has been very difficult for France.

Our country is small in comparison with yours, and we have a limited number of universities. At present, the number of young French scholars, I mean men of very great distinction, is great, so that the way to a career is very difficult. Therefore, not one refugee has been appointed in our public educational system and we cannot think of making any such appointments.

For about one year we had been able to support financially those who were on our soil. However, this has ceased since last summer as the funds were exhausted.[24]

Happily, at least as far as concerns mathematicians, practically all those who had come to France have found, or I hope are soon to find, employment abroad. I regret this for my own country. Some of them have already gone to America—in my opinion the better for you and the worse for us. It will be a great thing for your country and for civilization if you have full success in that direction. We can do nothing more as we are over-crowded by a mass of refugees of all sorts, not merely of intellectuals.

On the other hand, I have just written to Geneva to Dr. Kotschnig, who will inform you, if you do not know already, what has been done on the part of the High Commissioner appointed by the League of Nations.

Wishing you complete success, I beg you to believe me,

Yours truly, J. Hadamard.[25]

---

be noted that France, in June 1940, divided into a German-occupied northern and an officially unoccupied southern part, where the Vichy government collaborated with the Nazis. France played a particularly important role as a country of transit for refugees (among them L. Bers), many of whom left Marseille for America. Others, such as K. Grelling, were deported from camps in France to extermination camps.

[22]Gumbel to Einstein, January 10, 1935. Einstein Papers, Jerusalem, 50 134-2 (T).

[23]"Once More . . . the Refugee Problem Abroad," *Jewish Advocate*, February 5, 1935, p. 2. Copy in Wiener Papers, MITA, box 11, f. 543.

[24]Probably the "Caisse nationale," founded 1930, the precursor to the French Research Council CNRS, which temporarily went into financial problems during the 1930s.

[25]*Jewish Advocate*, February 5, 1935, p. 2.

Figure 23 *"Once More . . . the Refugee Problem Abroad." From Norbert Wiener's article in the* Jewish Advocate, *February 5, 1935, p. 2.*

Refugees not getting out of France in time before the German occupation were, even in the unoccupied south of France, threatened by deportation to Germany and to the extermination camps. A vivid description of the conditions in the French internment camps is given by the "emigrant of the second generation," Walter Rudin (born 1921) in his autobiography.[26]

---

[26]Rudin (1997), pp. 47–53. I thank Heinrich Wefelscheid (Essen) for this information.

Felix Pollaczek,[27] the noted applied mathematician and pioneer of queuing theory, who had already been recommended to Veblen by Hadamard in December 1938,[28] could not leave France before the German invasion on May 10, 1940. He wrote to Veblen six days after the French surrender, which had taken place on June 22.

> Actually I am *boursier de recherches* of the Ministry of National Education, but my Austrian nationality has caused me continuous trouble, and only due to the influence of M. Emile Borel as former Minister of the Navy have I remained at liberty during the greater part of the war. But since May 10 the situation for all foreigners, and my situation in particular, has again grown worse. It has become tragic and absolutely untenable since the recent complete reverses of the military and political situation. After seven hard years as an emigrant I find myself an exile again, and there is no other country of refuge open to me except the United States.[29]

Th. von Kármán and Veblen supported Pollaczek's placement in the United States and recommended him to the leading mathematician at Bell Laboratories, Thornton Fry (1892–1991), specifically because of the importance of Pollaczek's field for defense purposes.[30] Fry, however, was much more reluctant to engage him due to his "shy and retiring" character. The problems of employing foreigners in defense research (problems that were confirmed to Fry by W. Weaver of the Rockefeller Foundation), and the general difficulty of rescuing people even from the unoccupied Vichy part of France played a role as well.[31] In a discussion at the Rockefeller Foundation with Veblen and Weyl on October 7, 1940, it was stated:

> P. is an Austro-German, now in France, so WW [Warren Weaver] thinks there is little possibility of rescuing him.[32]

Pollaczek, however, survived by being hidden by a French family of peasants for a year in 1943.[33] But Kurt Grelling (see above for Belgium), who was interned in the camp Gurs in southern France, could not accept the appointment he received at the New School for Social Research in New

---

[27]Pollaczek was the first husband of Hilda Geiringer, who later married von Mises. His fate as an emigrant has been investigated by Schreiber and LeGall (1993). See also Cohen (1981).

[28]Hadamard to Veblen, December 24, 1938. OVP, cont. 32, f. Pollaczek, Felix, 1938–40.

[29]Ibid. Pollaczek to Veblen, June 28, 1940. Translation into English by H. Weyl.

[30]Ibid. Kármán to Veblen, July 24, 1940. Veblen passed the letter on to Fry, scribbling on it the following note: "It would seem to me that it would be a good idea to get together in this country a few of the men in applied math, who would be useful in connection with problems of national defense . . . and the antifascists who come here from Europe are extremely pro-American and ready to help."

[31]Ibid. Fry to Veblen, August 12, 1940.

[32]RAC, R.F. 1.1, 200, box 46, f. 532.

[33]Schreiber and LeGall (1993), p. 279.

York City anymore due to his papers having been processed very slowly by the American immigration authorities.[34]

Great Britain and Australia[35]

Norbert Wiener, in his article in the *Jewish Advocate* (Boston) of February 5, 1935, quoted above, wrote the following on the comparison between problems experienced by the English and the Americans regarding immigrants:

> As far as England is concerned we have had the privilege, during the past week, of a visit from Professor J. B. S. Haldane of the University of London. Professor Haldane is in the forefront of those who have labored to help academic refugees in England. We have talked over the English and the American problems, their similarities and differences; and Professor Haldane tells me that the English universities have been as little able to support refugees from their own funds as our universities are.
>
> England, with half the available population of the United States, has probably done at least as much as the United States in the care for these refugees; yet, every cent has come by voluntary contributions from the pockets of private individuals. The difficulties in the way of public support are the same as here. England can no more afford than we to displace young men from academic positions by the competition of foreign scholars. On the other hand the English have had an even livelier sense than ours of the need of keeping international scholarship intact and of offering an asylum for persecuted scholars.[36]

In mathematics, many algebraists, among them several members of the Schur School in Berlin, immigrated to Great Britain. While British algebra and number theory apparently profited from immigration, there are surprisingly few applied mathematicians among the refugees who finally settled in the United Kingdom. Among the thirty-seven refugees to the UK

---

[34]Peckhaus (1994).

[35]Great Britain was temporary or permanent home to R. Baer, H. G. Baerwald, F. A. Behrend, S. Bochner, H. von Caemmerer, R. Courant, H. Hamburger, H. O. Hirschfeld (Hartley), W. Hauser, H. Heilbronn, O. Helmer, G. Hermann, M. Herzberger, K. Hirsch, L. Hopf, H. Ille (Rothe), F. John, H. Kober, K. Kober (Silberberg), G. Kürti, W. Ledermann, V. Levin, K. Mahler, A. Mayer, P. Nemenyi, B. Neumann, A. Prag, R. Rado, W. Rogosinski, F. Rothberger, E. Rothe, O. Schilling, O. Taussky, S. Vajda, W. Wasow, A. Winternitz, and H. Zatzkis. In 1938, F. A. Behrend's father, Felix Wilhelm B. (1880–1957), who in 1933 was director of a Berlin high school (Gymnasium) and philologist, immigrated to Great Britain as well. Because he published a mathematical school textbook in 1932, he could be considered a mathematician as well. Kind communication by H. Begehr (Berlin). On Australia see below.

[36]"Once More . . . the Refugee Problem Abroad," *Jewish Advocate*, February 5, 1935, p. 2. See facsimile of this letter above, under "France."

mentioned in the footnote above, there were seven that could be called applied mathematicians in a sense relatively close to engineering or physics: H. G. Baerwald, R. Courant, M. Herzberger, L. Hopf, G. Kürti, P. Nemenyi, and W. Wasow. None of them stayed in Britain; all but one (Hopf to Ireland) proceeded later to the United States. Two more refugees to the UK, F. John and O. Helmer, went to the United States as pure mathematicians and became later engaged in applied domains. Only one refugee to England, H. O. Hirschfeld (Hartley), worked as an applied mathematician (statistician) in the UK before proceeding to the United States as well, if only after the war. Hartley, born in 1912, had to learn statistics in England from scratch. After receiving his doctoral degree on a topic in the calculus of variations from Berlin University in 1934, Hirschfeld-Hartley took an English PhD in statistics under John Wishart (1898–1956) in Cambridge, the former assistant to K. Pearson, in 1940. Wishart felt, however, that statisticians trained on the continent did not fit into the English system. In 1939 he wrote to the SPSL, in response to an inquiry about a possible position for Hilda Geiringer, the statistician trained by Richard von Mises in Berlin:

> A purely statistical post, for which there is from time to time a vacancy, would require a technical training in applied statistics, and a knowledge of methods developed in this country, particularly in relation to the biological sciences, or in the industrial sphere, which it is not clear that she at present possesses.[37]

So, if one wonders about the reasons for this apparent lack of demand for applied mathematics in the UK at that time one is tempted to point to the older and strong tradition in applied mathematics in the UK and its peculiar connections to mathematical physics and (more recently) mathematical biology, which differed from the continental(mostly German) tradition to institutionalize applied mathematics in special departments either at technical or (more recently) traditional universities. The United States, however, which finally profited greatly from the German immigration in applied mathematics (see chapter 10), had the younger system of science and mathematics compared to Britain. It apparently followed very much the German example in applied mathematics in the 1930s as it had followed the German role model in pure mathematics several decades before.

There are historians who conclude more generally that the "closed university system in Great Britain was not favorable to immigrants."[38] This

---

[37]SPSL, box 279, f. 3 (Pollaczek-Geiringer, H.), folio 87. Wishart to E. Simpson, June 2, 1939. R. A. Fisher of the Galton Laboratory replied in a similar vein to the SPSL on May 22, 1939. See ibid., folio 113.

[38]Kröner (1989), p. 22 (T).

resonates with Hardy's characterization of the extreme insularity of the British system at least in the 1920s, as mentioned in chapter 3.

In a similar vein the career patterns for scientists and mathematicians in Britain were apparently—unlike the situation in the United States—to a strong degree determined by tradition, which made it even difficult for some immigrants to get their academic degrees recognized in the host country. In this context a remark by Selig Brodetsky from Leeds seems revealing, which he made on July 10, 1938 in a letter to a "Professional Committee for German Jewish Refugees" in London about the topologist Boris Kaufmann from Heidelberg, who was at that time thirty-four years old:

> Dr. Kaufmann is undoubtedly one of the best of the younger mathematicians, but it is going to be very difficult for him to settle down in any sort of way unless he becomes a Cambridge man in a technical sense, by taking a Cambridge degree. Quite a few young mathematicians who have done this have since been able to find places at various Universities. I think that Kaufmann should work for the Ph.D. at Cambridge.[39]

Eventually, at least five among the immigrants (Bernhard Neumann, Richard Rado, Kurt Hirsch, Olaf Helmer, Hirschfeld-Hartley) took the English PhD, although they all had a German doctorate before. No case of a similar requirement in the United States is known to me.

Competition on the job market was an obstacle to immigrants too, as it is for instance revealed in a letter by the SPSL to the mathematician Mary Cartwright, who had expressed doubts whether forty-six-year-old Hilda Geiringer would be willing to accept a junior academic position. Esther Simpson of the SPSL wrote on May 22, 1939 to Cartwright:

> I do not doubt that Dr. Geiringer herself would be willing to take a junior position; the difficulty there would be to obtain permission from Home Office to employ her in a junior position. One would have to prove that no British candidate was available, and usually this is not the case where junior positions are concerned. Perhaps computing is an exception? I am afraid that her information about her chances in U.S.A. is too vague for this Society to take the responsibility of bringing her over.[40]

The economic situation of the immigrants was generally not without problems. Werner Rogosinski, who became known for his work on Fourier series, had to work under difficult conditions up until the war:

> In 1941 the war provided Rogosinski with the opportunity to obtain a teaching post when Aberdeen appointed him assistant at £300 p.a. It was not a great

---

[39]SPSL, box 281, f. 1 (Kaufmann), folio 202.
[40]SPSL, box 279, f. 3 (Hilda Pollaczek-Geiringer), folio 116.

deal for an established mathematician in his middle forties with a wife and son but many others were even less fortunate.[41]

A student of Landau in Göttingen and a specialist in analytic number theory, Hans Heilbronn—according to Taussky a "giant among mathematicians"[42]—was considerably better off than other immigrants due to his having been granted a five-year appointment in Cambridge.[43]

Research mathematicians coming from abroad received help in—above all—Cambridge from Godfrey H. Hardy and in Manchester from Louis J. Mordell (1888–1972).[44] However, the Academic Assistance Council (AAC, later SPSL), which was often approached by Hardy on their behalf, had only limited resources. Therefore the AAC asked, for example, Courant in 1935 whether or not he could help Fritz John to proceed to the United States, as the funds of the AAC were exhausted.[45] Indeed, invitations to Britain were often extended to the refugees on the understanding that these would lead to a chance in the United States later. Hardy, for one, wrote the following concerning Hans Hamburger to Oswald Veblen on December 6, 1940. The letter also reveals a certain degree of naïveté on the part of Hardy with respect to the American welfare system and the function of the privately secured American "affidavit," in reality a rather formal affair:

> I understand that he has an "affidavit" for the U.S., and may hope to get his visa inside a year. That means, doesn't it, that he is sure of <u>subsistence</u> if he goes (apart from a job)? The S.P.S.L. here (to whom Weyl wrote about him, rather discouragingly, a little while ago) certainly supposed, when they encouraged him to come, that their responsibility for him would be temporary.[46]

The physicist Victor Weisskopf (1908–2002), who was temporarily in Cambridge, England, and who later went on to the United States, considered the policies of the British university administration, which in his opinion did not provide long-term prospects to the immigrant, to be "very short-sighted."[47] There are also reports about resentment in Britain toward immigrants from middle and Eastern Europe[48] and an "undertone of

---

[41]Hayman (1965), p. 137.

[42]Taussky-Todd (1988b), p. 27.

[43]Ibid., p. 31.

[44]Rider (1984). R. Baer (temporarily before moving on to the United States) and K. Mahler stayed at the University of Manchester with Mordell, while H. Heilbronn preferred to accept the parallel offer from the more attractive Cambridge. See the Mordell Papers at St. John's College Library.

[45]CPP, March 22, 1935.

[46]OVP, cont. 6, f. Hardy, G. H., 1924–27. The affidavit played a similar role for Paul Kuhn's immigration to Norway, see below.

[47]Hoch (1983), p. 223.

[48]Ibid., p. 224.

Figure 24 *Godfrey H. Hardy (1877–1947). The leading English mathematician provided temporary jobs in Cambridge for eighteen expelled mathematicians. In the 1920s he had cooperated with H. Bohr, R. Courant, H. Weyl, O. Veblen, and the Rockefeller philanthropy for international communication.*

anti-Semitism,"[49] which was typical for many other European countries and some American quarters as well (see below). In May 1934 the applied mathematician from Oxford, E. A. Milne (1896–1950), responded with the following words to a request by A. V. Hill to support refugees:

> That there are two exiled German families in this road in very good houses & gardens, one with a car, may cause me a little irritation, but I have no knowledge whatever as to how they are supported. . . . I wonder how our doctors, or lawyers, or bricklayers would welcome an excess of exiled doctors, or lawyers, or bricklayers. Fortunately the academic struggle for existence is not so severe as in other professions, & I agree fully that we must do something to preserve academic freedom. . . . I have some very good friends amongst Jews, men I like very much, but I think anti-Semitism is partly connected with a certain ineradicable tactlessness amongst Jews. . . . At bottom, I don't understand how it's that the Jews, who are reputed to be the wealthiest community in the world, don't undertake the whole support of their nationals.[50]

Dahms writes with respect to the family of the logical empiricist and cofounder of the Vienna Circle, Friedrich Waismann:

> On the private level Waismann became very lonely; because of the anti-Semitism, which was also rampant in England, his wife and his only son committed suicide.[51]

One Jewish immigrant and Czechoslovakian citizen with a German education and limited knowledge of English, Artur Winternitz, made it to England simply because he happened to have been born in Oxford.[52]

There was also help coming from England specifically with "non-Aryans" in mind. Several Jewish organizations made sure that the refugees they supported were practicing Jews, as mentioned in a letter related to the case of Heilbronn:

> The Jews here do not want to give money to help someone who will not associate with them when he comes here.[53]

---

[49]Rider (1984), p. 133.

[50]Milne to A. V. Hill, Oxford, May (?), 1934. Milne Papers, Bodleian Library Oxford, f. D 65. I thank June Barrow-Green (London) for sharing this source with me.

[51]Dahms (1987), p. 107 (T). Also Rider (1984), p. 133, talks about "an undertone of anti-Semitism [that] ran through many of the nationalist assertions made by British and American scholars."

[52]OVP, cont. 33, f. Winternitz, Artur, 1938–40. The differential geometer and topologist Winternitz originally tried to immigrate to the United States and wrote a letter to this effect to Weyl on November 16, 1938. The letter is in the file together with an undated opinion by Weyl, who calls Winternitz "well versed in the foundations of mathematics."

[53]H. R. Hassé to L. Mordell, Bristol 6.12.33, Mordell Papers, Cambridge, 11.9.

The mathematics teacher from Freiburg, Wilhelm Hauser, finally received a position as a teacher in 1941 at the Royal Grammar School in Newcastle, then evacuated to Penrith.[54] Hauser had been taken by the Nazis into so-called protective custody (Schutzhaft) after the November pogrom of 1938 and temporarily sent to the concentration camp Dachau, before he reached England. He remembers his first contacts with Englishmen when he was temporarily in Paris:

> I tried to go on to the USA. But my attempts were not successful. Then my wife heard that a relief organization in Oxford was looking for a teacher who could take over a boarding school [Heim] for young Jewish people between 15 and 18. A former good acquaintance from the lodge in Freiburg, a Quaker, still had contacts with English Quakers, and he told my wife about the position. . . . The Quakers appointed me in spite of my bad (Baden) English.[55]

Hauser and Adolf Prag[56] are examples of refugees making it to the British school system, no easy task either—particularly not for research mathematicians from Germany such as Hamburger, as described in another passage of Hardy's letter to Veblen, dated December 6, 1940:

> We are hoping that some of the people may be able to get schoolmasterships for the duration.—There is beginning to be a real shortage, and, although there is a good deal of prejudice to be overcome . . . sooner or later headmasters will have to realize that they must take "aliens." Indeed there are already some signs that this is happening. Hamburger is not a very good case for this because (though extremely pleasant and cultivated—he is, for example, a good linguist) he has no experience at all of elementary teaching. He is excellent with a good mathematical audience (e.g., at my seminar).[57]

Even Prag, who had been a teacher in a private Jewish school in Germany before emigration, had in England to limit his research, which had been inspired by the Frankfurt seminar on the history of mathematics around Max Dehn.[58]

---

[54]Oswald (1983). Thanks to Jeremy Gray (London) for pointing out the evacuation.

[55]Quoted from Wirth (1982), p. 165 (T).

[56]Scriba (2004), p. 411.

[57]OVP. cont. 6, f. Hardy, G. H. 1924–27. From Hamburger's file kept at the British SPSL it becomes clear that Hamburger was generally admired but at the same time seen unfit for both teaching and the war effort. One has to assume that Hamburger felt embarrassed by the situation and used the first opportunity after the war to leave Britain, first going to Turkey then returning to Germany. See SPSL, box 279, f. 6, particularly the very friendly but also critical remarks on Hamburger by H. S. Ruse from the University College Southampton.

[58]Dehn wrote to Viggo Brun on September 10, 1939 from Oslo, reporting that Prag and his wife had visited him and that Prag had no time anymore for research in his teaching job in England. See Brun Papers, Oslo.

Conditions for immigration and for the refugees already in Great Britain changed with the outbreak of the war. One official from the SPLS wrote to John von Neumann in October 1939 concerning Karl Löwner, who had received a stipend for Cambridge (England) and an English visa in June 1939, but who had not by then been allowed to leave German-occupied Czechoslovakia:

> All visas for this country which were granted before the outbreak of war are automatically void and each case has to be considered anew on its own merits. The Home Office say that if a refugee has already reached a neutral country they will consider re-confirming the validity of the visa but I do not think they would take any steps to help a potential refugee who has not been able to leave Czechoslovakia.[59]

In Great Britain, the internment of "enemy aliens," primarily on the Isle of Man, affected even German immigrants coming as early as 1933, like Hans Heilbronn[60] and Adolf Prag. In April 1939 Heilbronn applied for British citizenship. Unfortunately it was a few days too late to enable the papers to be processed before the start of World War II. As a contribution to the war effort, Heilbronn organized the Trinity College A.R.P. Fire Service. Nonetheless, he was interned on the Isle of Man in 1940 as an enemy alien, something he regarded as highly unfair since nobody could have been more opposed to the Nazis than he was. The Academic Assistance Council, Hardy, and others made strong representations on his behalf, so he was released after several months of detention. He then served in the British forces with the Signals Corps and Military Intelligence until the autumn of 1945, on which Hardy wrote the following to Veblen on December 6, 1940:

> Heilbronn is now in the Army (Signals, a good regiment with many very intelligent people, and he seems pretty happy there).[61]

Reports on the conditions in the internment camps are varying and partly contradictory. A rather mild description is given by Stefan Vajda,[62] a less friendly one by Kurt Mahler.[63] Occasionally the internment camps have been described as a result of "the general hysteria of 1940."[64] Poorten (1991),

---

[59]OVP, cont. 31, f. Loewner, Charles. E. Simpson (SPSL) to von Neumann, October 5, 1939. On Löwner's inability to leave the country earlier, see his letter to von Neumann, dated June 14, 1939, in the same file. See also the similar case of W. Fröhlich below, who unlike Löwner did not, in the end, reach a safe haven abroad. Löwner managed to go to the United States directly.

[60]Cassels and Fröhlich (1977).

[61]OVP, cont. 6, f. Hardy, G. H., 1924–47.

[62]According to Bather (1996).

[63]According to Poorten (1991), p. 374.

[64]Kani and Smith (1988), p. 51.

however, reports on efforts in further education among the interned in a "university of the interned."

Lausch (1987) confirms this for Felix Adalbert Behrend, who was deported from Great Britain to Australia.[65] A long letter, written by Behrend in January 1941 to Thomas Mann in America, whom he admired, illustrates the hardships of deportation plus the fact that Behrend would have preferred to go to the United States instead, this despite his not being fully informed about immigration policies:

> In June of last year I was like many other refugees interned and—due to error or accidence—I was sent to Australia. This happened in a moment when I was trying to become familiar with voluntary military service which according to my nature is not very attractive to me. An application for release as a scientist has been granted meanwhile, but the release can only be effective in England, and I do not see any real point in returning to England where I had no occupation at all and where there is even less chance now for a reasonable position. . . . For this reason I now want to apply for a non-quota visa to the USA (I heard this opportunity was recently opened for scientists) and I take the liberty to immodestly ask for your support. I assume that Professor Weyl and Dr. A. Brauer in Princeton and other mathematicians who know my case will support me.[66]

Hermann Weyl in Princeton, who was informed about Behrend's wish to come to America, was not willing to help. In a letter to Dean Richardson at Brown University, Weyl wrote in June 1941 about Behrend and other émigrés who had already succeeded in leaving Germany:

> They are not in any serious danger, and may eventually find openings in the British Dominions.[67]

Behrend's biographers explain how his mathematical versatility (number theory, topology, geometry, foundations) coupled with the absence of any

---

[65]Australia became temporary or permanent residence to F. Behrend and H. Schwerdtfeger. H. Lausch (Clayton, Australia), who has worked systematically on immigration to Australia, has found another Vienna mathematician, Stefan Petö, who fled to Australia, whose mathematical training has not been verified yet (letter to me, February 11, 1998). H. Löwig, who because of erroneous information in Pinl and Dick (1974), p. 175, was included in the German version of this book as a refugee to Australia before 1945, came to that country apparently only in 1947, after somehow managing to survive in German-occupied Czechia. See his file (1939–47) in OVP, cont. 31.

[66]Thomas Mann Archive (ETH, Zurich), F. Behrend: "Die Fahrt zu den Vätern" [The journey to the fathers] (1961), 19 pp., p. 4, Behrend to Th. Mann, January 31, 1941 (T). Behrend remarks on the same page that, "probably due to censorship," he did not receive a reply. In another letter to Thomas Mann toward the end of the war, Behrend, who tried his hand at writing novels at that time, complained about the cultural isolation in Australia. Ibid., p. 8. Behrend to Th. Mann, November 25, 1944.

[67]Weyl to Richardson, June 18, 1941. OVP, cont. 30, f. Bers, L.

Figure 25 *Felix Adalbert Behrend (1911–1962). The versatile Berlin mathematician who had first reached England after stopping in Prague was finally deported to Australia. The repeated uprooting affected the quality of his mathematical work. In his literary contacts with Thomas Mann in the United States he showed his unabated closeness to the German language and culture.*

outstanding achievement in these fields were caused by the constant wanderings his fate as an emigrant imposed upon him, thereby leading to a great divergence of stimuli during what could have been his most productive years:

> It is possible that the deep disturbances of his life during the pre-war and war years—he was six times uprooted—may have prevented the concentration of his energies into one deep channel.[68]

Indeed the situation in Australia at that time was not stimulating for front-line research. In a December 1939 letter to the English Nobel Prize winner in physiology, A. V. Hill (1886–1977), who was active in placing refugees, G. H. Hardy in Cambridge commented with the following words on efforts to channel the refugees to Australia, based on grants from the Carnegie Foundation in the United States:

> I am writing to you on Rutherford's advice. I was talking at different times today, (a) to him and (b) to Adams at the R. S., about the Carnegie offer to the A. Assistance Council. In a way, of course, it is much the biggest thing that has come the Council's way. And the sum suggested (£400, if I remember) is so much more substantial than what most of the younger refugees are getting, that one's natural instinct would be to make a small selection of the very best people, and to push their claims very hard.
>
> But the "empire"—which apparently excludes England—is an awful snag. There may be one or two places—e.g., Sydney—where you can send a man without handicapping him too badly: but even Sydney is not London, Cambridge, or Princeton. Suppose X is the very best of the mathematical refugees under 30—I have in fact a definite man in mind. Have I to go to him and say "you have about £180 a year up to the end of next year: I can probably get you £400 for 3 terms: <u>but</u> for that you have got to leave Cambridge (or wherever you are) and go to Adelaide, where there are no libraries, no mathematical society, and nobody who can possibly care two damns about you or your work?" It would seem that human insanity can hardly go further: yet that is the advice one may be forced to give.
>
> <u>America</u> would be quite O.K.—that's another matter altogether. This not mere anti-imperialistic sentiment: I know Wilton of Adelaide, and the sort of handicaps he has fought against there.[69]

At the beginning of the war immigrants to Australia were required to report weekly to the police. The use of non-English languages in public was forbidden.[70] Hans Schwerdtfeger reached Australia, coming from

---

[68]Cherry/Neumann (1964), p. 266.

[69]Hardy to A. V. Hill, December 15, [1939]. Hill Papers at Churchill College Cambridge, AVHL II 4/33. I thank June Barrow-Green (London) for this information.

[70]Kind communication by Mrs. Hanna Schwerdtfeger to me, undated, received July 21, 1993.

Prague and going through Switzerland, due to help from Max Born shortly before the war broke out. As Max Born himself commented in 1965:

> After many vain attempts I eventually managed to place Schwerdtfeger in Australia, with the help of the great physicist Sir William Bragg, who came from there.[71]

Schwerdtfeger's immigration to Australia and not to the United States, which he would have preferred, is probably also due to "selection" by his colleagues such as Weyl, Busemann, and Courant who knew him from Göttingen and were not fully convinced of his capabilities.[72]

India[73]

In 1935 N. M. Basu (Calcutta) sent detailed offers to Richard Courant. The latter recommended Max Dehn instead. Dehn, who finally gained a minor position in the United States, was interested, but wrote skeptically to Courant in June 1935:

> It is very doubtful whether the board there will be interested to hire an old German geometer, given that Indian interests lean towards the opposite direction.[74]

Finally, Friedrich Levi got the position, a fact that Courant regretted.[75] At least at the beginning, Levi, the algebraist and topologist from Leipzig, who in his textbook of 1929 had presented combinatorial surface topology for the first time, experienced considerable difficulties adapting to India. This becomes clear in the following slightly condescending passage from a 1937 letter to Richard von Mises:

> The Bengali are very intelligent and lethargic people, they usually have an admirable memory. Real penetration of the subject is replaced by memorizing.[76]

---

[71]Born, ed. (2005), p. 125. See chapter 4 for the discussion between Born and Einstein on Schwerdtfeger.

[72]See CPP, in particular Courant's correspondence with H. Busemann. See also chapters 5 and 7 above and below.

[73]Along with F. Levi, Victor Levin was temporarily in India.

[74]Dehn to Courant, June 21, 1935, CPP (T). Dehn was apparently alluding to the strong Indian tradition in number theory, following Ramanujan.

[75]See Courant's correspondence with Basu in CPP. According to Kegel and Remmert (2003), p. 401, Levi was supported by H. Bohr, Hardy, and Weil. Levi stayed several times with Baer in the United States when he still hoped to find a position there. See OVP, cont. 31, f. Levi, Friedrich W., 1935–41.

[76]Levi to von Mises, May 8, 1937 (T). Mises Papers, Harvard, 4574.5, box 3, f. 1937.

After the outbreak of World War II, Levi even feared dismissal or internment, as he confided in a letter to Weyl in June 1940:

> Slowly the situation improved and my position grew steadily, even during the war, up to a few weeks ago. In consequence of the events in Holland, people have become very suspicious of the enemy subjects, making recently not much distinction between refugees and non-refugees. When the drive started, I was here in the Himalaya for vacations, and I have not been interned up to now. It is, however, doubtful if arrangements can be made allowing me to return to Calcutta. . . . My term of appointment expires on Jan. 12th 1941. . . . Thus I am again a refugee looking for a job, and at this time there is really no other place left but the United States.[77]

Weyl, in his response, showed more faith in India and proved to be right. Levi helped to build the Tata Institute of Fundamental Research in Bombay after the war, before returning to Germany in 1952.

Netherlands[78]

Van Dalen reports that "the Dutch were so much used to their traditional neutrality, that . . . they were inclined to consider the exiles from Germany as reasonably safe."[79] Unfortunately, they could not offer them jobs. Stefan Warschawski wrote from Utrecht to Courant on October 16, 1933:

> There is a huge number of younger Dutch mathematicians, who have no chance to become teachers or assistants. Therefore Prof. Wolff tries to find a position for me at some other place. He has written to his friend Prof. Struik. . . . The latter has now answered and reports that he has turned to Prof. Richardson (Brown University . . .) on my behalf.[80]

On June 20, 1933, Kurt Mahler had written to the English number theorist Louis J. Mordell:

> At present since some weeks I am in Amsterdam, where I have worked with my friends Koksma and Popken, for I could no longer endure the life in Germany.[81]

---

[77]OVP, cont. 31, f. Levi, Friedrich W., 1935–41. Levi to Weyl, Darjeeling, June 16, 1940.

[78]The Netherlands hosted temporarily O. Blumenthal, M. Herzberger, K. Mahler, R. Remak, A. Rosenthal, S. Warschawski, and A. Weinstein. L. Boll and K. Freudenberg survived the Nazi occupation there in hiding.

[79]Dalen (2005), p. 732.

[80]CPP (T). The Dutch mathematician Julius Wolff (1882–1944) later on was deported by the Germans and murdered in the concentration camp Bergen Belsen. See Barrau (1948).

[81]Mordell Papers, 17.6.

Norway

On immigration to Norway[82] there is the general description by Lorenz (1992). It contains only a few remarks specific to mathematics, primarily about Werner Romberg's influence on the development of numerical analysis in Norway, after the war. On his emigration from Munich through the Soviet Union to Norway, Romberg wrote the following in 1998:

> In 1937 my residence permit in the Soviet Union was not renewed, because I had a German passport. Since Poland still existed, I could travel through Warsaw to relatives in Prague. . . . I offered to professor E. A. Hylleräs in Oslo to work as his assistant. He succeeded in finding some money from the "Brøgger Committee." Thus I was able to fly to Oslo, after the occupation of the Sudetes but prior to the occupation of Prague.[83]

According to a note in the Oswald Veblen Papers, Max Dehn, who had been dismissed in Frankfurt in 1935, "was one of the fathers of modern topology." In 1938, another topologist, originally Danish but later Norwegian Poul Heegaard (1871–1948) wrote to Veblen, pointing out Dehn's predicament in Germany but without seeing how to get Dehn to Norway or another Scandinavian country: "The Scandinavian lands are too small, to give possibilities."[84] Dehn got the chance to substitute for Viggo Brun (1885–1978) in Trondheim (Norway), but Brun anticipated further danger for Dehn in a letter to Weyl written on June 6, 1940, after the German occupation of April 1940:

> My hope—and yours—that I could place Max for good in my country is now equal to zero.[85]

Dehn, who in late 1940 had escaped via Sweden and the Trans-Siberian Railway to the United States,[86] described the fate of another refugee to Norway, Ernst Jacobsthal, in a letter to Weyl in March 1943:

> We were already very worried after all the news about the jew-baiting in Norway. . . . When we left Trondheim he [Jacobsthal] had a somewhat regular

[82]Norway was temporary residence to M. Dehn, E. Jacobsthal, P. Kuhn, and W. Romberg. Jacobsthal and Romberg returned to Norway after the war, coming back from refuge in unoccupied Sweden.

[83]W. Romberg to R. Siegmund-Schultze, October 1, 1998 (T). See another quote from this letter in chapter 4. That Romberg indeed "flew" from Prague by plane and not just "fled" is clear from another, Norwegian, letter by Romberg to Per Christian Hemmer (Trondheim), dated October 3, 1991, where Romberg says that at the time of his "flight" the Czech borders were already regularly checked by German military.

[84]OVP, cont. 30, f. Dehn, M. Heegaard to Veblen, Helsingfors, undated, received August 26, 1938.

[85]Ibid.

[86]Dawson (2002).

Figure 26 *Werner Romberg (1909–2003). In 1933 Romberg wrote a physics dissertation with Arnold Sommerfeld in Munich and fled, because of his leftist political leanings, in 1934 to the Soviet Union. He had to leave in 1937 and went to Norway in 1938, where he became famous for his methods in numerical analysis after the war, returning from his asylum in Uppsala (Sweden) in 1944.*

source of income, opened by these nice Norwegian colleagues, and in the first line, of course, by Viggo Brun. Jacobsthal and a man from Prague, Pavel Kuhn, brought to Norway through the endeavors of Brun, were hired to produce a new collection of mathematical formulas, to be edited by the Norwegian Mathematical Society.[87]

Pavel (or Paul) Kuhn was originally a Czech citizen and a German-speaking Jew working as an actuary in Prague. He had been interested in number theory for many years and had therefore been in contact with Viggo Brun in Norway. In a letter to Weyl from Prague in August 1939, Kuhn described the Norwegian policies for immigration and how an American "affidavit" could solve problems in this context. In the English rephrasing (probably by Weyl) this reads as follows:

> Professor Brun is trying everything in his power to bring about Dr. Kuhn's immigration to Norway. The Norwegian government requires two guarantors ready to give a guarantee without time limit. One is Professor Brun himself. But he cannot find a second guarantor unless Dr. Kuhn can show an affidavit for the United States. . . . There is little likelihood that Dr. Kuhn will ever make use of his American affidavit (supposing he gets one); it would only help him in getting into Norway.[88]

A detailed description of his and Paul Kuhn's flight from Norway to Sweden is given by Ernst Jacobsthal in his letter to Weyl from Uppsala, dated June 19, 1943:

> We left Trondheim on January 6 [1943] in the morning and were on Swedish territory 2:30 am on January 9. Because of the deep snow in the high border mountains we had to take a detour through Oslo. 18 hours on the train without valid documents! Usually there is Gestapo control on the train. We were lucky, nobody asked for passports that day. In order to make the trip we even had to ask for a written permission. Why K. and I were not arrested in Trondheim I have no idea. By October no human being was left there who was a Jew according to the German law.[89]

[87]E. Jacobsthal to H. Weyl, March 20, 1943 (T). OVP, cont. 31, f. Jacobsthal, E., 1938–44.

[88]Note from November 13, 1939. OVP, cont. 31, f. Kuhn, Paul, 1939–44. The letter from Kuhn to Weyl is also in the files and dated August 10, 1939. The files also contain a letter of support written by Karl Löwner, dated January 1, 1940. The function of the American affidavit was the same here as in the case of Hans Hamburger in the UK. See above in this chapter.

[89]OVP, cont. 31, f. Jacobsthal, E., 1938–44. Handwritten (T).

Palestine[90]

The continuing financial problems of the Hebrew University in Jerusalem during the 1930s made subsidies necessary even for Abraham Adolf Fraenkel's professorship there (see chapter 8).[91] In general the work situation in Palestine held less appeal for theoretical scientists than for individuals in practical academic professions, such as medical doctors.[92] In 1935 Wolfgang Sternberg related in detail his problems finding a job and his permanent efforts to immigrate to another country such as the Soviet Union or the United States.[93] According to Sternberg,[94] and according to the file kept on Samson Breuer in the SPSL, Breuer, brother-in-law of Fraenkel, had no permanent position in Palestine either. Breuer, who had been both an algebraist and an actuarial mathematician before emigration, was appointed at the Migdal Insurance Co. Ltd. in Jerusalem. He tried repeatedly to proceed to another host country where "orthodox Jews" could live.[95] As late as July 1945 Breuer asked the SPSL for support in finding a position because he had "not been able to maintain my family out of my salary."[96] Sternberg also reported to Courant that there was a "special fund intended for German immigrants"[97] at the University of Jerusalem. The applied mathematician Michael Sadowsky did not find employment in Germany because his wife was Jewish. He went to Palestine with his wife in 1937, but in a letter to the British SPSL, he said:

> Referring to the question of establishing contact with the Hebrew University in Jerusalem as you suggested . . . I found out that this university is a National

---

[90]Palestine gave temporary or permanent residence to R. Artzy (Deutschländer), S. Breuer, A. Cohn, A. Fraenkel, G. Leibowitz, R. Peltesohn, M. Sadowsky, M. Schiffer, I. Schur, W. Sternberg, D. Tamari, and O. Toeplitz.

[91]In a memorandum by the American Friends of the Hebrew University in New York City, sent to Duggan's Emergency Committee on November 3, 1938, shortly before the Kristallnacht pogrom in Germany, it is emphasized that European countries such as Germany, Austria, Czechoslovakia, and Italy no longer subsidized the Hebrew University and that the Americans had to step in. EC, box 8, f. Fraenkel, A.

[92]Kröner (1989), p. 18. Emmy Noether had already tried in vain to obtain a position for her student J. Levitzki in the United States in the 1920s because he could not find anything in Palestine. See chapter 3.

[93]CPP. For instance on January 18, 1935: Sternberg was working at the university of Jerusalem without any remuneration. Sternberg finally reached the United States in 1939, after having stayed, according to Pinl (1969), p. 210, four years (!?) in Prague. See also Courant's letter to Berwald, quoted in the preface to this book.

[94]Sternberg to Courant, undated, after January 30, 1935, CPP.

[95]Breuer to SPSL, August 23, 1934, SPSL box 277, f. 9 (Breuer, S.), folio 435.

[96]Ibid., folio 451.

[97]Sternberg to Courant, March 27, 1935, CPP (T).

Jewish Institution of an extremely nationalistic trend in the Jewish-Orthodox sense. . . . My being a non-Jew (Russian by birth, Greek Orthodox by religion) prohibits the University of any kind of affiliation with me.[98]

Also Sternberg, of Jewish descent, reported about problems with Jewish nationalism in Palestine and with the New Hebrew "Iwrit" (which would later become the official language of Israel), of which he allegedly understood "only 6 words." He came to the conclusion: "There are too many intellectuals here."[99] Regarding the possibility of obtaining Palestinian citizenship, Sternberg wrote to Courant in 1936:

The value of this citizenship is, however, dubious. It is also dependent on whether the rather unpredictable English policies remain friendly to the Jews or not. I admit that an American citizenship would be much preferable to me.[100]

Issai Schur's student Menahem Max Schiffer was prevented from completing his doctoral dissertation due to the political events in Berlin in 1933 and the (at first only temporary) dismissal of his teacher.[101] Quite accidentally, Schiffer met a woman in Schur's apartment from the English-Jewish Emergency Council who offered him a stipend at the Hebrew University in Jerusalem. Schiffer, who later on changed his mathematical fields to Riemann surfaces and differential equations, went to the United States in 1946, where he became a very influential mathematician at Stanford University. Schiffer's teacher in Berlin, Issai Schur, who helped many émigrés with expert advice, was himself forced to go to Palestine in 1938, in order to save his life.

Unlike Sternberg and apparently also unlike Schiffer, Rafael Artzy (born Deutschländer) had already thought about immigration to Palestine before 1933 (see above). About the details of his emigration he remarked in a letter in 1998:

I was fortunate enough to have a cousin in Jerusalem who acted as guarantor. So I only had to wait for the immigration "certificate," which arrived already

[98]SPSL, box 284, f. 5 (Sadowsky, M.), folio 283. Sadowsky to SPSL, October 13, 1937. Sadowsky proceeded to the United States in 1938, although the files of the SPSL do not keep record of this fact, which is however sufficiently documented by the Poggendorff dictionary and the Web site of the Illinois Institute of Technology.

[99]Sternberg to Courant, July 1, 1935, CPP (T). He reported there was no chance of finding a job in an insurance company. He also had problems with the climate.

[100]Sternberg to Courant, November 8, 1936, CPP (T).

[101]Schiffer (1986/1998). Schiffer remarks here that Schur "liked much" his work, which he wanted to submit as a PhD dissertation (p. 178).

Figure 27 *Issai Schur (1875–1941). The algebraic work of the student of G. Frobenius focused on group representation theory, which gained importance also in quantum mechanics. Historically, the Berlin school around Schur has been somewhat overshadowed by the more spectacular "abstract" school around Emmy Noether in Göttingen. Schur could not make up his mind about emigration in spite of early offers from America and ended up going to Palestine in 1939 in order to save his life.*

towards the end of the summer semester of 1933. At the same time I received a questionnaire from the university about my race. I scribbled "Jew" on it with big letters and threw it on the table in the Rector's office.[102]

Artzy could not complete his PhD with Reidemeister who had just been dismissed. In Palestine Artzy wrote the "dissertation without any guidance because there was no geometer in Jerusalem."[103] Since Artzy was, in addition, very active within the "Haganah," the organization for Jewish self-protection, and since he also had to earn his living, he did not obtain his doctor's degree until 1945. His academic career did not really take off until 1951, when, at the age of thirty-nine, he accepted an assistantship at the Technion in Haifa. About the reasons for his further migration to the USA in 1960, Artzy wrote to me:

> In 1960 I accepted an assistant professorship at the University of North California. Why? Somehow the provincialism at that time in Haifa, in Israel in general, had become too boring to me, and I felt an urge to travel, also with regard to mathematics.[104]

South America

South America[105] had a prior history of scientific immigration.[106] Even before 1933, scientific isolation experienced in South America forced several immigrants to go on to North America. Given the conditions of 1933, however, immigration to South America remained an attractive option. This applied even to the well-known logician and Hilbert student Paul Bernays. However, the envoy from Ecuador to Switzerland, on interviewing Bernays there, had doubts about the latter's pedagogic capabilities.[107] Therefore the specialist in function theory and student of Heinrich Behnke (1898–1979), Peter Thullen, was appointed in Ecuador:

> In the beginning of 1935 I was appointed by the Ecuadorian government— through the mediation of the "Emergency Committee for German Scientists abroad" (Zurich, later London) [Notgemeinschaft für deutsche Wissenschaftler

---

[102]R. Artzy in a letter to me, January 11, 1998 (T).

[103]Ibid. (T).

[104]Ibid. (T).

[105]South America hosted temporarily or permanently R. Breusch (Chile), K. Freudenthal (Fulton, Colombia), R. Frucht (Chile), E. and W. Fanta (Brazil), and P. Thullen (Ecuador).

[106]For mathematics see the case of Karl Grandjot, as discussed in chapter 3.

[107]G. Pólya to H. Weyl, February 2, 1935. Weyl Papers, ETH Zurich, Hs 91: 420. In the same letter, Pólya mentions that Thullen had now been selected by the envoi. See also Bernay's letter to Courant, February 11, 1935, CPP.

im Ausland]—to Quito. I worked there together with other German and Swiss scientists for the foundation of the "Escuela Politécnica" of Ecuador.[108]

Robert Frucht would have liked to go from Chile to the United States. Unfortunately, he failed to gain a nonquota visa, not being eligible for one due to his temporary employment outside the university system as an insurance mathematician in Italy.[109] His wish for immigration to the United States after the war was not supported by the American statistician H. Hotelling.[110] Frucht's scientific isolation in Chile is documented by the fact that when he wanted to write an obituary of his teacher Schur in 1941, he found no mathematical journal in Chile where this could be published.[111] Curt Fulton (born Freudenthal) immigrated to Bogotá in Colombia due to C. Carathéodory's help, where he taught from 1938 until 1946, before proceeding to the United States.[112] The failure of Alfred Tauber to escape to Ecuador is mentioned in the last chapter.

Soviet Union (in Particular Russia)[113]

From the outset there were bureaucratic and political hurdles in the Soviet Union with respect to the appointment of foreigners. P. S. Aleksandrov describes this in his obituary of Emmy Noether, who had first tried to go to the Soviet Union after her dismissal.[114] Further mathematicians hoping to find a position in Soviet Russia but who did not make it primarily due to political obstacles included Kurt Mahler,[115] Felix Pollaczek, and

[108]Thullen's undated German vita [after 1944] (T). CIP, f. Thullen, 1944–48. Thullen was in Rome in 1933–34 with a grant of the German Research Association (former Notgemeinschaft Deutscher Wissenschaft, not to be confused with the Notgemeinschaft of refugees after 1933 mentioned above) and stayed there longer until his final immigration to South America in 1935. See also chapter 4 and Appendix 6 for Thullen's diaries in 1933. On the Emergency Committee see below under Switzerland and the article Erichsen (1994).

[109]H. Weyl to H. Shapley, January 27, 1939. Shapley Papers, HUA, box 6A, file: European Refugees F.

[110]Hotelling reflected on that occasion, in a letter to Weyl, dated February 10, 1947, rather arrogantly on European statisticians. OVP, cont. 30, f. Frucht, R. See chapter 10.

[111]R. Frucht to G. Szegö, May 5, 1941. Szegö Papers, Stanford, box 5, f. 17.

[112]Letter by C. Fulton to me, March 20, 1994.

[113]The Soviet Union gave temporary or permanent residence to H. G. Baerwald, S. Bergmann, S. Cohn-Vossen, E. Lasker, V. Levin, F. Pollaczek (?), W. Romberg, M. Sadowsky, and F. Noether. Baerwald is mentioned as temporary refugee in the Soviet Union in the following document: E. Simpson of the British SPSL to B. Drury (Emergency Committee N.Y.), January 12, 1939. EC, box 84, f. Noether, F.

[114]Aleksandrov (1935/1981). That Aleksandrov dared to report publicly on these problems is even more remarkable given the political situation in the Soviet Union in 1935.

[115]Temporarily Mahler had hopes for a position at the university of Saratov with the help of A. Khintchin, as he wrote to L. J. Mordell from Groningen (Netherlands) on November 8, 1935. See Mordell Papers, 17.20.

Wolfgang Sternberg. Pollaczek, who became known for his statistical work within telecommunications and for queuing theory, was nominated for two positions in Russia, but apparently to no avail.[116] The early immigrant Chaim (Hermann) Müntz, however, who had come to Russia from Berlin in 1929 and had an influential chair for analysis at Leningrad, claimed in 1937 that Pollaczek had received a temporary position in the Soviet Union due to his assistance:[117]

> The appointments of Cohn-Vossen, Walfisz, Pollaczek (the latter was not allowed to slip in again) were immediately influenced by myself, the ones for Plessner and Bergmann indirectly.[118]

The Polish mathematician Arnold Walfisz (1892–1962) was able to stay at the Georgian University of Tbilisi (Tiflis) during the war, and the other mathematicians mentioned by Müntz survived as well. The fate of the refugee to the Soviet Union from Köln, Hilbert's coauthor for the book *Geometry and the Imagination* (*Anschauliche Geometrie* 1932, English 1952), Stefan Cohn-Vossen, who died in 1936, still needs elucidating more fully.

Immediately after 1933 there existed somewhat naive attitudes by some Americans vis-à-vis the political conditions in the Soviet Union. Richard Brauer wrote from Princeton to Szegö in Königsberg in 1934, responding to Szegö's request to help Erich Rothe in his effort to immigrate to the United States. According to Brauer, the American Oswald Veblen had suggested the Soviet Union as a host country for Rothe instead:

> Veblen judges the possibility of absorbing German mathematicians [in America] rather pessimistically . . . in appointments nationalistic motives play a growing role. Veblen, of course, strongly regrets the latter. . . . Personally I understand the inhibitions which Rothe has vis-à-vis Russia. But it is difficult to talk about this here. All free-minded people [frei denkenden Menschen] look to Russia with a certain admiration and with much interest. The dangers that may be connected to a life in Russia seem small from a distance.[119]

Solomon Lefschetz, for instance, who had been born in Russia, was very impressed by modern Russian topological work, and, in 1934, still

---

[116]This according to information in the Oswald Veblen Papers, OVP, cont. 32, f. Pollaczek, F. On the personal sheet, kept for Pollaczek, a position for algebra in Tiflis in 1935 is mentioned, "which failed because of local political opposition," as well as a professorship for applied mathematics in Baku in 1936, "for which the required visa was refused."

[117]On Müntz, who went on to Sweden in 1937, see below under "Sweden" and in Ortiz and Pinkus (2005).

[118]SPSL, box 282, f. 7 (Müntz, Ch.), Müntz to Edmund Landau, December 12, 1937 (copy).

[119]R. Brauer to G. Szegö, November 20, 1934. Szegö Papers, Stanford, box 5, f. 20 (T).

believed in the possibility of "breaking down the present situation" by generous American grants to Russians.[120]

With respect to Wolfgang Sternberg's wish to immigrate to the Soviet Union, Aleksandrov remarked skeptically in a letter to Courant in 1936:

> The rush of scholars, in particular mathematicians, for a position in the USSR has increased greatly in recent years. Thus it is really only top people who have a chance of a position here. That is why I am dubious as to Sternberg's prospects. . . . I have become very cautious after being unable to do something for Lüneburg, Zorn, Aronszajn, and other outstanding young people.[121]

Sternberg reported 1935 in a letter from Jerusalem to Courant about his attempts to go to the Soviet Union, using the emigrated biologist Julius Schaxel (1887–1943) there as a contact person. However, he had already received information about "German scholars, who had settled in Russia, but who all of a sudden and without being given a reason were forced to give up their positions and leave Russia."[122] This also affected Müntz, who had remained without Soviet citizenship and who had to go in 1937 although he had once—as many others such as Albert Einstein—had illusions about the political system there.[123] At about the same time Michael Sadowsky, Russian born but apparently without Russian citizenship, who had been employed by the University of Leningrad since 1934, had to go as well, allegedly because he refused to make communist propaganda in his lectures: "I was asked by the G.P.U. to leave the country within 2 days."[124] Other emigrants such as Stefan Bergmann had to leave the Soviet Union as well, in view of the political situation there. It is known of at least one mathematician-emigrant, Emmy Noether's brother Fritz Noether, that he became a victim of Stalinist terror.[125]

---

[120]See the respective quotation by W. Weaver on his discussion with Lefschetz, as quoted in Siegmund-Schultze (2001), p. 133.

[121]P. S. Aleksandrov to R. Courant, CPP, November 17, 1936 (T).

[122]Sternberg to Courant, July 1, 1935, CPP (T). Sternberg writes in a very detailed fashion about his negotiations with the biologist-emigrant Julius Schaxel and with the mathematician in Leningrad V. I. Smirnov.

[123]See Ortiz and Pinkus (2005). Grundmann (2004), p. 254, quotes one article from *Das Neue Russland* (The New Russia) of 1931, where Müntz is called "one of Albert Einstein's closer scientific collaborators" and is indirectly quoted as saying that "Einstein . . . is keeping close track of the successful advances made in the direction of the socialist construction of the Soviet Union." See also Müntz's file in SPSL, box 282, where he describes himself as Einstein's collaborator from 1926 until 1929 and emphasizes that he was driven out of the Soviet Union despite "strict impartiality" (Müntz's vita, dated December 10, 1937, ibid., folio 336).

[124]SPSL, box 284, f. 5 (Sadowsky, M.), folio 274. Sadowsky to SPSL, May 17, 1937.

[125]He was arrested under false accusation and executed by the Stalinist secret police. See Schlote (1991) and Schappacher and Kneser (1990). See also remarks above in 5.D.

Finally, it should be mentioned that the Soviet Union served as a transit country (via the Trans-Siberian Railway) to America for the mathematicians Max Dehn and Kurt Gödel.[126]

Sweden[127]

In a 1934 letter from Stockholm, Willy Feller, who later became well known as a probabilist in the United States, sent the following sarcastic report to Courant:

> I never catch a glimpse of pure mathematics, partly because it virtually does not exist as a university institution—the students are learning on their own, the professors are only doing the exams—partly because Carleman is of the touching [rührend] opinion that one should execute [an die Wand stellen] all Jews and immigrants (which, however, he only tells his assistant after consuming a nonnegative [nichtnegativ] amount of alcohol).[128]

Five years later Feller wrote to Neugebauer, again from Stockholm:

> The emotions here are aggravating from day to day. There is a flood of student revolutions, Nazism at the universities is mushrooming.[129]

In spite of all these critical remarks by Feller, a man given to sharp utterances, a history of emigration cannot ignore Sweden's role as a haven for Jewish refugees, particularly from other Scandinavian countries during the time of the greatest territorial expansion of Hitler's Reich.[130] Nominally neutral,[131] Sweden was also the last resort for the early emigrant from Germany Chaim (Hermann) Müntz who had been expelled from the Soviet Union in 1937. From Stockholm, Müntz wrote to Courant in October 1941:

---

[126]Dawson (2002). For Dehn the result was political mistrust in the United States because of his flight route. See chapter 9. In 1941 Dehn reported in a talk about his flight. See Dehn Papers (AAM Austin).

[127]Neutral Sweden hosted temporarily or permanently W. Feller, W. Fenchel, E. Jacobsthal, P. Kuhn, Ch. H. Müntz, W. Romberg, F. Rothberger and K. Sperling. The only ones who stayed in Sweden after the war were Kuhn and Müntz.

[128]W. Feller to Courant, November 11, 1934, CPP (T). It is unclear to this reader whether Feller was using this bit of mathematical slang to express that Carleman would abstain from his anti-Semitic feelings only in moments of extraordinary soberness (negative amount of alcohol).

[129]Feller to O. Neugebauer, February 22, 1939 (copy [T]). Richardson Papers, BUA, correspondence, 88 F (Feller 1940–42).

[130]See Lorenz (1992) and Jessen (1993), p. 130, the latter in particular for Fenchel and his wife, K. Sperling.

[131]Sweden collaborated in various economic and political respects with Nazi Germany.

The Swedish scientific world, which is not capable of providing a regular position for me because the country is culturally saturated, supports me by allocating research assignments. For external reasons I chose hemodynamics, a border field toward physiology, an area which after Euler's initial research has barely been touched by mathematicians.[132]

Müntz felt threatened at that time without being captured by the Nazis; he wrote from Stockholm to Weyl in November 1941:

In the unfortunately no longer impossible event of an incorporation of the last small piece of a free north into the domain of the most modern zealotry, my work (I do not speak of the person anymore) would be uselessly sacrificed. I hope, however, to still be able to tell something.[133]

Weyl considered Müntz a "mathematician of very high rank," in particular for his work on boundary value problems of differential equations. However, at the same time, Weyl wrote:

Einstein, who is willing to help him in any other way, warns against bringing him to America because of his somewhat unbalanced personality.[134]

Feller proceeded to the United States in 1939. Five years later, the situation of several other mathematical refugees in Sweden was summarized in a letter written by the statistician Harald Cramér (1893–1985) to Weyl in April 1944:

Fenchel is in Lund, where he is in contact with the mathematicians at the University and is working as a teacher in a school for Danish refugees. He seems to be able to live decently on the payment he receives there. Jacobsthal and Kuhn are both in Uppsala, and I think that with respect to both of them any contribution that you might be able to offer would be very welcome. Jacobsthal receives a small monthly allowance from the Swedish Committee for Intellectual Refugees, while Kuhn is occupied with statistical work under Professor G. Dahlberg. . . . Harald Bohr is here. . . . We have now—partly through the kind help of the Rockefeller Foundation—been able to offer him a position as guest professor in the University of Stockholm, and he has settled down to live in Djursholm. . . . The mathematicians still left in Denmark and Norway seem to be fairly well. Some of them have been imprisoned, but as far as I know, none have yet been killed or sent to Germany.[135]

---

[132]Müntz to Courant, Stockholm, October 30, 1941. CIP Bobst Library New York City (T).

[133]Müntz to Weyl, Stockholm, November 25, 1941 (T). OVP, cont. 32, f. Muentz, Hermann.

[134]Ibid., undated sheet with information about Müntz.

[135]An extract from this letter, dated April 26, 1944, is in OVP, cont. 31, f. Jacobsthal, E., 1938–44.

Switzerland

Switzerland[136] had, in 1933, and still has, a certain reputation as a classic country of exile. Indeed, the foundation in Zurich in June 1933 of the "Emergency Committee for German Scientists abroad" (Notgemeinschaft Deutscher Wissenschaftler im Ausland, Zurich, later London) must be mentioned in this respect as it proved to be of great importance for Thullen, von Mises, and other emigrants immediately after 1933.[137] However, the following documents demonstrate the restrictions imposed by the Swiss immigration policies (case W. Fröhlich) plus the difficult economic conditions for immigrants such as Samelson and Bernays.[138] Being born in Switzerland did not automatically qualify one for citizenship, even if this fact could be used to get one on the Swiss immigration quota for the United States, as seen in the case of Wolfgang Wasow. Even Swiss citizens such as the prominent logician Paul Bernays encountered problems making ends meet. G. Pólya, a professor in Zurich, wrote to the American J. D. Tamarkin in 1934 that it was—given the problems of Bernays—even more difficult to do something for G. Szegö:

> I could not do anything for him [Szegö, RS] here in Switzerland. All we could do till now on the Technische Hochschule for the expelled mathematicians, is to give a job to Bernays for half a year (for the "Sommersemester" 1934) because he happens to be a Swiss citizen.[139]

Weyl and Pólya tried to do something for Bernays with a stipend at the IAS but apparently succeeded only once, for the year 1935/36.[140]

Before reaching his final host country before the war, Australia, Hans Schwerdtfeger was also temporarily in Zurich, where he was supported by G. Pólya and his wife, according to the following report by Hermann Weyl from March 1939:

> Professor G. Pólya of the Institute of Technology in Zürich, Switzerland, has given shelter to him and his family in Zürich since the end of January. At that time the situation in Prague had become too dangerous on account of the Gestapo. But the Swiss police will tolerate him in Switzerland at most for three

---

[136]Switzerland hosted temporarily or permanently F. Behrend, P. Bernays, W. Hauser, I. Heller, P. Hertz, A. Pringsheim, H. Samelson, H. Schwerdtfeger, and A. Weinstein.

[137]Erichsen (1994). The committee issued the two "Lists of Displaced Scholars" (LDS 1936–37) in London.

[138]For Bernays see also his failed attempt to migrate on to South America (see above).

[139]Askey (1982), p. 1. In a letter to Weyl on January 26, 1939, Pólya wrote that "doing something for him [Bernays] here for the year 1939/40 seems impossible" (T). IAS Archives, School of Mathematics, Member files: Bernays, Paul.

[140]See List of Mathematical Stipends, IAS Archives, Faculty, Veblen (box 33), f. School of Mathematics.

Figure 28 *Paul Bernays (1888–1977). The logician and important collaborator in Hilbert's program for the foundations of mathematics did not obtain a secure position in his place of refuge, Switzerland, although he had Swiss citizenship. The marginality of his research topic within mathematics and the extreme modesty of his character might not have helped.*

months, and he may even have to face deportation into Germany at the end of this month. What can one do to come to his rescue? Mrs. Pólya is active in some relief organization for refugees in Zürich.[141]

During the war years the physical security of immigrants to Switzerland and of Jews in that country gave great cause for concern to observers abroad. This included fears for G. Pólya and for the early immigrant to Switzerland from Germany, Heinz Hopf.[142] Pólya, the former collaborator of Weyl's in Zurich, was described by Weyl as "at least half-Jewish" and as a "top-notch mathematician, of great productivity" in a letter to the director of the IAS in July 1940. He continued in the letter:

> I do not think that Pólya is in any immediate danger or was in immediate danger of losing his job in Zürich, although it is unpredictable to which lengths the Swiss will go in order to placate the Nazis. Pólya abhors Nazism and will find it hard to breathe in the present European atmosphere. He obviously wants very much to have a chance to come to this country, yet it may well be that later he will decide to return to his position in Zürich.[143]

As is well known, in late 1940 Pólya went to the United States (first to Brown University, two years later to Stanford), where he remained for the rest of his life. In 1941, the topologist Heinz Hopf, who had been a Rockefeller fellow in Princeton in the 1920s, wrote to S. Lefschetz in Princeton:

> In one of your letters you ask whether I am interested in a position in the USA. . . . Of course my wife and I have been thinking for quite some time about that problem. But, although even in more quiet times America and its scientific life have much that is attractive, we nevertheless try to hold out here. Because, apart from personal obligations that keep us here rather strongly, we deem it right, for principal reasons, not to leave the ship as long there is, in spite of the tempest, a chance that it does not sink. The risk of falling into the water must be taken, but even this is easier when one knows that on the other shore there are friends and colleagues who would help to pull one out of it.[144]

[141]H. Weyl to Miss E. Daetsch, American Friends Service Committee, March 21, 1939. OVP, cont. 32, f. Schwerdtfeger, Hans, 1935–40.

[142]The fate of Pólya and Hopf, who were both threatened by Nazi racism, was often discussed by their friends in the United States. Wasow (1986), p. 271, also talks about "the threat of a German invasion" as a widespread feeling in Switzerland at that time.

[143]Weyl to F. Aydelotte, July 21, 1940. OVP, cont. 32, f. Pólya, George, 1940–41.

[144]H. Hopf to S. Lefschetz, copy. Weyl Papers, ETH, Hs 91:289 (T). Hopf was half-Jewish and potentially threatened by the Nazis.

The topologist Hans Samelson described in a letter from 1994 his stay at the ETH in Zurich, after having left Breslau as "half Jew" ["Halbjude"] with the following words:

> It was clear from the beginning that my only chance was eventual immigration into the USA; in Switzerland my stay was definitely temporary, but as long as I got enough money from home or otherwise to live on—but not from work—they let me stay there.[145]

Samelson also notes that he came to the United States in 1941 on the French immigration quota, since he had been born in Strasbourg in 1916, regarded by the Americans as French even though that town was only reincorporated into France after World War I. In Samelson's letter to me there is also a breathtaking description of his escape through Spain, an extremely dangerous undertaking since he was—in spite of his descent—liable for military service in Germany. Hermann Weyl wrote in 1941 to Heinz Hopf in Zurich, who had coauthored an article with Samelson on Lie Groups in 1940:

> The Samelsons arrived safely, though after quite an unpleasant journey. He makes an excellent impression on everybody around here. In one of your earlier letters you wrote that his probability of surviving in Switzerland was zero.[146]

In 1940, Heinz Hopf wrote to Hans Freudenthal in the Netherlands about Walter Fröhlich, Prague, who, although he had already received a stipend and visa for England, had to reach a neutral country first, in order to emigrate:

> I deem it, unfortunately, totally unrealistic to get him an immigration visa to Switzerland now. How about Holland. It is, by the way, a matter of Fr. with wife.[147]

In the end Fröhlich did not make it; he perished in a Nazi camp in 1942.

---

[145]H. Samelson to me, May 2, 1994. The fate of the Samelson brothers Hans and Klaus, the latter a pioneer of informatics in Germany after the war, is described in Tamari (2007), pp. 288–303. Particularly moving is the suicide of their Jewish father Siegfried Samelson in Breslau after the November pogrom in 1938, which enabled the rest of the family to survive under slightly better conditions. The great-uncle of the Samelson brothers was the famous German Jewish geometer Moritz Pasch (1843–1930).

[146]Weyl to H. Hopf, June 19, 1941. Weyl Papers, ETH, Hs 91:291. I thank Liliane Beaulieu (Nancy) for this information.

[147]Hopf to Freudenthal, February 19, 1940. Freudenthal Papers, ETH, Hs 1183:53 (T). Thanks to Liliane Beaulieu (Nancy) for pointing this letter out to me. In a letter to Weyl, dated Prague, December 8, 1939, Fröhlich reported that his British visa had become invalid due to the outbreak of the war. OVP, cont. 30, f. Fröhlich, W., 1939–42. See above on immigration to Great Britain.

Figure 29 *Hans Samelson (1916–2005). The noted topologist who was born in Strasbourg came through Switzerland (1936) to the United States, using the French immigration quota in 1941. He is the youngest refugee considered in this book and lived until 2005 in Stanford, California. His younger brother Klaus Samelson (1918–1980) could not emigrate, he started his career only after the war and became a pioneer of informatics in Germany.*

Turkey[148]

In 1933, Richard Courant turned down an offer for a position in Istanbul, which Richard von Mises, however, later accepted. In July 1933 Courant explained to the applied mathematician Wilhelm Prager (who would eventually go to Istanbul as professor for mechanics) the plans of the Turks and their contacts to a "Swiss Committee." Therefore Courant and the

[148]To Turkey fled temporarily or permanently H. Geiringer, R. von Mises, W. Prager, H. Reichenbach, and W. Ornstein. For the general situation of the emigrants see Neumark (1980) and Widmann (1973). See also Erichsen (1994).

Göttingen experimental physicist James Franck traveled to Zurich for negotiations and also visited Istanbul.[149] Courant gave a reason for his refusal to accept a position in Istanbul in a letter to Reidemeister:

> Franck and I have turned down the appointments, with the remark that for the scientific-pedagogic and organizational tasks there the younger scholars should be preferred.[150]

Instead, it was not a younger scholar, but the four-years'-older von Mises who did not shy away from taking over the burdens of starting from zero once again. In September 1933 Courant had written to von Mises that he had come to the conclusion that "the Turks have serious aspirations to build something solid out of the now existing nothing there."[151] Courant had apparently rejected the position for financial reasons, too, as indicated in October 1933, in a letter to Marianne Landau, the wife of the dismissed Göttingen number theorist Edmund Landau:

> As Mises (who as a bachelor with a considerable fortune in Austria has apparently also a position in Prague as a reserve) has offered his service in the meantime, the matter has become unrealistic for me anyway.[152]

For the immigrants to Turkey there were, of course, all kinds of problems in adapting to the rather foreign culture. Language posed a particular problem. Main languages in teaching were French and Turkish. Finally, von Mises and his former assistant from Berlin and future wife, Hilda Geiringer, who had come to Turkey one year later, gave their lectures partly in Turkish.[153]

The philosopher and probabilist Hans Reichenbach experienced considerable problems in overcoming the cultural shock of his immigration to Turkey. In 1936, he wrote the following letter to the physicist Alfred Landé (1888–1975) who had been in the United States since 1931:

> In intellectual respects it is really disappointing. The country here is still far from being a cultured people [Kulturvolk]. The government is open to reforms, but the people are not behind them, and also the university reform, which is, by the way, lavishly funded, lacks support from intellectually interested strata of society. . . . So far I know very little Turkish. We all teach with translators who translate every sentence separately into Turkish. . . . In addition there is a rather

---

[149]Courant to W. Prager, July 20, 1933, CPP (T). The "Swiss Committee" is probably the "Emergency Committee for German Scientists abroad" (Notgemeinschaft für deutsche Wissenschaftler im Ausland) mentioned by P. Thullen in his report to Courant quoted above.

[150]Courant to K. Reidemeister, September 7, 1933, CPP (T).

[151]Courant to R. v. Mises, September 13, 1933, CPP (T).

[152]Courant to M. Landau, CPP (T).

[153]On October 22, 1937, von Mises writes in his personal diaries: "For the first time lectured fully in Turkish." HUA, Richard von Mises Papers, HUG 4574.2.

nasty form of nationalism in the country, also among students, which contains an unpleasant element of anti-Semitism. . . . You write that you will soon achieve American citizenship. That is the great advantage with a country like the U.S.A. being able to adapt one's entire being to the new country. Here such naturalization is not attractive at all.[154]

To Einstein, Reichenbach wrote at the same time that "the organization of the university is such that the natural scientists are entirely cut off from philosophy."[155]

Economic problems exacerbated the difficulties. The increasing pressure to move on to another country is described in a letter of mid-1939 by Prager in Istanbul to Weyl:

Although Turkey has joined the English alliance, German influence in the economic domain is steadily growing. The fact that Germany is paying prices, if only for commodities, way above the world market level, does not make a move away from German influence any easier. . . . It cannot be ruled out that the repeated demand on the part of the Germans to get rid of the emigrants at the university here will be honored this time.[156]

Weyl replied in July 1939:

It should not pose a problem for an "applied" [mathematician] of your direction and caliber to find something acceptable [in the U.S.]. . . . It has become more and more complicated in this world to escape the Nazi polyp.[157]

But it turned out to be indeed difficult also for Prager, and he applied at about the same time in vain for Turkish citizenship, as he explained in April 1940 in a letter to another refugee, Hans Reissner in Chicago, when Prager was still in Istanbul:

15 months ago, asked by the rector of the university, I applied, together with several colleagues, for naturalization, because I realized that such a large number of foreigners in the faculty would sooner or later become unbearable for the government. . . . These applications were first supported by the government but shelved after the war started in Europe. So I am still a German citizen. The dangers which result from this fact for my position in a country allied with the Western powers are self-explanatory.[158]

[154]Reichenbach to A. Landé, December 27, 1934. Landé Papers 64, fol. 5, Manuscript Division, SBPK Berlin (T).

[155]Reichenbach to Einstein, April 12, 1936 (T). Einstein Papers, Jerusalem, 20 107-1.

[156]W. Prager to H. Weyl, June 28, 1939 (T). Veblen Papers, Library of Congress, cont. 32, f. Prager, Willy. I thank H. Mehrtens (Braunschweig) for pointing this letter out to me.

[157]Ibid. Weyl to Prager, July 17, 1939 (T).

[158]W. Prager to Hans Reissner, April 27, 1940. Reissner Papers, San Diego, box 1, f. 32.

In early 1939 Hilda Geiringer, who had a minor and temporary position at the university, was denied a continuation of her contract. Von Mises wrote in April 1939 to the British SPSL:

> Mrs. Hilda Geiringer . . . was until now employed as an extraordinary professor of mathematics at the University of Istanbul. Her contract with the Turkish government expires at the end of this term and will not be renewed because the Turkish authorities think that they can already advance Turks to places like this. After our general opinion this attempt of the government is considerably premature and we are all convinced that it can only be considered as the beginning of systematically replacing of all foreign professors by indigenous forces and explained by political intentions.[159]

Von Mises made the continuation for Geiringer's post a condition for renegotiations of his own position. Therefore he inquired in March 1939 with von Kármán in California about conditions for immigration to America. He wrote he did not have a pressing need for a remunerated position straight away, but: "The risk of being captured by the Third Reich is becoming too great."[160] Indeed the Jewish immigrants were without doubt in a much more precarious situation than other immigrants to Turkey.[161] In June that same year, Weyl wrote to von Mises's old acquaintance from Frankfurt, Ernst Hellinger:

> Von Mises is coming—without a salary, obviously he is prosperous enough to live this way in America—to Harvard. It looks like the whole affair in Istanbul is about to collapse.[162]

Hilda Geiringer, however, whose many applications for positions in America were turned down,[163] found a temporary and dangerous refuge Lisbon in September 1939, when the war had already broken out. Geiringer wrote, in September and October 1939, several emotional and desperate letters from Lisbon to her teacher and friend Richard von Mises, who was already in the

---

[159]SPSL, box 279, f. 3 (Pollaczek-Geiringer, H.), Mises to SPSL, April 26, 1939, folio 92.

[160]R. von Mises to Th. von Kármán, March 28, 1939. Kármán Papers, Caltech, Pasadena, 79.25 (T). See also Appendix 4.3 and the epigraph in the present chapter.

[161]The English mathematician Patrick du Val (1903–1987) went to Istanbul in 1941 and reported in a letter to Louis Mordell on June 4, 1942, that he was the only foreigner among the mathematics professors of the university but that in other faculties there were almost exclusively foreigners among the professors. See the Mordell Papers at St. John's College Library, Cambridge, 19.43.

[162]Weyl to Hellinger, June 30, 1939 (T). OVP, cont. 31, f. Hellinger, Ernst, 1938–41.

[163]According to von Mises's personal diaries, May 5, 1939, she was turned down by the Bryn Mawr College, which finally would accept her in September 1939, following von Mises's personal visit in Philadelphia.

Figure 30 *Hilda Geiringer (1893–1973). The former Berlin assistant and future wife of Richard von Mises achieved important results of her own (statistics, plasticity). In both the Turkish and the American emigrations she never received a position commensurate with her abilities. After von Mises's death in 1953 she devoted most of her time to the edition of his scientific papers.*

United States. On October 1, 1939 the long-term assistant and future wife (1943) of Richard von Mises wrote:

> Is there no way to marry pro cura? Here an emigrant who has a resident's permit has married his "bride" and she was then allowed to come to him straight from Vienna.[164]

Finally, in October frantic efforts on the part of von Mises, then in the United States, secured the life-saving visa.

Yugoslavia

Despite having been born in Munich, the student of A. Hammerstein in Berlin, Michael Golomb, had a Polish passport, due to his parents having once come to Germany from the Austrian part of Poland. Golomb had not succeeded, during the 1920s, in obtaining German citizenship. This was an additional handicap to his emigration in 1933:

> In 1933—time of deep [economic; R. S.] depression—I could not with a Polish passport get a visa to any of the countries with some significant mathematical activity that I would have preferred to go, not even a visitor's visa. By chance I received an offer for an assistantship from a professor of astronomy at the University of Belgrad and at the same time a 3-months visitor visum for the visit of my sister, who lived in Yugoslavia. The Yugoslav government would not permit me to accept the offered assistantship nor any other employment, and I lived there for five years under constant threat of deportation, supported only by charity and the help of friends, until I got the American immigration visum toward the end of 1938.[165]

In the same retrospective letter of 1993, Golomb made it clear that, originally, the United States was not his first choice of a host country:

> If I had had the choice of a country for building my career in mathematics, I would have preferred strictly on scientific criteria a dozen European countries to the USA. . . . I myself believed American mathematics and the American scientific culture in general to be mainly a European import. This belief was reinforced by the fact that the young Berliner Privatdozenten H. Hopf and v. Neumann were invited to the most prestigious American institutions.

---

[164]Richard von Mises Papers, HUA, 4574.5. box 3, f. 1939 (T). See also Appendix 4.3 for von Mises's successful efforts to secure a position for her in the United States.
[165]Golomb to me, July 19, 1993.

Other Places of Refuge (Poland, Czechoslovakia, Ireland, Italy, Egypt, South Africa)[166]

Although Mussolini's fascism was in power in Italy, the country offered relative protection to some refugees such as R. Frucht, P. Thullen, and W. Wasow immediately after 1933.[167] From 1934 on, however, restrictions on German immigrants applied. In 1936 the German-Italian "Axis" was formed. The Italian racial laws of 1938 finally forced R. Frucht out of the country.[168] In *Prague* Max Pinl shared a room with F. Behrend who was covering the cost during Behrend's stay as a refugee there.[169] The efforts of Heinrich Löwig to emigrate after his dismissal as a high school teacher from German-occupied Neutitschein (Moravia) in 1938 and as a Privatdozent from the German University in Prague in 1939 were not successful. Although Jewish, Löwig somehow survived German occupation and immigrated to Australia in 1947.[170] In the case of Peter Scherk, the Nazi authorities used his regular visits to Prague in 1935–36 on the invitation of Karl Löwner as a pretext to get rid of him when he wanted to return to Berlin in 1936. This provoked Weyl to the response that he had not yet experienced such an extreme case of arbitrary Nazi policies.[171] In 1935, W. Sternberg went to Prague from Palestine. Courant's suggestion to look for a place in Egypt was declined by Sternberg due to his fear of nationalistic, anti-Semitic sentiments there.[172]

---

[166]These "other places," which have been barely investigated so far in the literature, gave temporary or permanent residence to L. Hopf (Ireland); R. Frucht, P. Thullen, and W. Wasow (Italy); L. Lichtenstein and F. Rothberger (Poland); A. Marx (South Africa); W. Ornstein (Poland, Egypt); F. Behrend, W. Dubislav, P. Hertz, F. Pollaczek, W. Romberg, P. Scherk, H. Schwerdtfeger, W. Sternberg (Czechoslovakia).

[167]See Wasow (1986). For the general situation of Jewish refugees in Italy see Voigt (1988).

[168]Frucht (1982). He hoped temporarily to get help from the IAS in Princeton and wrote in a letter to a Miss G. Blake of the IAS that he had become stateless due to the anti-Semitic laws in Italy, and "no other European country permits the immigration of persons without citizenship" (OVP, cont. 30, f. Frucht, R. Frucht to Blake, Trieste, September 16, 1938).

[169]Pinl (1969), p. 174. See also Pinl's report on the situation after the Munich Dictate of 1938, in Appendix 3.5.

[170]See Löwig's file in SPSL, box 282, f. 1, where there is among other things a report, "Details of national and racial politics in Czechoslovakia," both during German occupation and after the war (ibid., folio 110/111).

[171]Scherk to Weyl, Prague, October 28, 1936, and Weyl to Scherk, November 17, 1936. OVP, cont. 32, f. Scherk, P. Scherk wrote in his letter: "I have never been active in politics, neither in Prague nor elsewhere" (T).

[172]Sternberg to Courant, undated, after January 30, 1935, CPP. According to this letter the established mathematician in Jerusalem, the Hungarian M. Fekete (1886–1957), did not see a chance for Sternberg in Cairo, where Fekete himself had once been denied an invitation.

Ludwig Hopf in Aachen escaped only at the last minute.[173] The chapter 5 epigraph is from the gloomy prophecy by James Franck about Hopf's emigration, which fortunately did not turn out to be true in his case. However it is unclear to what extent Hopf's early death in 1939 in Dublin, where he fled at the last moment via Cambridge, is connected to his sufferings during the time of his persecution or not. He had been threatened by incarceration, but it was his son Arnold Hopf who was taken instead of him to a concentration camp before he could immigrate to Kenya. Hopf's wife wrote in a letter to von Kármán about her son and the "three weeks concentration camp Buchenwald, the hardest thing my husband and I had to suffer in recent years."[174]

[173]On Hopf's emigration see Eckert (1993) and Sommerfeld and Seewald (1952/53).

[174]Mrs. Alice Hopf to Th. v. Kármán, May 19, 1940. Kármán Papers, Caltech, Pasadena, 13.31 (T).

SEVEN

Diminishing Ties with Germany and Self-Image of the Refugees

> We German mathematicians and physicists abroad have no reason for drawing such a sharp line between ourselves and our colleagues who have remained in Germany. We know that by far the greater part of them inwardly think as we think.
> —Hermann Weyl 1936[1]

> Your friends in America, for example, could not understand why you as a Dutchman chose to stay with the Nazis.
> —Richard Courant 1945[2]

In 1953, the social scientist Franz L. Neumann (1900–1956), himself an immigrant to America, distinguished between three types of immigrants: those desperately clinging to their pasts; others embracing, often without a backward glance, the culture of their new home; and, finally, those attempting to combine the two.[3] This differentiation, probably most applicable to the social sciences, has to be modified regarding the natural sciences and mathematics by taking into account the bigger division of scientific and political opinions in these fields. As to their scientific leanings almost all of the mathematical emigrants were of the "Neumannian type 3": in order to remain successful in their work they had to adapt to the new environment and also retain old "European" traditions in their research areas. On political and philosophical levels, however, concrete and individual circumstances of emigration played a deciding role in whether or not the emigrant in mathematics accepted the new environment and to what extent he maintained personal and emotional ties with his/her past in Germany. These "concrete circumstances" include both expulsion and reception, as discussed in the previous chapters.

One major concern for many emigrants was, of course, the fate of their relatives left behind in Germany. This affected all emigrants whether or not

---

[1] Hermann Weyl (Princeton) to Emil Julius Gumbel (Lyon) on March 19, 1936. OVP, cont. 31, f. Gumbel, Emil, 1936–44. Original English quotation.

[2] Richard Courant (New York) to B. L. van der Waerden, December 13, 1945. OVP, cont. 4, f. Courant, 1939–48. Original English quotation.

[3] Quoted from Greenberg (1996), p. 273.

they were mathematicians, and therefore this concern will be given only a fleeting glance in this chapter (D). Instead I will concentrate on discussing what effect concrete conditions of expulsion had on the emigrants' professional and political "self-image" and on their relations to the *mathematicians* remaining in Germany. In the first chapter the difference between "early, relatively voluntary" and "forced" emigration has been examined in detail. Further differentiations within "forced" emigration are, however, necessary. It becomes clear that circumstances surrounding the emigration, the "reasons" and pretexts for expulsions created by the Nazis, are important in understanding the nuances and facets revealed in the varying self-images among the emigrants.[4] Available documents point out differences in the political positions of the emigrants and in their relations to Nazi Germany in general, to the mathematicians remaining there, and to their fellow emigrants. However, the documents also reveal strong similarities in the attitudes of emigrants, based on the common experience of persecution and on shared mathematical interests. Among these similarities were strong feelings of responsibility for the further development of German mathematics even under the conditions imposed by the Nazis, and interest in maintaining of scientific contact with Germany. That feeling of responsibility seems remarkable, seen from today's viewpoint. More surprising is the fact that it also existed among Jewish mathematicians who had suffered the most arbitrary and unexplainable persecution. Many of them did not hold a grudge against those non-Jewish mathematicians remaining in Germany, ones who had often taken jobs previously held by the emigrants (D). At the same time, emigrants' first reactions were frequently coupled with a considerable amount of incredulity and with desperate, irrational hopes for a change in the conditions in Germany for the better, especially hopes for the quick fall of the regime. The emigrants often showed understanding for those German mathematicians who came to terms with the Nazi regime, since they did not envisage realistic alternatives for them. Less understanding was shown for foreign mathematicians choosing to stay in Germany, such as Bartel Leendert van der Waerden (1903–1996). Similarly, mathematicians choosing to return to Germany in spite of having positions or opportunities abroad, such as Eberhard Hopf and (temporarily) Carl Ludwig Siegel, were met with criticism by emigrants. The same applied to mathematicians such as the Austrian K. Gödel,[5] who hesitated in accepting offers

---

[4]The mathematician W. Magnus, in his critique of Pinl (1969), points to the importance of taking into account the concrete "reasons" of expulsion. See the epilogue (chapter 11) below. At times, Hermann Weyl considered himself a "voluntary emigrant," not least in order to avoid too strong a separation from the mathematicians in Germany (D). That Weyl's idea was marked by self-deception, given the concrete threats against his family in 1933, has been pointed out above.

[5]See the criticism by Gödel's countryman, Karl Menger, in Menger (1994), p. 224.

from abroad. Home-comers to Germany such as W. Maier and Walter Tollmien (1900–1968), who did not have real alternatives in the United States, were apparently perceived with more leniency by the emigrants. Even in these cases, however, the attitudes of the emigrants were not necessarily united or politically unequivocal and were often influenced by the dominating interest in maintaining international communication in mathematics. Contemporary opinions (of both emigrants and foreigners) about van der Waerden and E. Hopf became even more critical after World War II, influenced by the defeat of Hitler's Germany.

In fact, the concrete point of time is very important in judging emigrants' attitudes. Geographic and temporal distance, coupled with their journey through several transitional countries of refuge, caused the emigrants to distance themselves more and more both emotionally and politically from Hitler's Germany. Correspondingly, scientific communication (publication in German journals, letters) weakened. Disillusionment set in. As will be shown in chapter 9, conditions during emigration often led to a certain "politicization" even of previously "apolitical" mathematicians, based on the relatively new experience of dependence on political developments and decisions both in Germany and in the host countries. The still existing relations to mathematicians in Germany parallel with the requirements of adaptation to the new political environment in the United States imposed considerable restraint on the immigrants, at least regarding *public* political statements. A striking example was the hesitant way in which the prominent emigrants Weyl and Courant reacted to Emil Julius Gumbel's plan for publishing a book containing political statements by emigrants.[6] The book *Free Science: A Collection from German Emigration* (*Freie Wissenschaft. Ein Sammelbuch der deutschen Emigration*) was finally published in German in 1938 in Strasbourg (S). The discussion about Gumbel's book on emigration also pointed out continuing political divides prevalent before 1933 between persons, who later became emigrants, especially Gumbel and Weyl. These aftereffects were also very visible in a controversy between Weyl and Busemann on the one side and Lüneburg and Courant on the other, and played an active part in Lüneburg failing to gain an academic appointment in the United States (S).

The fate of the emigrant Emil Julius Gumbel, who produced work of considerable significance in mathematical statistics[7] but whose political

---

[6]Einstein's refusal to publish in this book may have had other reasons. He certainly was not politically cautious.

[7]Gumbel's specialty was the "statistics of extremes" (rare events, floods, metal fatigue), a field that owes him its first monograph in 1958. See Hertz (1997). On Gumbel's politics see Mehrtens (1990a), Jansen (1991), and Gumbel (1991). See also the case study in chapter 9 on Gumbel as a political emigrant.

activity and personality had caused a division in the community of German mathematicians in the 1920s,[8] shows the complexity of the scientific and political factors influencing the ideology of the immigrants, who, often together with Americans, had a large say in positions for their fellow immigrants (D). Among the factors preventing a lasting academic appointment for Gumbel in the United States were prejudices against certain "marginal areas" of research like parts of mathematical statistics, not cultivated at the main centers of pure mathematics such as the Institute for Advanced Study in Princeton.[9] The initial policies of Veblen, Weyl, and others to only support excellent "pure" mathematicians were later on corrected or complemented by efforts from other quarters, when the war created an increasing demand for trained mathematicians. Also, the concern for pensions for older mathematicians led to a stronger consideration of social aspects in the appointments of immigrants.[10]

7.D. Documents

7.D.1. Concern for the Fate of Relatives Left Behind

Artur Rosenthal, who had arrived in the United States in 1940 from Holland, wrote in November that year from Ann Arbor (Michigan) to Hermann Weyl about the fate of his mother in Heidelberg:

> My beloved mother has been carried off by the unscrupulous Nazi criminals to a concentration camp in Southern France (Camp de Gurs)—in spite of her 75 years! The same fate has apparently affected nearly all Baden and Palatine Jews. According to everything one hears . . . it is horrible there. During the summer there was already a lack of food and the most primitive things there. . . . Of

[8]Because of his antimilitaristic and antinationalistic politics, Gumbel was dismissed in Heidelberg in 1932, even before the Nazis came to power. The protest against that dismissal by a few colleagues had later consequences for their careers, such as for H. Rademacher, who was dismissed in 1933. Another non-Jewish mathematician who had signed for Gumbel was Gustav Doetsch (1892–1977), who later tried to compensate for his antimilitaristic stance and his joint publications with the Jewish emigrant F. Bernstein by pleasing the Nazis and proposing an institute for war research. See Remmert (2004), pp. 234ff., and Segal (2003), pp. 98–105.

[9]Only in exceptional cases, persons who were not outstanding pure mathematicians could gain access to the IAS, such as Alfred Brauer, who worked mainly as a librarian there. The statistician Felix Bernstein had no chance for a position at the IAS, not even after Einstein's intervention. See below chapter 9. The same applies to Eugen Lukacs, who was also recommended by Einstein but whom Veblen considered "a young mathematician of only moderate distinction . . . not a man we would offer a stipend or any other financial assistance" (O. Veblen to A. Flexner, November 22, 1938, IAS Archives, Faculty, Veblen (box 32), f. 1938–39).

[10]See chapter 9.

course I am trying everything to get her out of the camp, which will hopefully be possible by regular money transfers.[11]

Rosenthal succeeded in getting his mother out of Gurs and to America, while Kurt Grelling had not been able to escape with his wife from the same camp.[12]

7.D.2. The Emotional Ties to Germany and to German Mathematics on the Part of the Emigrants

In September 1933, H. F. Blichfeldt in California wrote to Veblen about Issai Schur in Berlin:

> The best way for him would be a pension in Germany, even a modest one. He is perhaps too old to be transplanted to a country whose language and customs are so different from those of Germany.[13]

Schur received a pension, but this was hardly a solution for the then fifty-eight-year-old scholar. He was finally even denied the use of the mathematical institute's library.[14] In his memorial address over thirty years after Schur's death, his student, Alfred Brauer, had this to say about the situation of his teacher in 1933:

> No sooner had his dismissal been announced than he got an offer from the University of Wisconsin in Madison. He turned the offer down since he did not feel capable enough anymore of teaching in another language.[15]

On the same occasion in 1973 Brauer, once responsible as a mathematician for the mathematical libraries both in Berlin and in the IAS at Princeton, had the following to say about Schur's emotional ties to the German language and German mathematics after his dismissal in 1935:

> When Schur could not sleep at night, he read in the *Jahrbuch über die Fortschritte der Mathematik* [the traditional German mathematical reviewing journal; R. S.]. When later he was forced to sell his library from Israel [obviously Palestine; R. S.], and the Institute for Advanced Study in Princeton was interested in the *Jahrbuch*, Schur sent a telegram, as late as a few weeks before his death, refusing to sell the *Jahrbuch*. The copy was acquired by the Institute only after his death.[16]

[11]Rosenthal to Weyl, November 30, 1940 (T). OVP, cont. 32, f. Rosenthal, Artur, 1938–42.
[12]Ibid. Rosenthal to Weyl, July 12, 1941. On Grelling's fate see chapter 6.
[13]H. F. Blichfeldt to O. Veblen, September 4, 1933, Richardson Papers, BUA, Correspondence 1933 (German-Jewish Situation), f. Blichfeldt.
[14]See Brüning, Ferus, and Siegmund-Schultze (1998), pp. 25ff.
[15]Brauer (1973), vi (T).
[16]Ibid., p. xiii (T).

Figure 31 *Tombstone of Issai Schur. Tombstone of Issai Schur (1875–1941) and his wife in Tel Aviv with the following inscription:*

Yeshajahu Schur
Professor der Mathematik
4 Schwat 5635
12 Tevet 5701

Regina Schur
8 Schwat 5641
20 Adar A 5725

Mathematicians generally, regardless of age and nationality, often held strong emotional attachments to Germany. In many letters written by Willy Feller during his emigration, even in discussions about his former friend and later Nazi activist E. Tornier (in correspondence with E. J. Gumbel), Feller's longing becomes clear. Feller's American student Gian-Carlo Rota remarked in 1989 on the strong ties with Germany on the part of Feller (originally from Croatia) who had changed his name when studying in Göttingen:

> He did not like to be reminded of his Balkan origins, and I had the impression that in America he wanted to be taken for a German who had Anglicized his name.[17]

Again, according to Rota, Feller possessed a "boundless admiration for . . . German mathematics."[18] It is probably no coincidence that Feller, unlike many other emigrants, published in German in the *Mathematische Annalen* as late as 1937.[19] In his letters Feller often alludes to his Jewish descent, showing to which degree he was hurt.[20] In 1934 he wrote to Courant from Stockholm, apparently mixing sarcasm with some racist prejudice of his own:

> The prospects here are of course ε/3, but the Assistant Council tries to sell me off [verschachern] to a Negro college in Atlanta (Georgia) according to the maxim "the inferior races to the inferior races."[21]

John von Neumann, who had been teaching semester-wise in Princeton since 1930, maintained his strong emotional attachment to the European and in particular German mathematical culture. Even after the Nazis came to power in 1933 he indicated in letters to America that he was finding it hard to be cut off from the German culture and that given a choice he would have preferred to teach in Berlin rather than in Princeton also during

[17]Rota (1989), p. 227.

[18]Ibid., p. 229.

[19]Feller (1937).

[20]In SPSL, box 278, f. 7, folio 368, Feller says: "Non-arian descent. My grand-father was a Jew." Richard von Mises, who apparently sensed a certain competition with Feller in probability theory during American emigration, mocked Feller for allegedly having said to be "75% Aryan." Von Mises to von Kármán, September 2, 1942, Kármán Papers 20.37.

[21]W. Feller to R. Courant, November 11, 1934, CPP (T). Maybe it was the same "Negro college" in Atlanta that was discussed in the Emergency Committee as a possible place of employment for Gumbel in 1934. Gumbel, who was then still in France, was concerned whether "the fact of having been in a Negro College [is] absolutely prohibitive so that there is no possibility to return to a 'decent' University?" (Dunn Papers, American Philosophical Society, Gumbel to Dunn, Lyon, August 22, 1934).

the impending summer term. He wrote to Veblen on March 19, 1933 from Budapest, in his still rudimentary English:

> On the way to Budapest, we stopped on Febr. 7/8 for 36 hours in Berlin. I met there E. Schmidt and I. Schur. The effects of the political changes where [*sic*] not yet to be felt explicitly then, f. i. the new "Nazi" minister of education (Mr. Rust) was only 24 hours in office, and nobody new [*sic*] exactly, what he would do. People felt very uncertain then, but not too pessimistic. So I decided, to act as I intended, when we talked over it in Princeton: to resign immediately, but to give lectures in this summer-term. The newer German developments make me doubt, wether [*sic*] this is reasonable, and if [it] would not be better to cancel my German lectures altogether.[22]

7.D.3. Maintenance and Gradual Restriction of the Emigrants' Personal and Scientific Relations to Germany

It would have been surprising and even more painful for all parts involved if the mathematical collaboration between the emigrants and the mathematicians remaining in Germany had ended immediately with the dismissals. The close cooperation between Courant (New York City) and K. Friedrichs (still in Brunswick) in volume 2 of the "Courant-Hilbert" (*Methoden der Mathematischen Physik*, 1937) has often been mentioned in the literature.[23] Theodor von Kármán maintained close contacts with the Kaiser Wilhelm Institute for Fluid Dynamics in Göttingen and even with the head of the research department in the Reich Aviation Ministry (Reichsluftfahrtministerium), Adolf Baeumker (1891–1976).[24] Specific problems in the cooperation between emigrants and nonemigrants in mathematical publishing are illustrated by the controversy between B. L. van der Waerden (who remained in Germany) and the emigrant Richard Brauer over the—in the end—nonpublication of Brauer's textbook on algebra.[25] This affair will be discussed in some detail in chapter 10. In mathematical reviewing (*Zentralblatt für Mathematik*), organized by Otto Neugebauer

[22]Veblen Papers, Library of Congress, cont. 15, f. J. von Neumann, 1929–56. In von Neumann's personal file at the archives of Humboldt University Berlin, one finds a copy of his resignation (February 12, 1933) with the expressed option of an unpaid teaching assignment in the summer of 1933. UAB, UK von Neumann 44, fol. 19.

[23]Reid (1976) and (1983). The cooperation can be followed in its mathematical detail in Courant's private correspondence CPP.

[24]After the war, Adolf Baeumker worked for the American Air Force and then again for German rearmament.

[25]Similar conflicts arose over the cooperation of emigrated mathematicians in the second edition of the *Encyclopädie der Mathematischen Wissenschaften*, which was edited by Erich Hecke und Helmut Hasse. They are documented in Hecke's papers in Hamburg. See also Feit (1979) and Tobies (1994).

from Copenhagen after his emigration in 1933, there was a rather sober and objective cooperation between Germans, emigrants, and foreigners, until in 1938 anti-Semitic repressions in the *Zentralblatt* triggered the foundation of the *Mathematical Reviews* in the United States in 1940.[26]

Personal contacts were also maintained as much as possible. The physicist Maria Göppert (1906–1972) reported to Courant on April 29, 1935, about her visit to the seventy-three-year-old ailing David Hilbert:

> We all noticed his increasing loss of memory, which makes one sad sometimes. He does not want to concern himself with the new times; he deliberately fends off everything and does not read newspapers anymore. (CPP [T])

In 1935, Richard Courant turned to Richard Brauer with the proposal that his brother Alfred Brauer, who was still in Germany, should help sell in America the mathematical books owned by dismissed colleagues, in order to finance their emigration.[27]

Hermann Weyl's preparations for his obituaries of Emmy Noether (1935) and David Hilbert (1944) show on the one hand his unabated attachment to German mathematics, but, on the other hand, his dwindling contacts with Germany.[28]

Outstanding results in mathematical research were, of course, observed attentively from both sides. For instance, Erich Bessel-Hagen (1898–1946) in Bonn wrote on January 3, 1935 to Courant in New York:

> In our science I was extremely excited by Heilbronn's discovery (proof of the Gauss conjecture) with everything connected to it, in particular the class number approximation [Classenanzahlabschätzung] by Siegel. This is a <u>very</u> great source of pleasure. (CPP [T])

7.D.4. Conflicting Opinions on Mathematicians Remaining in Germany and on Those Who Returned in Spite of Chances Abroad

In 1935 the number theorist Wilhelm Maier decided to return to Germany. At that time he held a permanent position at Purdue University in Lafayette, Indiana, one he had received after his Rockefeller stipend in 1929–30. His efforts, however, to build a research group in Indiana had

---

[26]Siegmund-Schultze (1994).

[27]Courant to R. Brauer, December 3, 1935, CPP. The IAS, as was seen above, bought I. Schur's mathematical library with A. Brauer's help.

[28]In the Noether obituary, Weyl deplored that he could not get exact biographical details from Germany. Richardson supported Weyl in the preparation of his Hilbert obituary, which finally appeared in the *Bulletin of the AMS*. See Richardson Papers, Correspondence, BUA, f. 125 W (1943–45).

failed. The head of the mathematical department at Purdue wrote about this in October 1935:

> We regret very much Professor Maier's resignation but on the whole perhaps it was the best thing which could have happened. He was not particularly happy here, largely because he did not have students enough and felt that he was not appreciated. He came some five years ago and we hoped at that time that we could gradually build up a modest graduate school in Mathematics.[29]

The return of Maier to a rather insecure teaching assignment in Freiburg was an admission of defeat. The fact that Maier nevertheless obtained a full professorship in Germany (Greifswald) somewhat later in 1938 and in the traditionally strong German domain of number theory probably shows us something about diverging tendencies in international research and partly about the diminishing international competitiveness of German mathematics.[30]

Maier's disillusioned return to Germany could not easily be criticized by the emigrants. However, other mathematicians with good chances abroad but nonetheless choosing to remain in or return to Germany received less understanding abroad. The relatively young Dutchman B. L. van der Waerden, who had become the most prominent proponent of abstract algebra due the publication of his book *Moderne Algebra* in 1930/31 and who chose to stay in Leipzig, wrote to Courant in 1933 (undated) about an invitation from Princeton University:

> I think I will suggest to the Americans that they can use their money more wisely in these times than in getting me out, one who still has a position. (CPP [T])

Van der Waerden, who published a courageous obituary of his teacher Emmy Noether in the *Mathematische Annalen* of 1935, remained in Germany and had to make numerous compromises with the regime, including reverting to Nazi vocabulary on some occasions.[31] In certain respects the emigration of colleagues caused distinct advantages for him in his career,[32]

---

[29]W. Marshall to Richardson, October 15, 1935. Richardson Papers, Correspondence, box German-Jewish Situation, f. O. Szasz.

[30]On the biography of Maier, who did not flourish very much internationally, see Böhm (1965). H. Wefelscheid (Essen) informs me (letter, April 2, 1998) that Siegel continued to appreciate his erstwhile student Maier. The lack of international appreciation is probably connected to the fact that Maier's work was not in the soon dominating trend of "Bourbakism." The latter trend, however, was observed critically by Siegel as well (see chapter 10).

[31]See Soifer (2004/5). In a letter to Courant on September 28, 1935, van der Waerden was surprised about the reactions abroad over the dismissal of A. Berliner as editor of Springer's *Die Naturwissenschaften*. He stressed that Berliner was already seventy-five years old (CPP).

[32]Van der Waerden became coeditor of the *Grundlehren* series of the Springer publishers on the proposal of Courant (Courant to van der Waerden, October 10, 1933, CPP).

even though he certainly deserved every promotion given his outstanding results. The consequences of his stay in Germany became visible and palpable for van der Waerden after the war. When he reported to Courant, in a letter of November 20, 1945, about his problems in the Netherlands, where he was accused of collaboration with the Germans, Courant replied rather coolly:

> Your friends in America, for example, could not understand why you as a Dutchman chose to stay with the Nazis.[33]

In August 1945, Otto Neugebauer wrote to Heinz Hopf in Zurich about van der Waerden:

> I do not mind his remaining a German Professor until the end—I do mind his remaining a German Professor at the beginning.[34]

The return to Germany of Carl Ludwig Siegel and of Eberhard Hopf[35] provoked some discussion among emigrants and Americans. In a consultation with the Rockefeller Foundation in October 1934, Richard Courant called Siegel "the most able and promising mathematician alive."[36] On December 6 that same year, Courant asked Max Dehn in Frankfurt—then not yet dismissed—to write a letter of support for Siegel's stay at Princeton, suggesting a suitable wording for the Nazi authorities:

> For the moment one is immensely interested in a contact to Siegel's works. . . . It is not possible to think of a better cultural propaganda for Germany than to let such a personality work here [in America] for some time. (CPP [T])

However, Siegel, who held himself somewhat aloof from the world and who suffered strong psychological problems in adapting to American society, soon wished to return to Germany. On April 20, 1935, he asked Courant to write a diplomatic letter of resignation in English for him, of which Courant, in Siegel's opinion, had a better command. In his letter to Courant, Siegel stressed his concern about the announced replacement of the dismissed E. Hellinger by the Nazi activist W. Weber:[37]

> In any case there is a real reason for my return to Frankfurt: the expulsion of Weber amicably or by force. (CPP [T])

Siegel, who in his final letter to the mathematicians in Princeton asked for a conferral of his personal invitation to Dehn, made it clear to Courant

---

[33]Courant to van der Waerden, December 13, 1945 (T), Veblen Papers, OVP, cont.4, f. Courant, 1939–48. Also quoted as an epigraph to this chapter.

[34]Neugebauer to H. Hopf, August 15,1945, Hopf Papers, ETH Hs 621:1041 (T).

[35]On Hopf, more below in a case study (S).

[36]Note from October 16, 1934. RAC, RF 1.1., 200 D, box 143, f. 1769.

[37]On Weber's role in Göttingen see Schappacher and Kneser (1990).

that the real reason for his resignation was the atmosphere of prudishness in Princeton.[38] Harald Bohr, in a letter to Courant dated August 27, 1935, deemed it "unbelievably foolish of Siegel" to return to Frankfurt (CPP [T]).

7.D.5. Political Information, Caution, and Self-Censorship in the Contact between Emigrants and Mathematicians Remaining in Germany

Information on the political developments in Germany was provided to the emigrants and to the Americans by Harald Bohr (Copenhagen) and by the function theorist Karl Löwner in Prague, as well as in various anonymous reports.[39] This later worked in Löwner's favor, when seeking an appointment as a refugee in the United States:

> Professor Loewner was able to help American mathematicians keep informed regarding the growth of the Nazi mathematicians in Germany. He and Professor Harald Bohr of Copenhagen were the two persons who kept English and American mathematicians informed concerning the movement. They made trips to Germany and then reported to people on the outside. . . . I have every reason to believe that Loewner was strongly Anti-Nazi.[40]

Mathematicians remaining in Germany had to be cautious in their foreign contacts and political statements.

B. L. van der Waerden to Courant in his letter about his brother-in-law, Franz Rellich (1906–1955), who was at that time a mathematician in Marburg:

> The fact that Franzi due to his heart condition was prevented from visiting you in America or making a trip with Busemann in a collapsible boat is probably not damaging to his career.[41]

Courant replied to van der Waerden who was then vacationing in the Netherlands:

> What you are writing about personal things, also on Franzi, and what I hear from my present guest, Kurt Fr.[iedrichs], confirms the sad picture, which one gets from the newspapers. The pressure on dependent people becomes so strong that they are compelled to end relationships to undesirable [unerwünschten]

---

[38]See chapter 9.

[39]See Appendices 3.1 and 3.2.

[40]Undated note by R. G. D. Richardson. Richardson Papers, BUA, Correspondence, f. 117 L (Ch. Loewner 1943–45).

[41]Van der Waerden to Courant, August 10, 1935, CPP (T).

friends for the sake of their own careers. Nobody can be chided for that. . . .
I wish everybody could get out of this stuffy [stickig] atmosphere. I must admit I cannot understand those who remain in Germany, unless they do it out
of conviction or strong patriotism or from a willingness to fight. It seems to
me more and more that remaining there as a civil servant is impossible without compromises.[42]

Richard Courant wrote on August 29, 1935 to Alwin Walther in Darmstadt, indicating that he wanted to visit Walther's Institute for Practical
Mathematics:

I hope that the rumors which have been spread out in American newspapers
with respect to difficulties for Jews in finding rooms in Hotels are wrong. (CPP)

The letter is contained in its original English form in Courant's correspondence with the German word for "Not" [Nicht] written in large letters crossing the text out. This seems to indicate that Courant found it too
dangerous for Walther to receive such a letter even though it was written
in English. It is unclear whether and somewhat unlikely that Courant
visited Darmstadt nevertheless.[43]

On January 3, 1935 Courant sent a letter to Max M. Warburg prior to
Warburg's trip to Germany, containing information on the political behavior of the physicist Max von Laue (1879–1960), of the mathematicians Max Dehn, F. K. Schmidt (1901–1977) and others. Courant con-
cluded the letter with the words:

You will certainly understand that I am asking you to destroy the letter before
entering German territory. (CPP [T])

In a letter dated December 3, 1934, Hermann Weyl indicated to Courant
that he did not intend to write to Siegel about possible problems arising
from living with his (unmarried) girlfriend in Princeton, because

communications on that in a letter, which might be opened, could do him harm
in Germany (if we recall the problems which Gerlach had from the accusation
of concubinage). (CPP [T])

In the correspondences of mathematicians in those years there are many examples of the use of a "secret mathematical language" in order to conceal

---

[42]Courant to van der Waerden, August 20, 1935, CPP (T). This is most likely an indirect
encouragement to van der Waerden either to fight or to leave Germany. The reference to "patriotism," which van der Waerden could not claim, seems to suggest "to leave" rather than
"to stay."

[43]Reid (1976), p. 179, refers to the same letter without quoting it. Reid does not say
whether Courant in the end went to Darmstadt.

political information. Courant, for one, wrote from Göttingen to Harald Bohr in Copenhagen, in June 1933:

> It seems as though Lüneburg will unfortunately be the first element to be transformed into the purely imaginary axis.[44]

To Otto Neugebauer, meanwhile also in Copenhagen, Courant wrote on September 12, 1933, obviously recommending him to stay abroad:

> Generally Harald [Bohr] tends to emphasize the transformation theory instead of fixed point theorems. That method is also more rewarding for your work. (CPP [T])

7.D.6. Condemnation of Former Colleagues' Commitment to the Nazis by Emigrants

The feeling of abhorrence, incredulity, and condemnation vis-à-vis the most prominent Nazis among the German mathematicians, such as Bieberbach and Vahlen, who came forward with racist theories like "Deutsche Mathematik," was naturally rather common among the emigrants. Nevertheless most emigrants left it to their foreign colleagues like Harald Bohr to take a firm stance on these acts of Nazism among German mathematicians. Even from these foreign mathematicians, however, relatively little open criticism appeared in public or in print. First reactions by Harald Bohr and the English mathematician G. H. Hardy had triggered a crisis in the German Mathematicians' Association (DMV) in 1934.[45] The reluctance of the emigrants to criticize Bieberbach, Vahlen, and the like in public is partly related to the discouragement of political actions in the host countries.[46] Richard von Mises did not, at that time, publish his manuscript on mathematics in the Third Reich,[47] although he wrote a statistical article in 1934 with a clear satirical undertone directed against Nazi racism.[48] Although Americans such as AMS-secretary Richardson were upset about Bieberbach's racist theories, some of them felt that "the public protests against the Nazi regime in Germany seem to us to have done more harm than good to the oppressed people in Germany."[49] Richardson was even slightly irritated when his translation of an article in the German journal *Deutsche Zukunft*, critical and ironical about Bieberbach's racism

---

[44]Bohr Archives, Copenhagen, partial estate Harald Bohr (T).

[45]Mehrtens (1989).

[46]See chapter 9.

[47]Sheynin (2003).

[48]See Siegmund-Schultze and Zabell (2007).

[49]R. G. D. Richardson to Harald Bohr, June 26, 1934. Richardson Papers, box: correspondence (German-Jewish situation), f. H. Bohr.

as it was, appeared in the American journal *Science* without his knowledge and without mentioning his name, because he was "not at all convinced that the publication of this review will be of any service to the cause of freedom."[50] There was more openness of opinion in the private letters of various emigrants.

In March 1936 Weyl wrote to Heinz Hopf in Zurich about the algebraist and Noether student Heinrich Grell, who had meanwhile been dismissed in Halle although he had been active in the Nazi movement shortly after Hitler's seizure of power:

> In Grell's case I am unfortunately rather pessimistic. Personally I am inclined to see him as a victim of the Nazis like the others. . . . Among my German colleagues and here, however, I find the opinion prevailing that he chose the other party in the spring of 1933 and that he must now accept the consequences. Scientifically he is also behind those for whom America should care in the first place.[51]

In December 1938, Otto Neugebauer, who by then had already moved from Copenhagen to Brown University in Providence, received a letter from Oskar Becker (1889–1964) in Bonn. In it Becker demanded the resignation of Otto Toeplitz who was editor (along with Neugebauer) of the historical journal *Quellen und Studien zur Geschichte der Mathematik* because of his Jewish descent, and he proposed E. Bessel-Hagen as a replacement.[52] Neugebauer responded with the announcement of his own resignation and proposed Becker, who had not been editor before, and Bessel-Hagen as the new editors. He then wrote in long-winded sentences filled with icy sarcasm and contempt to Becker:

> I have inferred from your letter that also scientists in Germany consider the fact of extensive pogroms a sufficient reason for making it for a scholar of outstanding merit impossible even to work scientifically. . . . You are writing that you as a National Socialist apparently have an opinion differing from mine, in spite of your personal respect for Mr. Toeplitz. I can only reply that I am not in the happy possession of any "Weltanschauung" and I am therefore in need to consider in each case individually, what to do, without being able to retire to a

---

[50]Ibid. Richardson to the Editor of Science, J. McKeen Cattell, July 20, 1934. The correspondence revealed that O. Veblen had sent Richardson's translation without the Richardson's knowledge to *Science*, where it appeared in volume 80 (1934), pp. 35–36 signed by the original German initials P. S. without giving away the translator, Richardson.

[51]Weyl to H. Hopf, March 17, 1936. Weyl Papers, ETH Zurich, Hs 91:282 (T). Weyl seems to allude to Neugebauer's opinion in particular. See also Grell to Neugebauer, December 10, 1935, CPP.

[52]O. Becker to Neugebauer, December 3,1938. Courant Institute Papers CIP, file: O. Neugebauer. The reply by Neugebauer quoted in English translation below is dated December 13, 1938.

Figure 32 *Otto Neugebauer (1899–1990). The mathematician and historian from Göttingen, who became famous for deciphering the mathematical Babylonian Cuneiform Tablets, left Germany in 1933 due to political opposition. He moved the publishing offices of the abstracting journal* Zentralblatt der Mathematik *to Copenhagen and later (1940) founded the* Mathematical Reviews *in the United States.*

previously given dogma. This disadvantage in practical life is perhaps made good by sparing me to separate from people for whom I have respect simply because they are unhappy enough to be tortured by other people.

7.D.7. Self-Selection by Emigrants

In collaboration with Americans, German immigrants such as Weyl and Courant attempted to build a network of support for their persecuted colleagues. This will be discussed in detail in the next chapter. The other side of solidarity was, however, self-selection among the mathematicians. Generally the mathematical accomplishments and the positive social characteristics (age, social manners) of the refugees were mentioned in the expert opinions written for the various American hosts, clearly conversant with the interests and preferences of the latter.

Courant wrote to Carl Ludwig Siegel on October 16, 1934, about his success in winning over a functionary of the Rockefeller Foundation, who was probably not very competent in modern number theory:

> I can only tell you that the Rockefeller people are of the opinion that of all mathematical problems none is as important and urgent as the one which is connected to quadratic forms. (CPP [T])

In decisions on the direction of emigration, mathematical accomplishments as revealed in publications were decisive.

The mathematician from Göttingen, Hans Schwerdtfeger, received no support in his efforts to immigrate to the United States and went to Australia in 1939. Weyl wrote to Courant in May 1936:

> In the case of Schwerdtfeger it is important to note that he—this is at least the impression I had from him in Göttingen—should not really be a mathematician. And this less than ever under the current aggravating circumstances.[53]

Einstein responded in an undated letter from Princeton to a letter by Max Born from January 1937:

> Schwerdtfeger. Nothing can be done for him here. This is because Weyl and Courant are up in arms at the suggestion. Besides, I have had a closer look at one of his papers, and have the impression of a lack of really profound questioning. Because of the widespread unemployment amongst local people, it is, in any case, very difficult to place anyone here, and if one does succeed, it usually means a lowering of status. But in this case I cannot even attempt it with a clear conscience.[54]

[53]Weyl to Courant, May 2, 1936. Weyl Papers, ETH Zurich, Hs 91:79 (T).
[54]Born ed. (2005), p. 126. Born's letter is quoted above in chapter 4. On Schwerdtfeger's eventual immigration to Australia with the help of Born see chapter 6.

The Russian topologist P. S. Aleksandrov wrote to Courant on August 4, 1935 about the relatively small chances of success for Rudolf Lüneburg in the Soviet Union:

> Although Lüneburg is an outstanding mathematician, he has published much less (purely quantitatively) than some of his companions in misfortune (Cohn-Vossen, even Mahler); that is why our efforts are mounting to little for the time being. (CPP [T])

Courant tried to help his former assistant Lüneburg wherever he could. He found that social questions, such as the previous working conditions of the refugees, should not be disregarded. With respect to Erich Rothe he wrote to Weyl in January 1938:

> It looks like there is a decided aversion against doing something for people who are not wholeheartedly advertised as "first rate." This applies also to the Rockefeller Foundation which asked me about Rothe with the explicit remark that help can only be secured for people who are "absolutely outstanding." What can one do? I cannot reply disingenuously, although I find such schematic judgments meaningless and unjust.[55]

Also Einstein, who was perhaps not in a position to judge Emil Julius Gumbel's mathematical accomplishments and who seemed to support Weyl and others (see below) in their mathematical judgment of him, nevertheless saw the one-sidedness of that approach. To Gumbel he wrote in May 1933:

> I would be glad if you received a position. Accomplishments of character are as important as scientific ones; you must not let yourself be put in the shade.[56]

Gumbel's political activity before January 30, 1933, however, was apparently too much at odds with the positions of the average academics. Even the rather liberal-minded[57] Courant had refused to support a declaration in favor of Gumbel when replying to the prominent statistician and demographer Robert René Kuczynski (1876–1947) on January 18, 1933, even though Kuczynski's declaration had stressed Gumbel's scientific merits (and not his political activities against German nationalist and militaristic

---

[55]Courant to Weyl, January 4, 1938. Weyl Papers, ETH Zurich, Hs 91:88 (T).

[56]Einstein to Gumbel, May 16, 1933. Einstein Papers, Jerusalem, 38 615 (T).

[57]At least Courant was not prone to exaggerated German nationalism, which was the main source of the anti-Gumbel sentiments among many German scholars. When Courant evaluated his student Lüneburg, the latter's former communist sympathies were no issue. Lüneburg's relation to the Göttingen philosopher L. Nelson, however, was rather a recommendation in Courant's eyes (kind communication by N. Schappacher, Strasbourg). See below the case study 7.S.5 on Lüneburg.

organizations) and had called him a "representative of a discipline rarely cultivated in Germany, namely mathematical statistics."[58] Weyl, however, responded to a letter of recommendation in 1944, stressing Gumbel's political merits, with the clear opinion that nothing could be done for him at the IAS in Princeton. According to Weyl it was the dominant opinion among statisticians that Gumbel was competent but not outstanding and that he had made little progress in the last decade[59]—the fact that this had been a decade of hardship and emigration being totally disregarded.[60]

7.S. Case Studies

7.S.1. Richard Courant's Gradual Estrangement from Germany

Shortly before his immigration to England but still hoping for a reversion of that move, Courant wrote in June 1933 from Göttingen to his friend Harald Bohr in Copenhagen:

> Personally we are all quite well. One lives much more quietly here than you abroad with your tendentious newspapers might imagine. Of course you cannot make an omelette without breaking eggs [wo gehobelt wird, fallen Späne], and it is sometimes regrettable that among the eggs there are also Lüneburgs. But all in all I believe that after the situation has cleared and the parties are abolished there will arise—also psychologically—a more stable situation.[61]

On May 3, 1933, Courant had written to Hellmuth Kneser in Greifswald, who had responded to an earlier desperate letter of Courant's[62] with a less than tactful description of his national enthusiasm.[63] Courant replied with stunning tolerance and in desperation:

> Of course I can and must ignore [abstrahieren von] my personal situation and have the general perspective; and in this respect I recognize as you do the positive

---

[58]Kuczynski's declaration is attached to his letter to Courant, dated December 21, 1932, CPP.

[59]Weyl to H. M. Kallen (New School for Social Research, New York City), September 30, 1944.

[60]See also below in chapter 9 more on Gumbel's problems as a political refugee to France and to the United States.

[61]Courant to H. Bohr, June 30, 1933 (T). Niels Bohr Archives, Copenhagen, partial estate Harald Bohr. On R. Lüneburg see the case study below.

[62]Probably Courant's letter of April 28, 1933, as quoted in Beyerchen (1977), p. 22, and mentioned in chapter 4.

[63]A similarly tasteless reaction of H. Hasse's to Courant is reported below in this case study.

accomplishments of the government and see the inner unification of such a huge part of the people as an enormous source of potential strength. To be excluded from participation is something I regret even more.[64]

It sounded more ironical and ambiguous in December 1933, when Courant advised Helmut Ulm (1908–1975), not yet as academically established as Kneser and less willing to adapt politically:

In the current state of mobilization of the German youth the individual [Einzelne] cannot keep away for "individualistic" reasons, if he is not forcibly kept away. Rellich, for example, feels this way and goes to a military camp with great pleasure.[65]

Writing on February 21, 1934 from Cambridge (England) to his former student Fritz Schultz, who had shown himself an ardent supporter of Hitler in a previous letter, Courant still seemed to harbor hopes for a return:

It is very remarkable how quickly Germany regains its reputation here [wieder Boden gewinnt]. . . . Anyway, seen through German eyes, the main reason for our stay abroad is the purely objective dissemination of the German cultural values. (CPP [T])

One has to take into consideration that, at that time, Courant was still applying in Germany for a dispensation from the Reich Flight Tax, and political caution was influencing any letter's wording anyhow. That he had not yet given up hope of returning at some point in time, is revealed in a letter to the industrialist Carl Still (1868–1951) in Recklinghausen (Germany). Facing his imminent move to America, Courant wrote on May 6, 1934:

I had and I have the wish to maintain the connection with Germany and Göttingen as close as possible during my activity in New York. I do not want to live there as an embittered emigrant but rather as a representative of German science. I simply cannot give up hope for a course of events which, after several years, will allow me factually and morally to return to my activity in the homeland and give my children the chance to develop freely. (CPP [T])

In the same letter to Still, Courant remarked that he had not applied to the ministry for a pension but only for leave of absence for two years. However, Courant also reported that the number theorist from Marburg Helmut Hasse (1898–1979) had entered into negotiations with the Nazi

---

[64]Courant to H. Kneser, May 3, 1933, CPP (T). In a later letter, dated June 1, 1933, Courant advised Kneser to join the Nazi party NSDAP.

[65]Courant to H. Ulm, December 22, 1933, CPP (T). With "forcibly kept away," Courant obviously had in mind the exclusion of the Jews. Rellich, as an Austrian citizen, was not admitted to the camp when he arrived. See Segal (2003), p. 128.

ministry for a call to Göttingen and had in these discussions not objected to sending Courant into retirement "in order not to forfeit [the ministry's] confidence."[66] Courant, in his letter to Still, wrote bitterly:

> It is very clear that it is much easier for him and for others to have a free hand in G., to appoint people to their own choice and point of view, without being afraid of possible disturbing home-comers. (CPP [T])

Even as late as in fall 1934 Courant had still not accepted his exile as final. On November 28, 1934 Courant received a letter from Weyl in Princeton communicating to him the opinion of Courant's friend, the Göttingen experimental physicist James Franck:

> Franck is decidedly of the opinion that you should show the courage to put a definite end to your relation with Göttingen and to no longer see yourself as being here on vacation. I understand how severe this decision would be for you.[67]

Weyl's letter may well have had an effect. Courant finally accepted the proposal, drafted by Hasse in collaboration with the ministry, whereby he should apply for "voluntary" retirement.[68] This led to an end of any further contact with Hasse.[69]

In early 1935 Courant resigned as a member of the German Mathematicians' Association DMV.[70] A letter, dated March 6, 1935, to the physicist Arnold Berliner (1862–1942), shortly to be dismissed as editor of *Die Naturwissenschaften*, illustrates Courant's growing emotional estrangement from Germany, also due to increased geographic distance:

> It is amazing how quickly contact with current events and developments in Germany fades. While I thought about these things almost exclusively day and night when I was in England, one feels much more relieved from the pressure here. (CPP [T])

Nevertheless, Courant continued to feel a responsibility toward German science. Importantly, he maintained his contacts with the Springer publishing

[66]Courant is here indirectly quoting Hasse's letter to him, which was dated April 24, 1934. CPP (T).

[67]Weyl to Courant, November 28, 1934 (T). Weyl Papers, ETH Zurich, Hs 91:57. I thank Liliane Beaulieu (Nancy) for this information. Franck had, as mentioned before, publicly resigned his post in Göttingen. The reaction of the mathematicians to Franck's resignation has been discussed in chapter 4.

[68]Schappacher (1987), p. 351.

[69]Courant's justified grudge against Hasse became obvious particularly after the war. See below chapter 11.

[70]Courant's resignation from the DMV, dated February 19, 1935, is in copy in CPP and reproduced in Appendix 4.2.

firm. On January 16, 1935 he sent a report to the German General Consulate in New York City on "The Question of the German Scientific Literature," the original of which he had presented to the cultural department of the Foreign Office [Auswärtiges Amt] in Berlin (CPP). It is very likely that Courant's cooperation with Charles H. Brown (head of the American Library Association), resulting in that report, played a decisive role in a decision being made by the Nazi authorities to subsidize the export of German scientific journals.[71] Springer, with his many journals, obviously profited from that decision. Courant even advised Gabor Szegö in February 1936 to publish his book on orthogonal polynomials with Springer, contrary to Szegö's intent:

> Meanwhile it has become very clear to me that Springer is personally and with extraordinary energy following an objective policy, and I think one has to support him vigorously in that effort. Therefore I have maintained my relations to him as an editor.[72]

Courant, however, knew that political pressure and intrigues were under way in Germany to expel him from his position as an editor of the *Grundlehren* series.[73] He also knew that he had to adapt gradually to the American mathematical community in his publication routines. Thus Courant wrote to Otto Blumenthal, then still managing editor of Springer's *Mathematische Annalen*, on May 22, 1936:

> Regarding the Annalen: I am as always prepared to acquire material for the Annalen. For my part I am preparing a longer article on the Plateau Problem and on conformal mapping, where I have made some definitive progress. Under the prevailing circumstances, however, I have to publish these things here in American journals.[74]

Richard Courant, closely connected to the Berlin Springer publishing house, suffered, with respect to mathematical publication a certain conflict of loyalty. This, however, apparently did not damage his position with the

---

[71]Courant's correspondence with Charles Brown is contained in CPP. On the Nazi measure mentioned, in which also Goebbels's ministry of propaganda was involved, and which was partly related to the notorious lack of foreign currency in Nazi Germany, see Knoche (1990).

[72]Courant to Szegö, February 13, 1936, CPP (T). Szegö nevertheless chose to publish his book in the United States. See also chapter 9.

[73]There are letters to this effect by W. Blaschke and H. Bohr (October 8, 1936) in CPP. Blaschke wrote to Courant (March 12, 1936) that he should not be the only editor of the series because this could produce political problems in Germany. Blaschke proposed F. K. Schmidt as joint main editor.

[74]CPP (T). Blumenthal was managing editor for the *Annalen* for the last time in volume 115 (1937/38).

Americans; in particular it did not affect relations with the Rockefeller Foundation and to Warren Weaver.[75]

7.S.2. Concern for the Future of German Applied Mathematics and the Young Generation: Richard von Mises and Theodor von Kármán Supporting Walter Tollmien's Return to Germany

In spite of his forced emigration from Berlin on which he reflected with bitterness in his diary,[76] Richard von Mises felt a responsibility for the future of his Institute of Applied Mathematics there. In a letter to Erhard Schmidt on October 21, 1933, he even recommended the sixty-four-year-old Nazi mathematician Theodor Vahlen as his successor, because he obviously hoped this move would provide political stability for his former institute:

> As far as the final arrangement is concerned, the only successor who can be taken into consideration is Prof. Erich Trefftz in Dresden. . . . However it appears possible and beneficial to find an interim solution for the coming years, i.e., to find someone who would be capable of safeguarding the existence of the Institute and this even in a direction which conforms with the prevailing trends [heutige Zeitströmung]. I suggest that Prof. Th. Vahlen, who is active for the time being in the ministry for education, but who is evidently not happy there, should take over the chair and the Institute for Applied Mathematics.[77]

Von Mises, who took part as late as October 30, 1933 in the faculty meeting about his successor, had obviously realized that Trefftz, due to his political attitudes, did not have a chance at that time.[78] Vahlen's work in ballistics was—in addition to his political stance—in tune with the "prevailing trends." However, von Mises's recommendation could be interpreted as ironic in some passages. In any case, von Mises's long-term goals were not achieved. The institute would play but a marginal role within German applied mathematics after 1933. The nomination of the astronomer Alfred Klose (1895–1953) as successor to Vahlen in 1935 showed the relative neglect of the institute by the authorities. Insofar as von Mises might have hoped to secure, by his political move, a pension from Germany during his exile in Turkey, that hope was shattered as well.[79]

---

[75]See Siegmund-Schultze (2001). During the war, Weaver was also head of the "Applied Mathematics Panel." See below.

[76]See Appendix 3.3.

[77]Mises Papers, HUA 4574.5, correspondence, box 2, f. 1933 (T).

[78]It turned out that Trefftz would not have a chance two years later either. See Siegmund-Schultze (1984). One may add that Eberhard Hopf was ruled out as a successor at that time as well, as will be seen from a letter by Alfred Brauer, quoted below.

[79]See above chapter 4, and for the aftereffects of this affair chapter 11 below.

Figure 33 *Richard von Mises (1883–1953). The very versatile mathematician (stochastics, mechanics) built up mathematics in Istanbul (Turkey) after his expulsion from Berlin. He came to the United States relatively late (1939), where he finally became a professor at Harvard University in 1945. In his diary, von Mises reflected on his repeated emigration from Germany and Turkey. See Appendices 3.3 and 4.3.*

Von Mises, with his old acquaintance from Vienna, Theodor von Kármán, at the California Institute of Technology in Pasadena, was also concerned about a career for Walter Tollmien in Germany. They felt that as
an "Aryan" his chances of obtaining a position were strong. On June 10,
1933, von Mises wrote to von Kármán about the thirty-three-year-old
Tollmien, who was at that time at Pasadena and had no permanent position there:

> I'd like to return to the matter of Dr. Tollmien. I do not know if he still desires
> to come to Germany under the present circumstances. In any case, I have to tell
> you that the irrevocable prerequisite for any kind of employment or scholarship
> or suchlike is to make a statement on his honor [ehrenwörtliche Erklärung]
> that his four grandparents are of "Aryan, in particular non-Jewish descent." As
> long as I do not know whether Mr. Tollmien can or will make such a statement,
> it is impossible for me to do anything. Besides I believe that in the positive case
> the prospects are not bad as indeed a large part of all the previous candidates
> has to be disregarded now. I ask you [Dich] to give me the relevant information
> as soon as possible.

Von Kármán wrote on the reverse side of this letter:

> Dear Mr. Tollmien,
> enclosed a letter from H. v. Mises. Indeed a "document of our time." Please let
> me know if I should transmit the written evidence of your racial purity to Berlin
> or whether you want to write to Mises yourself. Please return the letter after
> having enjoyed it.[80]

Tollmien returned to Germany and found a position rather soon. In
March 1934 he received the following letter from von Kármán:

> Due to the great number of immigrating scientists, there must be a great de-
> mand for re-immigrating scientists. Wieselsberger wrote me that Tietjens re-
> turned to Germany and now has a position with the DVL.[81]

In a similar vein, Kurt Hohenemser—who as a "half-Jew" according to
the Nazi definition, had problems finding a position in Germany—wrote
to von Kármán in May 1935:

> It should not be very difficult in the near future for a German with flawless [ein-
> wandfrei] grandparents to find an adequate position in Germany. Perhaps one

---

[80]Theodor von Kármán Papers, Caltech, Pasadena, 20.37 (T). Written by von Kármán on
the reverse of a letter by von Mises, dated June 10, 1933, which is here also translated.

[81]Von Kármán to Tollmien, March 24, 1934. Von Kármán Papers, Caltech, Pasadena,
30.15. DVL is the German Air Traffic Proving Ground [Deutsche Versuchsanstalt für Luft-
fahrt] in Berlin-Adlershof.

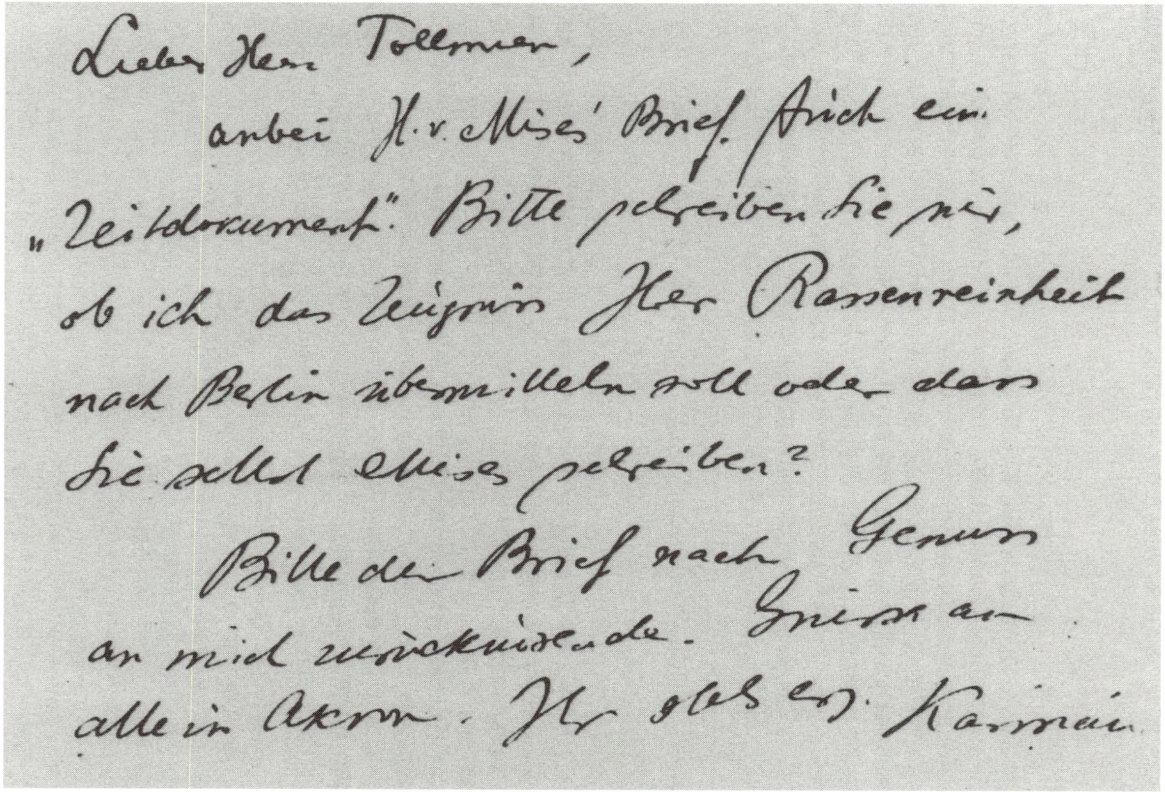

Lieber Herr Tollmien,

anbei H. v. Mises' Brief. Wirklich ein „Zeitdokument". Bitte schreiben Sie mir, ob ich das Zeugnis Ihrer Rassenreinheit nach Berlin übermitteln soll oder dass Sie selbst etwas schreiben?

Bitte den Brief nach Genuss an mich zurücksenden. Grüsse an alle in Akron. Ihr stets erg. Kármán

Figure 34 *Letter to Tollmien. Von Kármán to W. Tollmien in June 1933, about his "racial purity" as a prerequisite to return to Germany.*

should make Germans abroad aware of this fact—something which would in turn open up chances for us abroad.[82]

7.S.3. Controversial Judgments about the Return of an Established Mathematician to Germany: Eberhard Hopf

Eberhard Hopf had come to America through a Rockefeller fellowship, similar to the way in which A. Wintner and W. Maier came. He had received a permanent position at the prestigious Massachusetts Institute of Technology (MIT) in Cambridge/Boston. Hopf became a noted researcher on statistical mechanics and dynamical systems. Upon his return to Germany in 1936, Hopf published the first monograph in ergodic theory in the literature, namely (Hopf 1937). This booklet, which appeared in German, was based largely on stimuli from America (Birkhoff, Shapley, Wiener).

Long before 1936, efforts had been under way to call Hopf back to Germany, but most of them seem to have failed due to political reasons. This is documented in a letter by Alfred Brauer (Berlin) to Otto Toeplitz (Bonn) from May 1935:

> My brother [the emigrant Richard Brauer; R. S.] has met Hopf and his wife at a congress [in America]. They have been extremely nice to him. My brother also reported that Hopf has a high reputation in America. Thus his accepting appointment in Bonn seems doubtful. At the same time it seems unlikely that he would receive a call in the first place, because he has enemies in the Dozentenschaft [the Nazi faculty organization; R. S.]. There are accusations that he was friends with an alleged communist during his university studies. Schmidt has tried hard to get Hopf as a successor to v. Mises and now to Vahlen here [in Berlin]. But his efforts have failed, as did the attempt to appoint him in Göttingen because of the resistance by the Dozentenschaft, although the political accusation is ridiculous. I am convinced he would behave decently towards you, and he could probably be more courageous than others because he would always have the chance to go back to America.[83]

Hopf's colleague at MIT, Norbert Wiener, writes very sensitively on Hopf in his autobiography and gives also interesting information on the positions of some emigrants vis-à-vis Hopf's return to Germany in 1936.

> He was a German of sufficiently correct racial origin to be acceptable even in Nazi Germany. Originally he was hostile to Hitler, or at least sympathetic to

---

[82]K. Hohenemser to von Kármán, May 19, 1935 (T). Von Kármán Papers, Caltech, Pasadena, 79.15.

[83]Alfred Brauer to Toeplitz, Berlin, May 11, 1935 (T). Toeplitz Papers, University Library Bonn, Manuscript Division, Otto Toeplitz Teilnachlass (1999), letters, no. 10.

those on whom Hitler had wreaked his ill will. However, there were strong family influences pulling him to the Nazi side.

When my cousin, Leon Lichtenstein, had died as an indirect result of the coming of Hitler, the Germans looked around for a successor. At that time good mathematicians were leaving Germany en masse, and a successor was not easy to find. Finally Hopf's name came up, and he was offered the position.

It must be borne in mind that a university position in the Germany of the good old days had a prestige both social and intellectual beyond any comparison with that of a similar position in America. . . . I will say that Hopf consulted with a number of German refugees from Hitlerism and that they did not oppose his acceptance of the offer as vehemently as one might have expected. . . . Many of the German refugees believed that Germany would either be defeated or by an intrinsic revulsion would sooner or later cast off Nazism, and all their opposition to Nazism had not affected their pride in Germany as such. Hopf would form part of an element in the new Germany which would be at least a possible basis for the re-establishment of academic sanity after the war.

The authorities at M.I.T. did not like to have a pistol presented to their heads in the form of Hopf's claims to an immediate promotion over the heads of older men. . . . Hopf accepted the German offer. In his delight at his sudden rise, he was most condescending to his colleagues at M.I.T. To me he expressed his feelings that I was not getting my full deserts, and he wished that I could find such an advancement as he had found in Germany. I need not say that this condescension was not welcomed.[84]

After the war, in a letter to Courant, Hopf regretted his "lack of political insight," which led him to accept the position formerly occupied by Leon Lichtenstein, who had died shortly after his dismissal from Leipzig. Hopf was also frustrated about postwar conditions in Germany and about a lack of scientific contact, and wanted to return to the United States, an effort in which he finally succeeded.[85]

7.S.4. The Lack of Demarcation toward Mathematicians Remaining in Germany: The Example of Gumbel's Only Partially Successful Book Free Science (1938)

Emil Julius Gumbel, who was in exile in France after having been expelled by his nationalistic colleagues in Heidelberg in 1932, published his resignation from the German Mathematicians' Association (DMV) on April 19, 1934 in the *Pariser Tageblatt*. His reason was that the DMV "had, against its statutes, failed to oppose the destruction of German

[84]Wiener (1956), pp. 209–11.
[85]E. Hopf to Courant, June 23, 1945 (T). OVP, cont. 2, f. Courant, 1939–48. In 1956 Hopf refused to accept another offer to go to Germany because he did not want to repeat the same mistake (Courant Institute Papers CIP, f. Eberhard Hopf, 1945–59).

Science."[86] In 1935 and 1936, even before his move to the United States and when still in Lyon, Gumbel appealed to prominent emigrants to collaborate on the collection *German Science in Emigration*, later to be called *Free Science (Freie Wissenschaft)*.[87] The plan of the book begins with the sentence: "The chairs of the German universities are deserted or occupied with pseudo-scientists." The exposé then continues:

> The book aims to give a concise overview of the scientific achievements of expelled professors abroad. . . . Balanced polemics against national-socialist imaginary grandees [Scheingrössen] are welcome. . . . Special grotesques such as . . . the "Aryan" mathematics of Herr Bieberbach shall be exposed.

Weyl remarked in his reply that manifestations of National Socialist ideology in mathematics such as L. Bieberbach's "Deutsche Mathematik" were discrediting themselves and did not require special exposure. Weyl also held that the majority of mathematicians remaining in Germany were of the same opinion as the emigrants and that Gumbel's claim of the chairs being deserted or occupied by pseudo-scientists was an exaggeration.[88] In his letter to Gumbel (written in English) Weyl revealed his efforts to avoid drawing too strict a line of delimitation regarding scientists in Germany and claimed status as a voluntary emigrant:

> I must add that I am not an "emigrant" in the sense that I was dismissed from my position in Germany; I resigned of my own free will, and accepted the offer from America. There is no doubt, however, on which side my sympathies are, and if occasion arose I would not hesitate for a moment to testify to it in public as clearly as Thomas Mann did it.

It must be assumed that Weyl did not see this "occasion" for a public testimony in Gumbel's book. Gumbel on his part did not make use of Weyl's offer to reprint Weyl's purely scientific obituary of Emmy Noether from 1935, apparently because it would not have fitted into the political orientation of the collection.[89] Even Einstein, who had been approached by Gumbel for a preface and who was generally not hesitant with political statements, considered "the planned publication ill-conceived [verfehlt]."[90]

---

[86]Quoted from Mehrtens (1990a), p. 37 (T).

[87]Veblen Papers, Library of Congress, cont. 31, f. Gumbel, Gumbel's correspondence with H. Weyl, 1936–42. The plan of Gumbel's book (2 pp.), which is here quoted in translation from German, is enclosed in the files. It is also in Gumbel's file in SPSL and in the L.C. Dunn Papers in the library of the American Philosophical Society in Philadelphia, which contains correspondence between the geneticist Dunn and Gumbel on the same project.

[88]Ibid. Weyl to Gumbel, March 19, 1936. See also the epigraph to the present chapter.

[89]Ibid. Gumbel to Weyl, June 28, 1938.

[90]Einstein to Gumbel, July 9, 1936, Einstein Papers, Jerusalem 53 265 (T). Einstein wrote furthermore: "Such a publication which contains very mixed contributions can be neither effective nor financially successful."

Courant, who had been asked as well, found the "mixture of balanced report and propaganda" problematic. As to the possible effects of the publication, he wrote to Gumbel on December 10, 1935:

> I principally approve the idea of presenting a systematic evaluation of the favorable and stimulating intellectual forces expelled from Germany into other countries. However, I doubt that these favorable effects are clearly visible yet and that the time is ripe for a balanced report on them. A premature and polemic publication, originating from the participating persons themselves could possibly in my opinion do more harm than good. (CPP [T])

From Gumbel's file kept at the British SPSL it is likely to conclude that he unsuccessfully approached E. Schrödinger, H. Rademacher, P. Bernays and others as well.[91]

One may wonder why there is no evidence that Gumbel approached Richard von Mises for a contribution as well. Gumbel was close to von Mises in a mathematical, if not fully in a political respect and had even studied under him for a short period in Berlin in the early 1920s. Von Mises was certainly reflecting on a historical and philosophical level about the Nazi regime. In fact, he wrote on October 6, 1935 in his diary: "Historical questions, also on the topic Mathematics in the Third Reich."[92] This in all likelihood alludes to a handwritten manuscript of Richard von Mises on "Mathematics in the Third Reich," which was recently (2003) edited from von Mises's papers at Harvard.[93] Interestingly, about half of this text is identical to Gumbel's publication in 1936 in the Paris German-speaking journal for emigrants *Das Neue Tage-Buch,* titled "The Squaring of the Circle." This publication comments ironically on efforts by mathematical laymen in Nazi Germany to square the circle via an ideological and racist reinterpretation of that classical problem from Greek antiquity.[94] The surprising fact is that Gumbel in his publication refrains from mentioning von Mises at all. There is another interesting entry by von Mises in his diary. On April 13, 1934, von Mises wrote: "Article sent to the Neue Tagebuch." I am not aware of any publication coming out under von Mises's name in the *Neue Tagebuch.* Anyway, both the entry in the diary and the publication by Gumbel make it very likely that von Mises had second thoughts about publishing on the Third Reich, and that he

---

[91]SPSL box 279, f. 5. The SPSL gave him in the beginning some support by providing addresses of refugees.

[92]Mises Papers, HUA, HUG 4574.2 Diaries 1903–52. Translated from German Gabelsberger Shorthand. See also the Appendix 3.3.

[93]Sheynin (2003). I had known about von Mises's manuscript for over a decade, but abstained from publication, aware of the need for more historical background such as the facts that follow, particularly the connection to Gumbel (1936).

[94]Gumbel (1936).

FREIE
WISSENSCHAFT

EIN SAMMELBUCH
AUS DER DEUTSCHEN EMIGRATION

herausgegeben von

E. J. GUMBEL

39.97-2.

1938

SEBASTIAN BRANT/VERLAG

Figure 35 Free Science. *Title page of the book on emigration from Germany, edited by the statistician Emil Julius Gumbel (1891–1966), who had been expelled from Heidelberg in 1932, even before the Nazis came to power. Courant, Einstein, Weyl, and (probably) von Mises did not want to contribute for various reasons; several of them shied away from a public discussion of the situation in Germany.*

allowed Gumbel to publish some of his work under Gumbel's own name. Certainly von Mises was not prepared to contribute to Gumbel's collection *Freie Wissenschaft* either. He seems to have preferred more sophisticated and indirect modes of criticism such as his ironic attack by means of mathematical statistics on the Nazi race doctrine in an article published in Moscow in 1934.[95] In any case, many scholars (probably also von Mises) seem to have had qualms about classing themselves publicly as refugees. In 1936, the geneticist at Columbia University, Leslie C. Dunn (1893–1974), who was an active participant of the Emergency Committee, had also warned Gumbel about his project:

> I think your experience with Lederer will probably be repeated. It is quite true that most displaced scholars do not wish to consider themselves refugees. Many of them, as you know, are ardent German nationalists and some of them would be Nazis if they were not Jews. . . .
>
> I have forwarded your letter to Goldschmidt. . . . I don't think he will cooperate in your scheme. He is one of the type of Lederer and wishes to avoid all political activity or involvement.[96]

Gumbel's collection finally appeared in 1938 in German lacking contributions by prominent mathematicians or physicists, but with two articles by well-known biologists.[97] Two articles, titled "On the Co-Ordination of the German Universities" and "Aryan Science," were written by Gumbel himself. A change in the title of the eventual publication might be due to warnings from emigrants about placing too much emphasize on the political context.

7.S.5. The Aftereffects of Previous Political Conflicts in Emigration: The Case of Rudolf Lüneburg

An old political argument between two talented refugees and former young mathematicians in Göttingen resulted in difficulties for one of them in emigration. It was between the differential geometer Herbert Busemann,[98] the son of an industrialist, and the topologist Rudolf Lüneburg, the latter being close to the circle of the philosopher L. Nelson in Göttingen and at times even showing sympathies for communism.

In American emigration, Busemann became known for his development of Finsler geometry with purely synthetic geometrical methods. Unlike

---

[95]That article, Mises (1934), is analyzed in Siegmund-Schultze/Zabell (2007).

[96]Dunn to Gumbel, April 24, 1936. Dunn Papers, American Philosophical Society.

[97]Walter Landauer: "Mutation und Merkmal," pp. 214–28; and Julius Schaxel: "Faschistische Verfälschung der Biologie," pp. 229–45.

[98]On Busemann see also chapter 5.

Figure 36 *Rudolf Lüneburg (1903–1949). The topologist and assistant to Courant was dismissed in 1933 for political reasons. Political conflicts with consequent aftereffects in the American emigration contributed to the fact that Lüneburg did not find a position in academia but went into industry instead. There he worked on the mathematical foundations of optics.*

Busemann, Lüneburg did not find a position in American academia. He went into the optical industry and showed that mathematical optics can be developed systematically from the Maxwell equations of electromagnetism.[99] Due to his early demise (1949) and because Lüneburg's accomplishments in applied mathematics were less visible to "pure" mathematicians, the course of his emigration has remained in the dark until recently.

Lüneburg, born in Volkersheim in 1903, was, in 1933, dismissed in Göttingen for political reasons and went to the Netherlands (Utrecht), where he received a modest Dutch grant.[100] Lüneburg and Busemann had both been assistants to Richard Courant, Busemann, however, receiving no salary since he was well to do and did not need it. According to Courant, "Lüneburg left Holland in 1934 in order to do anti-Nazi underground work in Germany."[101] He was unemployed then and was literally starving in Braunschweig (Brunswick) in 1935. Therefore Courant wrote a letter dated July 26, 1935 to Busemann, who was at that time in Copenhagen.[102] Courant asked whether Busemann's father would consider paying for Lüneburg's transatlantic travel. Busemann replied on September 11, 1935:

> When back in 1926 I learned about his insufficient means I asked my father to help him. But Lüneburg rudely refused, he was not prepared to be helped by a capitalist. . . . This strength of character was of course recognized by my father but I cannot expect him now to help such an explicit opponent. . . . By the way we recently discussed in a bigger group (in Germany) the question of whom to help and all agreed that given the terrible consequences which any kind of dictatorship implies, nobody should be helped who is seen to have even the slightest sympathy with a dictatorship. (CPP [T])

Courant replied on September 26, 1935:

> It would be remiss of me to conclude this letter without mentioning that I have read your remarks about Lüneburg with the greatest consternation. If you refuse to help Lüneburg this is your right as a private person and I do not hold this against you. . . . But I hope you won't seriously maintain the reasons for your refusal on the basis of an ethical system of axioms. I believe that a principled tendency to cling to whichever abstract formulas instead of looking at the individual himself is in the end as dangerous as the position of fanatics in favor of a dictatorship. (CPP [T])

[99]Lüneburg (1964), p. v. In this volume, which was edited after Lüneburg's untimely death, the preface by E. Wolf and supplements by M. Herzberger give an appreciation of Lüneburg's life and work.

[100]Schappacher (1987), p. 364, and SPSL, box 282, f. 4.

[101]R. Courant to SPSL, November 20, 1942, SPSL, box 282, f. 4.

[102]Busemann was the son of a Krupp director. See Reid (1976), pp. 106, 133, 153.

Figure 37 *Herbert Busemann (1905–1994). The son of an industrialist had to leave Göttingen for political "reasons" in 1933. Compared to other emigrants, he was comfortably situated for starting again in exile. He was in an ongoing political conflict with another émigré, R. Lüneburg. Since 1947 Busemann worked in Los Angeles where he wrote important papers on the foundations of geometry (photo from about 1934).*

Busemann replied on October 6, 1935, that in agreement with Fenchel, Bohr, and Neugebauer in Copenhagen he did not want to let "people reach influence . . . who would use the latter to curtail my resp. our freedom." He then wrote to Courant in New York:

> I believe the situation is this. In America these problems are not acute, thus one indulges, as previously in Germany, in judging applicants for appointment according to their purely human qualities <u>alone</u>. But here in Europe one no longer thinks this way, because danger is imminent. Bohr's formulation was: we continue of course to be liberals and allow everybody to fight for his conviction, but it is legitimate to impede that fight insofar as it is directed against our ideals. Would you still help Teichmüller today?[103]

Courant replied in a letter of October 17 that Lüneburg was not comparable with Teichmüller and Hitler. Somewhat later he wrote to Hermann Weyl with regard to Lüneburg, who meanwhile was unemployed in New York:

> In any case he is firmly determined not to be tiresome as a competitor in the academic job market by applying there. He has eked out a wretched existence recently in Germany as an ordinary laborer [Erdarbeiter].[104]

Lüneburg had, according to Courant, sent in applications in academia before, to Johns Hopkins University in Baltimore, as Aurel Wintner wished to have him there "in support of his probabilistic work."[105] However, the stipend was not granted and Courant had to step in privately:

> When I came back it became evident that Lüneburg was in a state of starvation. Therefore I have now hired him as my private assistant.[106]

Wintner wrote to Courant about the reasons for Lüneburg's failure in Baltimore and alluded at the same time to the then still dominating ideal of pure mathematical research among American and German mathematicians:

> Unfortunately, in a letter to the commission, L. mentioned his plans about a job in industry, and that damaged his case considerably.[107]

In a following letter Wintner informed Courant confidentially that Lüneburg had received a negative expert evaluation by Weyl. Wintner did not agree with the tone of that evaluation and assumed that Lüneburg "had

---

[103]CPP (T). Oswald Teichmüller was the young and brilliant mathematician and Nazi activist who had organized the boycott against E. Landau in Göttingen in 1933. See Schappacher and Scholz (1992) and 4.D.2 above.

[104]Courant to H. Weyl, November 16, 1935, CPP (T).

[105]Courant to H. Weyl, October 13, 1935, CPP (T).

[106]Ibid.

[107]A. Wintner to R. Courant, March 10, 1936, CPP (T).

once trod on Weyl's toes in Göttingen."[108] As Weyl was intimately connected with Busemann at the Institute for Advanced Study in Princeton it must be assumed that political and private motives played a role here as well.[109]

As the example of Busemann and Lüneburg shows, former political positions and economic conditions in Germany could also influence the fate of the emigrants after 1933.

[108] A. Wintner to R. Courant, May 29, 1936, CPP (T).

[109] Also Weyl's slight verbal political dissociation from Lüneburg in the connection with his dismissal from Göttingen in May 1933 points to this conjecture. See Schappacher (1987), p. 364. In the 1920s Weyl did not apparently have much sympathy for critics of time-honored philosophical and political systems such as for Lüneburg's philosophical teacher L. Nelson in Göttingen and H. Reichenbach in Berlin. For Weyl's relation to Reichenbach see Hecht and Hoffmann (1982).

EIGHT

The American Reaction to Immigration

HELP AND XENOPHOBIA

> If our story has a hero it was certainly Veblen. But there was also a collective hero: this generation of American mathematicians who, at the very beginning of their careers, experienced the influx of Europeans and who reacted to this influx with so much grace and so much cordiality.
>
> —Lipman Bers 1988[1]

> With this eminent group among us, there inevitably arises a sense of increased duty toward our own promising younger American mathematicians. . . . with the attendant probability that some of them will be forced to become "hewers of wood and drawers of water." I believe we have reached a point of saturation, where we must definitely avoid this danger.
>
> —George D. Birkhoff 1938[2]

> We must not forget that we once deceived ourselves about the safety of the ground we were living on. So let us stick together and show extreme caution.[3]
>
> —James Franck 1935[3]

8.1. General Trends in American Immigration Policies

As mentioned in chapter 2, the United States became the final host country for over half of the German-speaking mathematicians who emigrated after 1933. The refugees were usually neither scientifically nor emotionally (given the fact that in many cases relatives were left behind) attracted to the geographically distant United States, although political developments quickly revealed the extraordinary level of protection provided by the strong American society and even emphasized the desirability of obtaining American citi-

---

[1]Lipman Bers in Bers (1988), p. 242.

[2]George D. Birkhoff on the occasion of the fiftieth anniversary of the AMS in 1938 in Birkhoff (1938), p. 277. See below for a more complete quote.

[3]James Franck to R. Courant in a letter on November 1, 1935, CPP (Miscellaneous [T]).

zenship (D). This predominant direction of the emigration resulted from the political developments within Germany and Europe, from the course of the war, but was also due to the huge absorbing capacity of the widely branching American system of universities.[4] Further factors contributing to the orientation of the refugees toward North America were the stronger economical basis of the American system (in spite of the temporary economic crisis), considerable foresight of American science functionaries (Rockefeller), and the experiences of the 1920s. In third place one can mention the relatively liberal American immigration policies compared with other countries.

These policies were, however, in 1933, considerably more restrictive compared to the turn of the century, a development partly caused by improvements in the welfare system, particularly during Roosevelt's New Deal.[5] In the 1930s American immigration policies were marked by a certain bureaucratic "governmental apathy"[6] with regard to the problems of the refugees. It would be an illusion to believe that the United States would have generally and unconditionally opened its gates to the refugees, in particular to the Jews among them, after the events of 1933. The restrictive immigration policies are sufficiently documented in the works of Pross (1955), Friedman (1973), and Daniels (1983). In the case of at least one mathematician, the logician Kurt Grelling,[7] but in many more cases of ordinary, nonacademic refugees, too, the bureaucratic decelerating policies of the Americans, often combined with political mistrust, were partly responsible for the death of the refugee. Again this does not exculpate the German authorities in any way.

It is emblematic that the American immigration quota granted to German refugees of 25,000 people per year was only once filled during the 1930s, in 1939, while hundreds of thousands waited without a chance for a visa during all these years. According to the, very likely, too low figures presented by Pross, about 104,000 immigrants came from Germany/Austria to the United States up until the outbreak of the war in 1939.[8] Almost all

---

[4]See Pross (1955), pp. 37 and 45ff., who stresses the gaps in the American system that had to be filled.

[5]Daniels (1983) mentions the "likely to become a public charge" (l.p.c.) clauses in the American immigration laws, which "probably kept out more otherwise qualified immigrants than did any other" (Daniels 1983, p. 63) and which became even more crucial during the Roosevelt era. In January 1939, Hermann Weyl wrote to one émigré, Artur Rosenthal: "The other side of the coin of the social welfare which is now being introduced in America on a grand scale as well is that it automatically leads to a hermetic sealing of the borders" (OVP, cont. 32, f. Rosenthal, Artur, 1938–42, Weyl to Rosenthal, January 25, 1939 [T]).

[6]Krohn (1987), p. 32.

[7]Peckhaus (1994), p. 66.

[8]See Daniels (1983), p. 66, and Pross (1955), p. 45. Strauss (1991), p. 10, however, talks about half a million German-speaking immigrants into *all host countries*, which indicates a much higher figure than 104,000 to the United States alone. But Strauss also restricts the total

of these immigrants were expelled by the Nazi regime, among them 7.3 percent from academic professions, but only 1,000 scientists. Some German-speaking scientists were able to make use of the quotas of other countries such as France and Switzerland, even if officially they were not citizens of these countries. This was made possible, for example, for H. Samelson and W. Wasow due to their having accidentally been born in these countries (D). The proportion of non-Jewish emigrants within academic emigration was probably higher than that within the total emigration,[9] even though it most likely did not exceed 10 percent.

Although far from all scientists wanting to emigrate succeeded in doing so, they were—on average and excluding the individual sufferings from the callous and objective historical perspective—clearly "privileged victims" of National Socialism within the total emigration.[10] If they could present an offer from an American university and fulfilled an additional condition of uninterrupted academic activity, they came under the exemption clause of the Immigration Act of 1924. This clause accepted an immigrant independent of the existing immigration quota for the respective country "who continuously for at least two years immediately preceding the time of his application for his admission to the United States has been, and who seeks to enter the United States solely for the purpose of, carrying on the vocation of minister of any religious denomination, or professor of a college, academy, seminary or university; and his wife, and his unmarried children under 18 years of age, if accompanying or following to join him."[11] This exception to the rule, clearly in the interest of the United States, too, was usually handled much less bureaucratically.[12] There were further exemption clauses particularly for students wishing to continue their studies abroad ("student visa").[13]

In spite of the exemption clauses of the Immigration Act, admission was very selective, being not just conditioned by the arbitrariness of some American political authorities but also influenced by specific scientific needs. The clause was limited in its effect anyway, since refugees, at the time of their application to the U.S. embassies, had often been driven out of their academic jobs for longer than two years (D).

---

number of emigrated professors to about 1,000 to 1,500. A recent Web site gives a total of 130,000 Jewish emigrants from Germany/Austria to the United States and of 50,000 to Great Britain, and an overall emigration of about 300,000, which is less than half of the Jewish population of 1933 in Germany/Austria. At http://holocaust.juden-in-europa.de/shoah/pogrom/pogrom-1.htm. This again is a much lower percentage of Jewish victims than in Poland and other Eastern European countries.

[9]Strauss (1991), p. 10.
[10]Underlined also by Strauss (1991), p. 15.
[11]Quoted from Daniels (1983), p. 68.
[12]Pross (1955), p. 46.
[13]This clause enabled the rescue of F. Herzog and E. Reissner.

In 1940, one observer discussed the benefits of early immigration in the nineteenth century to the United States, but added cautiously:

> This, however, is not made the basis of a plea for the unrestricted reception of European scientists by the United States at the present time. Conditions are now vastly different from those existing ninety years ago. The frontiers now have all been occupied and openings for employment are so few that immigration, once unchecked, has had to be curtailed. The advantages of scientific and technological superiority, once held by the Europeans, no longer exist and foreigners have not the opportunities to make themselves useful that were enjoyed by refugees two and three generations ago.[14]

The same observer encouraged historical analyses of previous immigration and, apparently, the consequences of more selectivity in the present immigration policies:

> A detailed study of the race, age, occupation, religion and other available data pertaining to refugees who fled to America in previous migrations, with brief accounts of their successes and failures, would no doubt help clarify certain problems concerning the large number of exiles who have been banished from Germany since the adoption of the present national socialistic regime.

If one talks about "privileged victims" of National Socialism, although the plans for the emigrants' life were in many cases destroyed,[15] the suffering of millions of *other* victims of the terror regime becomes even more palpable. Therefore the present book does take into account—at least in the listings of the appendices[16]—the fates of those mathematicians who did not make it to a host country abroad, several of them being murdered by the regime.

8.2. Consequences for the Immigration of Scholars

The American society imposed restrictions on immigration not just for ordinary people but for scholars as well, among them mathematicians. Generally preferred were the most talented scientifically or who were otherwise useful. Parallel to the tightening of the immigration law in 1924, important American universities had introduced ethnical quotas and restrictions for admission.[17] Well into the 1950s, the predominantly private elitist universities of

[14]Browne (1940), p. 207.

[15]Most visible in the cases of less successful emigrants such as Felix Bernstein, as documented in chapter 9.

[16]See some documents in this chapter and Appendices 1.2 and 1.3.

[17]See Synott (1979) and, recently, Karabel (2005). As late as 1935 there were only five Jews among 635 students registered at Princeton University. Johnson (1952), p. 392, reports that the movement for a state university in New York in 1946 was inspired by feelings

the United States sealed themselves off from the infiltration of new social and ethnic groups and cultural trends,[18] although many of the scientists coming from abroad met most of the desired social standards.

The variety of possible reactions by the Americans to immigration is described in the very different epigraphs by the immigrant Bers and the conservative and influential American mathematician Birkhoff. The awareness on the part of the immigrants of the existing problems is exemplified in the letter of physicist James Franck to his friend Richard Courant, both refugees from Göttingen.

The aggravating impact of the economic depression around 1930 on the American academic job market and therefore also on immigration has been described in detail by Robin Rider (1984, 123–29). There was an argument among American mathematicians after 1933—an argument that had been raised to some extent during the 1920s—about whether the *new chances* for mathematics in the United States resulting from immigration would even out and possibly eliminate[19] the *disadvantages* connected to the diminishing prospects for young Americans on the academic job market in the United States. Concerns about possible disadvantages were apparently formulated even more clearly by leading American mathematicians than by the "affected" younger ones;[20] at least they attracted more attention. Moreover, in the policies of the emergency organizations priority was given to younger scholars. Mathematical results of older scholars,

---

against the anti-Semitic *numerus clausus*. Until the mid-1920s no Jews were appointed to permanent professorships at Princeton. The mathematician S. Lefschetz (appointed 1924) was an early exception. A topic of its own is the discrimination against Afro-American citizens who were accepted late in American academia, basically only after World War II. The Afro-American statistician David Blackwell (born 1919) reported in his interview with D. J. Albers about discrimination even during the war, and about the long-term effects on the career paths of other Afro-Americans. Cf. Albers and Alexanderson, eds. (1985), pp. 17–32.

[18]Ash (1996), p. 41. See also Porter (1988), p. 370. This included discrimination against women scientists, of which there were only few among the immigrants. See the complaint by Hilda Geiringer in her letter to the British SPSL on June 4, 1947 (D) and the case study on Emmy Noether below.

[19]Veblen, as quoted in Reingold (1981), p. 324, and A. Flexner (see below) hoped that the upswing of mathematics due to immigration would, in the end, create new positions. In the Documents part (D) to the present chapter the skeptical voices of American mathematicians toward immigration will be emphasized relatively more than the ones voicing encouragement. Also, the help Americans gave to the emigrants will probably be mentioned somewhat less than deserved, since it is already widely known from existing literature.

[20]This is the impression of L. Bers (in the epigraph) and of M. Golomb (D), while Courant (D) held the opposite opinion. The latter's view was probably fueled by utterances like the one in an article signed "Ph.D." and titled "Aid for American Scholars" in the *New York Times*, November 6, 1940, p. 22, in which the author complains about his insecure situation eight years after he acquired the PhD and says that "the European scholar . . . is at least in part responsible for the collapse of his world." See below for the reaction to this by A. Johnson.

produced at an early stage in their careers, were only of relative and limited value to host countries such as the United States.[21]

The attitudes of the political authorities were on average much more discouraging and obstructive to immigration than those of the scientific community.[22] Xenophobic sentiments were at any rate bound to be stronger in other professional groups not as used to contact on an international level as scientists were.[23] In the end, mainly due to the atrocities in Germany becoming public knowledge, solidarity in American society with the immigrants prevailed. With the outbreak of the war in Europe in 1939, which soon resulted in extensive preparations on the American side ("war preparedness" programs) and in war research, the sentiment grew that immigration was beneficial to American science.[24]

The main problem of immigration—apart from the feared implications for the American academic job market—was financing. Only a few of the immigrants had relatives with sufficient means in the United States[25] or came with money of their own.[26] Basically, the money had to come from private sources such as private universities, special funds, and industry, not, however (at least not before the United States entered the war), from the American state. Salaries apart, as a step prior to immigration, affidavits had to be secured from private individuals, who on average certainly hoped that their financial guarantee would never be claimed.[27]

The universities, where most of the refugee-mathematicians went, welcomed additional funds from relief organizations and, in some cases, financed the entire salary from such sources. Supplementary funds the

[21]See D and chapter 10 below.

[22]Abikoff (1995), p. 9, also articulates this opinion.

[23]As argued in Pross (1953), pp. 48–49. The fact, to be mentioned below, that Jewish scientific immigrants were not on average subject to stronger discrimination than non-Jewish ones supports this evaluation.

[24]See chapters 9 and 10.

[25]However, such cases did exist. The sister of Ernst Hellinger in Chicago offered to the IAS $1,200, using H. Weyl as mediator: "I feel that for obvious reasons it would make life easier for my brother than the consciousness of his being dependent on me" (OVP, cont. 31, f. Hellinger, Ernst, 1939–42, Hanna Hellinger to Weyl, January 7, 1939).

[26]Among the latter I have already mentioned Courant, von Neumann, Busemann, and von Mises, of whom at least von Neumann was secured anyway by his position at the IAS. Wolfgang Sternberg came to the United States with several thousand dollars that, in his case, however, were soon depleted without sufficient replenishment by a job. OVP, cont. 33, f. Sternberg, W.

[27]Weyl wrote to Ernst Jacobsthal in Berlin, on February 24, 1939, that he had received an affidavit for him from one Louis Bernstein in Maine. He added: "Mr. Bernstein is a perfect stranger to me as well as to you and I have assured him that it is pretty certain that you will never call upon him to give the financial help promised" (OVP, cont. 31, f. Jacobsthal, E.).

universities received for immigrants from private organizations and through special initiatives were often crucial to alleviating or avoiding existing concerns. Indeed they often enabled the creation of additional positions or the assignment of immigrants to pure research instead of subjecting freshmen to foreign accents and teaching methodology.

8.3. The Relief Organizations, Particularly in the United States

Looking—with the knowledge of hindsight—at the documents and files of the most important, mainly American and British, emergency organizations for refugees, the widespread lack of political acumen among the acting persons both with respect to the life-threatening situation of the refugees and the promise immigration held for the scientific development in the host countries appears somewhat surprising. However, it is a historically established fact that the horror of Auschwitz was not foreseeable for most contemporaries, at least before the war broke out.

Arguably the most important American relief organizations were—at least within mathematical immigration—the following two:

The "Emergency Committee in Aid of Displaced German (from 1939: Foreign) Scholars" (EC), under Stephen Duggan, founded in 1933 at Duggan's Institute of International Education (IIE) in New York City. The Institute had already played a certain role in the early immigration of the 1920s, as discussed in chapter 3.[28]

The Rockefeller Foundation (RF) in New York City with its emergency program for scientific immigrants.[29]

The amount of money allocated to refugees was usually the absolute minimum amount according to immigration laws and was way below the regular income of American professors, let alone the salaries of the permanent

---

[28]See Duggan/Drury (1948) and Duggan (1943). In the first years after 1933, Duggan was assisted by Edward R. Murrow (1908–1965), who later became a prominent journalist reporting from Nazi Germany and fighting McCarthyism in the United States. Joseph McCarthy would use Murrow's international connections at the Emergency Committee as ammunition for labeling him a communist. See the recent film *Good Night, and Good Luck* (2005), directed by George Clooney. Some historical controversy still exists about the role of Stephen Duggan's son Lawrence, a good friend of Murrow's, who took over as director of the IIE after his father and who committed suicide after being accused by McCarthy of being a Soviet spy.

[29]See Siegmund-Schultze (2001, pp. 187–215). In many cases the RF not only matched funds from either the EC or the universities, it also created certain special fellowships that were used for instance to support C. L. Siegel and H. Rademacher. See Berndt (1992), p. 209. The latter, however, were reserved for exceptionally prominent mathematicians. On the effects of the RF support for applied mathematics see chapter 10.

© Bachrach

DR. STEPHEN DUGGAN
Author of "A Professor at Large"

Dr. Duggan, now Director of the Institute of International Education, was formerly Professor of Political Science at the College of the City of New York and Lecturer on International Relations at Columbia University.

Figure 38 *Stephen Duggan (1870–1950). The liberal political scientist was the first director of the Institute of International Education, founded after World War I in New York City. After 1933 he organized through these offices the Emergency Committee in Aid of Displaced German (from 1939: Foreign) Scholars. He cooperated closely with O. Veblen, H. Weyl, and the Rockefeller Foundation.*

members of the IAS.[30] Once the immigrants were in America with a permanent visa, however, those stipulations were often disregarded, and immigrants such as Sternberg and Rothe had to survive for a longer period of

---

[30]The latter, like Weyl and Einstein, had fixed incomes of $15,000 per year, while "$2,500 a year [was] the minimum that the consuls will recognize as sufficient support to

time on less than $100 per month. This also implies that some American institutions[31] and individuals[32] took advantage of the cheap intellectual labor offered by the immigrants in need (D). As often in the history of science many practitioners were willing to accept material setbacks if only they could continue in their beloved occupation.[33]

Nevertheless, the historical perspective is somewhat blurred when focusing on those immigrants failing to find a position in the United States on their own initiative and on the basis of a "regular" appointment at a university or in industry, and for whom therefore extra efforts and concerted actions had to be taken. While the majority of the eighty-three or so mathematical immigrants to the United States appeared, at one time or another, on lists of refugees for whom support was sought, only about a third of them finally received financial support from special organizations and funds created for the immigrants.[34] Moreover, these organizations gave preference (at least with the renewal of stipends) to candidates for whom there was a reasonable chance of reaching a permanent position in the long run.[35] We must also consider that help through the relief organizations was not restricted to money and to funding, but was frequently given in the form of making connections to universities and governmental authorities, advising in visa matters, and such.

---

justify a non-quota visa," according to the director of the New School for Social Research in New York City, Alvin Johnson, who responded in the *New York Times*, November 17, 1940, p. 78, under the headline "The Refugee Scholars" to criticisms by "Ph.D." (see above). Twenty-five hundred dollars was indeed the sum Johnson paid his faculty at the "University in Exile." Some immigrants, such as Hilda Geiringer, received even less per annum, namely $2,300 at Bryn Mawr. See EC, box 10, f. Geiringer, H., 1933–40.

[31]M. Lotkin and E. Rothe were temporarily employed as schoolteachers with basically only room and board and a small allowance of $300 (Lotkin) and $500 (Rothe) per year (!). See for Lotkin, EC, box 79, f. Lotkin, Michael (1937–43), and for Rothe OVP, cont. 32, f. Rothe, Erich, 1934–44. Under these conditions Rothe lived in Iowa with his young son and his wife Hildegard, also a refugee mathematician. The latter died from cancer in 1942; Rothe was unable to afford a nurse for her.

[32]One individual who shows up several times in Hermann Weyl's correspondence as a man who took advantage of the immigrants was Malti, an American engineer of Armenian origin at Cornell University. See below the case study.

[33]This is pointedly expressed in a letter by Peter Scherk to Weyl of April 6, 1939, which was quoted in chapter 6.

[34]Rider (1984), p. 142, has already pointed out that those special funds mediated by the Emergency Committee and the Rockefeller Foundation generally supported only about 20 percent of the refugees.

[35]For instance, the Rockefeller Foundation stopped its help for Felix Bernstein in 1935, when Columbia University made it clear that there was no chance of taking Bernstein on permanently. Frank D. Fackenthal (Columbia) to E. Murrow, EC, box 2, f. Bernstein, F., (1933–35). In 1952, at the age of seventy-three and threatened by the McCarran Act of losing his American citizenship, Felix Bernstein tried to get some compensation or even a pension from Columbia, stressing that the funding of his position there between 1933 and 1936

Studying the cases of refugees experiencing difficulties in getting a position does not only complete the picture; in a way, it also reveals something about the general situation at the universities, with trustees and administrators frequently having to take into account the interests of the students and of the scientific staff. This is not to say that at the level of the universities efforts were not made to capitalize structurally and politically on the accommodation of the immigrants. In fact, in mathematics the immigration of foreigners coincided with and partly triggered the creation of mostly privately funded research institutes and strong graduate schools at universities, these being less exposed to the needs of teaching. Among these establishments the most important ones for the immigration of mathematicians were

> the "Institute for Advanced Study" in Princeton (New Jersey) with its strong school for mathematics, which had been founded in 1932 by the Bamberger-Fuld family
>
> the Graduate School of Mathematics at New York University under Richard Courant
>
> the Graduate School of Mathematics at Brown University under R. G. D. Richardson

The latter two institutions were oriented toward applications and gained momentum particularly during the war, when at Brown the "Summer School for Applied Mechanics" was organized.[36] There were efforts under way "to demonstrate to the nation at large the significance of all our colleges and universities"[37] in this context.

In addition, the New School for Social Research (NSSR) in New York City under the direction of Alvin Johnson (1874–1971), with its special division called University in Exile, gave support to some mathematicians but declined help for others as it was primarily oriented toward the social sciences.[38] The University in Exile cooperated with Duggan's EC in particular by providing nonquota immigration to refugees while other universities were less flexible and quick in their decisions.[39] Finally, several smaller

---

by the EC hadonly been possible on the grounds of some promise of permanency by Columbia. But the geneticist L. C. Dunn of Columbia University, who had been involved as a member of the EC as well, flatly denied that claim of earlier promise. Dunn Papers, APS, file: Bernstein, Dunn to G. B. Pegram, June 24, 1952. The existing files confirm Dunn's view. For Bernstein see also chapter 9.

[36]This was complemented in 1943 by the start of the *Quarterly of Applied Mathematics* at the same university.

[37]Harvard President Conant 1936, as quoted in Jones (1984), p. 214.

[38]Krohn (1987). The NSSR gave temporary support also to the mathematicians E. J. Gumbel and the non-German-speaking André Weil.

[39]Johnson (1952), p. 345.

1937

Figure 39 *Roland G. D. Richardson (1878–1949). The secretary of the American Mathematical Society between 1921 and 1940 had been a post-graduate student with Hilbert at Göttingen in 1907–8. He built a center for applied mathematics and mechanics at Brown University (Providence) in 1941. In this sphere of activity he placed great reliance on many immigrants, among them H. Lewy, K. Löwner (C. Loewner), and W. Prager, and he was in competition with R.Courant's institute in New York City.*

Jewish and non-Jewish relief organizations supported the victims of the racial and political purge. Among the correspondents of the Emergency Committee, the following organizations and committees that helped in supporting one or another of the refugees appeared: the British Academic Assistance Council (AAC), later called the Society for the Protection of Science and Learning (SPSL),[40] the American Committee for Christian German Refugees, the American Friends of the Hebrew University,[41] the American Friends Service Committee,[42] the Carl Schurz Memorial Foundation, the Carnegie Corporation, the Committee on Catholic Refugees from Germany, the Notgemeinschaft Deutscher Wissenschaftler im Ausland (UK, Switzerland),[43] and the Oberlaender Trust.[44] The Weltstudentenwerk in Geneva (continued by the International Student Service from 1935 onward) helped students to emigrate.[45] There were more, smaller relief organizations such as YIVO (Yiddish Scientific Institute Vilno, later New York).[46]

In mathematics the German Mathematicians' Relief Fund, founded by Hermann Weyl and Emmy Noether in 1934, had a certain marginal but—for the individuals concerned—often existential importance.[47] However, due to the small sums available from this fund, the personal efforts of mathematicians such as Weyl and Veblen in finding positions for the refugees and in obtaining funds from EC and the Rockefeller Foundation mattered even more. In the institutional triangle consisting of Princeton (IAS with American mathematician Oswald Veblen and émigré Weyl); New York City (EC, RF, New York University with Courant); and Providence, Rhode Island (with the secretary of the AMS, Roland G. D. Richardson, at Brown University), relief work was organized with the Rockefeller Foundation acting mostly behind the scenes and providing matching funds for appointments at various colleges and universities. Somewhat later, in 1938, that triangle was extended to a quadrangle including Cambridge, Massachusetts,

---

[40]See for instance Rider (1984).

[41]They helped to secure the professorship for A. Fraenkel in Jerusalem. See EC, box 8, f. Fraenkel, A.

[42]They supported for instance Peter Scherk. See EC, box 92, f. Scherk, Peter.

[43]This organization was active in the years immediately after 1933 and supported for instance the immigration of R. von Mises to Turkey.

[44]The Trust paid for instance part of Hilda Geiringer's salary at Bryn Mawr College and parts of the salaries for E. Helly and A. Basch at Paterson College in 1942. For the latter see OVP, cont. 30, f. Basch, A.

[45]It played a decisive role in the immigration of Walter Ledermann to Great Britain. Letter to me, December 29, 1997.

[46]Abikoff (1995), p. 10, reports on the support of this institute for Lipman Bers, who came from Lithuania.

[47]Rider (1984), p. 150. The Relief Fund supported among others E. Jacobsthal, F. John, E. Hellinger, P. Kuhn, E. Rothe, H. Schwerdtfeger, W. Sternberg, and M. Zorn. For Hellinger see Rovnyak (1990), p. 22. More on the Relief Fund below in this chapter.

Figure 40 *Harlow Shapley (1885–1972). The astronomer, famous for his calculation of the size of the Galaxy, was director of the Harvard College Observatory from 1921. In 1938 he initiated—with H. Weyl and O. Veblen— the "Asylum Fellowship Plan," the focus of which was on the accommodation of senior refugees.*

where the astronomer Harlow Shapley (1885–1972) of Harvard University organized the Asylum Fellowship Plan.[48]

It was certainly not the original desire of American mathematicians to make the United States the main country of immigration for refugees (D). In the same vein, the policies of the Rockefeller Foundation were originally directed toward providing support for the refugees in their various European host countries (for Gumbel in France, Neugebauer in Denmark, etc.). Immediately after the promulgation of the Civil Service law by the Nazis on April 7, 1933, the Americans received the first calls for help from Germany.[49] On May 9, 1933, Princeton topologist Oswald Veblen, perhaps the most influential American helping refugees, had contacted the Rockefeller Foundation. Having been informed about the Rockefeller plans, namely the creation of the first Rockefeller emergency program for deposed European scholars to be decided on May 12, 1933, Veblen organized a meeting with Duggan's EC at the end of May 1933. At that meeting, in which the American mathematicians Louis L. Silverman (1884–1967), Gilbert A. Bliss (1876–1951), O. Veblen, and Roland G. D. Richardson (1878–1949) took part, support for approximately twenty-five mathematicians was discussed. Matching funds came from RF and additional support was given "by a group of wealthy Jews."[50] The meeting focused on those potential immigrants deemed the most attractive and well known. However, social conditions restricting possible help for them were immediately mentioned. Schur was considered as possibly the best mathematician among the candidates, but as probably too old; Courant was seen to be an able administrator but probably unwilling to accept a subordinate position; and Landau had, in the opinion of the assembled Americans, enough means of his own. Interestingly enough, approximately the same number of twenty-five mathematicians was finally supported by the EC, partly through assistance from the RF. However, the idea that was also drafted, namely to provide for the majority of refugees at *some prominent universities,* had to be gradually abandoned and opportunities had to be sought for them at smaller institutions.

Both in the Emergency Committee and within the Emergency Fund of the Rockefeller Foundation, the immigration of natural scientists and mathematicians was given no priority, which again was probably partly linked to the unemployment of young American scientists competing for the same positions. Priority was given by the RF to the social sciences, and,

---

[48]Jones (1984).

[49]W. Blaschke (Hamburg) recommended in a letter to N. Wiener on April 12, 1933, W. Prager who "has at least temporarily lost his position in Göttingen" (Wiener Papers, MIT, box 1, f. 37). For further examples see chapter 4 and appendices 3.1 and 3.2.

[50]For details discussed at this meeting see the letter by R. G. D. Richardson to H. Bohr, June 29, 1933, in Richardson Papers, Correspondence, BUA (Providence), box: Correspondence 1933 (German-Jewish Situation), f. H. Bohr.

at least after war broke out in 1939, money was often funneled to the refugees via the New School for Social Research in New York City. The RF under its president, Max Mason (1877–1961), who had obtained his doctor's degree in mathematics in Göttingen under Hilbert in 1903, supported approximately three hundred immigrants in all the sciences. However, according to available records, only fourteen German-speaking mathematicians were supported by the RF Emergency Fund: F. Alt, F. Bernstein, R. Courant, W. Feller, K. Friedrichs, E. J. Gumbel, F. John, H. Lewy, O. Neugebauer, E. Noether, H. Rademacher, O. Szász, G. Szegö, and A. Weinstein.[51] While mathematics and physics had been the center of attention of the Rockefeller philanthropies in the 1920s, biology was now more considered within the natural sciences. But the social sciences received even more support by the RF, while Duggan's Emergency Committee focused on the humanities. Only eighty-one natural scientists and mathematicians were to be found among the 335 scientists supported with funds by the EC. They included the following twenty-three German-speaking mathematicians:[52] G. Bergmann, F. Bernstein, A. Brauer, R. Brauer, R. Courant, M. Dehn, A. Fraenkel, H. Fried, K. Friedrichs, H. Geiringer, K. Gödel, E. J. Gumbel, E. Hellinger, F. John, H. Lewy, K. Löwner, O. Neugebauer, E. Noether, H. Rademacher, A. Rosenthal, C. L. Siegel, O. Szász, and G. Szegö. All of these—with the exception of Fraenkel[53]—went to the United States. From a total of eighty-three German-speaking immigrant-mathematicians to the United States, twenty-six received financial support from either the EC or the RF (or both), while the remaining fifty-seven were forced to accept help from other sources. Several of them are listed as "non-grantees" in the EC files. The online list of nongrantees contains 1,073 names of people on whom folders exist.[54] This EC list consists, ex-

---

[51]See Siegmund-Schultze (2001), pp. 304–5. Seven non-German-speaking mathematicians were supported as well. C. L. Siegel, who technically speaking was not forced out of Germany but went "voluntarily," received support from other Rockefeller funds.

[52]Duggan and Drury (1948), pp. 204–8. List of grantees. The 335 included forty-seven "Rosenwald Fellows" for humanities and J. Hadamard, A. Tarski, and A. Zygmund, who were non-German speaking. The printed list is almost identical with the online list (1982) of "Grantees and Fellows," which refers to thirty-eight boxes with folders of the Emergency Committee, kept today at the New York Public Library. See the inventory, which was compiled in 1982, at http://www.nypl.org/research/chss/spe/rbk/faids/emergency.pdf.

[53]The printed list of grantees in Duggan and Drury (1948) includes A. Fraenkel from Kiel, who is erroneously not on the Web site list of the New York Public Library. Nonetheless, this library, which keeps the files of the Emergency Committee, has a folder on Fraenkel in box 8. However, the case of Fraenkel is exceptional, because he went to Palestine in 1933 where he was supported by funds collected by the EC and provided to the Hebrew University in Jerusalem.

[54]See http://www.nypl.org/research/chss/spe/rbk/faids/emergency.pdf, boxes 39–106. Thus there were about three times as many nongrantees as grantees among the refugees. To

cept for a small number of nonrefugees (for example American correspondents of the EC), basically of refugees, among which are found the following twenty-eight expelled German-speaking mathematicians: F. Alt, R. Baer, A. Basch, H. Baerwald, S. Bergmann, S. Bochner, W. Feller, C. Froehlich, H. Hamburger, H. Heilbronn, E. Helly, P. Hertz, A. Korn, F. Levi, M. Lotkin, E. Lukacs, P. Nemenyi, F. Noether, F. Pollaczek, E. Rothe, M. Sadowsky, P. Scherk, I. Schur, H. Schwerdtfeger, W. Sternberg, S. Warschawski, A. Weinstein, and H. Weyl. Excluding Heilbronn (to Great Britain), Levi (to India), Noether (to Russia), Pollaczek (to France), Scherk (to Canada), Schur (to Palestine), and Schwerdtfeger (to Australia), the rest of them ended up in American emigration before 1945. Several of those named, such as Bochner, Lotkin, and Scherk, received a lot of assistance from the EC in mediating visa and academic connections in the United States; others, such as Alt, Feller, and Weinstein, received Rockefeller grants.

Alongside reasons for xenophobia already mentioned (unemployment, financing, anti-Semitism), further causes such as professional jealousy and political mistrust also figured. Some scientists of German origin in the United States (particularly Germanists) conducted propaganda against emigrants as an alleged "fifth column" of Socialism.[55] Doubts about the capability of foreigners to adapt to American conditions were not only directed against immigrants from Germany.[56] However, there was, generally, a suspiciousness toward the alleged peculiarity of the "German national character." This was naturally also ascribed to the Jewish immigrants and was a frequent point of discussion among Americans.

From the outset, those active in integrating foreigners had to reckon with xenophobic sentiment and had to adapt their strategies accordingly. In 1933 Richardson of Brown University—himself not above all doubt with respect to possible anti-Semitic sentiment—tried to discourage hopes for immigration in almost every letter he wrote, fearing the consequences for young Americans.[57] It was therefore all the more important to stress the absolute priority of research as opposed to teaching jobs when arguing in favor of immigrants. In the beginning, however, even the people at the purely research-oriented Institute for Advanced Study in Princeton (an institution that later on would become a symbol for the rescue of the

---

be sure, among the former there were also several nonrefugees. In Appendix 1 (1.1), EC-N means there is a file on the refugee in EC, however as a nongrantee.

[55]Krohn (1987), p. 29. Veblen, on his part, emphasized the "anti-Fascism" of many immigrants, as mentioned in chapter 6 in his remark on F. Pollaczek.

[56]See the Documents part of this chapter with examples for mistrust against Russians, Frenchmen, and even Britons.

[57]Richardson wrote for instance on Courant: "America has enough organizers and Palestine perhaps not enough" (Richardson to H. Blichfeldt, August 8, 1933, Richardson Papers, BUA, Correspondence 1933 [German-Jewish Situation], f. H. Blichfeldt).

immigrant-mathematicians) voiced concern about the rush of immigrants they felt unable to cope with. Veblen wrote to AMS secretary Richardson on July 29, 1933: "Of course, the Institute is already pretty heavily involved with foreigners and Flexner is anxious to keep it primarily American."[58] A not insignificant part of xenophobia was an expression of academic anti-Semitism just as present in American society as in European countries.[59] Reingold illustrates this for American mathematics in the 1920s and 1930s.[60] Reingold quotes many documents concerning in particular George David Birkhoff, the mathematician at Harvard University. However, in order to keep matters in the proper perspective, it is imperative to stress that this kind of anti-Semitism cannot be compared, let alone put on an equal level, with the criminal, institutionally legalized, and incited anti-Semitism in Germany *after* 1933. Moreover, Birkhoff, acting under quite different societal conditions, cannot be compared to the Nazi Bieberbach in Germany, even though Courant once called him a "Nazi" in an emotional letter (D). There was a connection—though a complicated one—between anti-Semitism and still very virulent tendencies of American isolationism with which the American government under Franklin Delano Roosevelt—which was not unambiguous in its attitudes either—had to contend.[61] Nevertheless, because of Birkhoff's indisputable and great influence on the community of American mathematicians, the charge of anti-Semitism against him, even today a matter of controversial discussion in the American mathematical community,[62] shall be pondered in a separate

---

[58]Richardson Papers, Correspondence, box: German-Jewish Situation, f. O. Veblen. This somewhat unrealistic intent on the part of A. Flexner's who had just hired three leading European immigrants (Weyl, Einstein, von Neumann) is also reported in Porter (1988), p. 368, and is documented in Flexner's correspondence in the IAS Archives.

[59]For mathematics see Reingold (1981), Bers (1988), and Niven (1988). For the general problem of the relation to Jewish immigrants cf. Friedman (1973). The political scientist Franz Neumann deemed anti-Semitism in Germany before 1933 lower than in the United States (Krohn 1987, p. 30). Of the same opinion were Wasow (1986), p. 48, and Bers (1988), as well as J. Franck, as reported in Hoch (1983), p. 241. See Hoch (1983), pp. 239ff.: "Special difficulties for Jewish theorists."

[60]Reingold (1981).

[61]This connection is basic to the recently published counterfactual novel by Philip Roth, *The Plot against America* (2004). Roth describes the fear on the part of American Jews around 1940 of the consequences of an isolationist government for the cause of the Jews both in America and in Europe. Roth imagines a presidency held by the pioneer of aviation, Charles Lindbergh, who had strong connections to the German Nazis and who at times pursued anti-Semitic policies.

[62]See the letter to the editor by D. Gale in the *Notices of the AMS* 41 (1994): pp. 1099–1100, which raises that reproach and argues against an article by S. MacLane that had trivialized Birkhoff's anti-Semitism as unremarkable given the circumstances prevailing at the time.

case study below. Concessions to existing anti-Semitism were manifold even among scientists *not* prejudiced politically or by their religious leanings. Even Oswald Veblen, who definitely belonged to the last category of unprejudiced Americans, felt compelled to recommend Kurt Reidemeister for immigration above all others since he deviated least from the approved American political and racial standard.[63] Even among functionaries and in the official policies of the emergency organizations, academic anti-Semitism was present at least for a short while. In the Rockefeller Foundation there were tendencies, at least immediately after 1933, to belittle the effects of Nazi rule, and to consider some Nazi measures against the Jews as inevitable and as a reaction to the alleged previous favoritism in Germany of the "Jewish liberal element."[64] Duggan's Emergency Committee regularly gave information about the "racial" descent and religious denomination of the immigrants when trying to secure jobs for them at American universities (D). By and large and on average, however, Jewish immigrants were not treated significantly worse or better by American rescue initiatives than non-Jewish emigrants,[65] especially if one takes into account the balancing influence of Jewish support organizations.[66] With war looming and the resulting easing of restrictions on the job market in the movement for "war preparedness," anti-Semitic sentiment was bound to diminish. In hindsight some people even regretted not having attracted more immigrants to the country, which was now badly in need of mathematicians in war research.[67] Historical judgments state concurrently a decrease in academic anti-Semitism in the United States after the war at the latest. Reasons for that were the course of the war, the horror of the Holocaust, the foundation of Israel in 1948, and, finally, a further increase in international scientific communication, the last-mentioned having come to a war-related temporary standstill.

The son of George David Birkhoff, Garrett, an influential mathematician in his own right, described in 1977 a decisive new level of internationalization within the American mathematical community resulting from the developments of the 1930s and the war. Birkhoff claims that at least some Americans viewed the impoverishment of the European scientific cultures around 1940 with mixed feelings and as potentially dangerous for world science ("overkill"). Thus Birkhoff stressed in retrospect

[63]See chapter 4 and a case study below.

[64]A quote by a leading RF functionary of the Paris office in May 1933. See Siegmund-Schultze (2001), p. 200, for the full quote.

[65]This is also confirmed in Fischer (1991), p. 31, for immigration in physics.

[66]Among the latter the conflict between more liberal and more orthodox currents of Judaism further influenced the choice of immigrants to be supported.

[67]Duggan and Drury (1948), p. 68. However, secrecy regulations and requirements of citizenship restricted the possibility of employment at least for the more recent immigrants.

the far-sighted tendencies within American science policies (as represented for instance by the Rockefeller Foundation) of the 1930s and 1940s. He refrained from emphasizing the narrow and xenophobic currents that needed to be discussed in this chapter in order to be able to convey the fuller historical picture:

> Our once closely knit and patriotic mathematical community became internationalized, diversified, and even fragmented in the post-war years. . . . The [American mathematical] leaders . . . had achieved by 1938 their mission of matching Europe's mathematical culture . . . their success had become an "overkill" by 1941, and . . . it was taken for granted in a very changed world by 1950.[68]

8.D. Documents

8.D.1. Competition on the American Job Market and Attempts to Keep the Immigrants away from America

In December 1933, AMS secretary R. G. D. Richardson said the following, based allegedly on information collected from Duggan's Emergency Committee: "Owing to the depression in this country, more professors have been released than have been dismissed in Germany."[69] Oswald Veblen wrote to Richardson in May 1933, with respect to the preparation, together with the Rockefeller Foundation, of a "general committee for the relief of German scientists who are dispossessed":

> The idea would be not to keep a concentration of refugees in any one country but to distribute them in such a way as to arouse a minimum of objection from other unemployed scientists.[70]

Applying this to the situation in the United States, Richard Courant, in a letter to Wolfgang Sternberg of March 30, 1935, even talked about the possibility "that several of the foreigners now in the country will have to leave" (CPP [T]).

To Gabor Szegö, Courant wrote in 1935 that there existed "among younger people in America increasing resistance towards jobs for refugees."[71]

---

[68]Birkhoff (1977), p. 77.

[69]Richardson Papers, BUA, box Correspondence, 1933 (German-Jewish Situation), f. Miscellaneous Letters, R. to E. A. Adams, December 26, 1933. This information given by the EC in 1933 is put into doubt by later data on the dismissals from American universities, presented in Rider (1984), p. 125.

[70]Veblen to Richardson, May 9, 1933. Richardson Papers, Correspondence, box: German-Jewish Situation, f. O. Veblen.

[71]March 31, 1935, Szegö Papers, Stanford, box 5, f. 15.

To the English Academic Assistance Council, Courant wrote on April 23, 1935 with regard to Fritz John:

> The unfortunate thing here just now is the growing resistance among younger academic people towards immigrants who might occupy jobs. For this reason I would not deem it wise for John to immigrate into America just now. (CPP)

The Chancellor of New York University, H. W. Chase, reported to Richardson in 1935 on Courant's success at the university, but added: "We do not feel that we should at the same time add to our staff a second displaced German scholar . . . without hurting the morale among the other men."[72] Courant himself tried to distribute the fellowships at his disposal in New York evenly to Americans and immigrants. He wrote on July 10, 1935 to the former immigrant from Russia, J. Tamarkin:

> Since there is already one scholarship intended for a Jewish student and another for a German-born American citizen, it seems desirable to me (and to others) that the remaining scholarship be given to a more typical American, and the University authorities wish to get somebody from outside. (CPP [T])

8.D.2. "Selection" of Immigrants to Be Promoted and Bureaucratic Obstacles on the Part of the Americans

In a 1936 letter of recommendation, written to O. Veblen in favor of Olaf Helmer-Hirschberg, J. Whyte admitted that the latter, who had just received his PhD in Berlin, was not, originally, "a displaced German scholar in our restricted use of that term."[73]

Support for the former Vienna high school teacher in mathematics Hans Fried received from the Emergency Committee in New York was afterward expressly declared to be an exception: "Our Committee has declined to consider any more applications for men and women who were not actually professors or Privatdozenten in foreign universities."[74] In 1938, Heinz Hopf wrote to Richard Brauer, who was already in America, about the latter's brother Alfred Brauer, who had been waiting for a chance to emigrate since his dismissal in Berlin in 1935:

[72]Richardson Papers, BUA, Correspondence, box: German-Jewish Situation, f. H. Lewy, Chase to Richardson, April 11, 1935.
[73]Whyte to Veblen, December 7, 1936, EC, box 151, f. O. Veblen. Whyte acknowledged that Helmer-Hirschberg had not been a docent in Germany, the usual prerequisite for preferential immigration.
[74]S. Duggan to Richardson, October 15, 1941. Richardson Papers, Correspondence, f. 87 D (Duggan 1940–42). Habilitation as a prerequisite for help is also mentioned in Duggan and Drury (1948), p. 186. However, in the case of Fried, there were strong arguments for support because Fried was not only an excellent teacher but also had managed to publish five papers on functions of real variables in good journals during the 1930s, four of them in *Fundamenta Mathematicae*. See OVP, cont. 30, f. Fried, Hans, 1940–41.

The Berlin USA Consulate declared that he could not receive a "professor's visa" (which entitled one to <u>immediate</u> entry): according to the law, this is only for people who have been in an academic position during the last 2 years. As this obviously does not apply to him!!! He has to register as an immigrant and wait for a few years.[75]

Similarly, Arthur Korn did not immediately (1938) receive a nonquota visa, since he had not held a teaching position in the years after his dismissal. It was only through circumventing the system by taking a position in a laboratory in New York (judged equivalent to a teaching position, thereby fulfilling the conditions for a temporary visa) that Korn eventually qualified for a permanent immigration visa.[76]

For Paul Epstein of Frankfurt, who later committed suicide, the obstacles proved to be insurmountable. He tried to go to England in 1939 since his "waiting number" [Wartenummer] on the immigration quota to the United States was too high. Max Dehn wrote to Mordell from Oslo:

It is not about giving him a position in England, but only to allow him to stay there temporarily with the help of the Aid Committee. He has an affidavit from his sister who is an American citizen, but has unfortunately a very high waiting number.[77]

Wolfgang Wasow had been born in Switzerland and was therefore allowed to use the Swiss immigration quota. In 1986 he described the difficulties getting his wife with him into emigration on the same quota. After the American vice-consul in London had turned down the application, a female sympathizer at the same office granted it a few days later, when Wasow accidentally showed up in the consulate again:

I wonder what was behind this affair. It is well known that many Americans, including many officials in the diplomatic service, did not like the large influx of refugees, particularly Jewish ones, and I believe that the first vice-consul who dealt with our case was one of them. His interpretation of the immigration law was so artificial as to indicate his bad faith."[78]

The case of Carl Ludwig Siegel shows that "voluntary" immigration did not qualify one for support from the American emergency organizations. On November 26, 1934, Courant wrote in a letter to his friend Harald Bohr in Denmark:

The formal difficulty in the whole matter with Rockefeller and the Emergency Fund is that Siegel is not officially regarded as threatened in his position and

[75]H. Hopf to R. Brauer, November 21, 1938, Hopf Papers, ETH, Hs 621:306c (T).
[76]Shapley Papers, Harvard, box 6B, file: Korn. The same problem had also other applicants for immigration (Weyl to Shapley, December 21, 1938, Shapley Papers, box 6F, file: IAS).
[77]Dehn to Mordell, April 27, 1939, Mordell Papers, Cambridge, 1941 (T).
[78]Wasow (1986), p. 205.

can therefore not be supported on the basis of the funds agreed on for displaced scholars. Rockefeller would have to give him a Traveling Professorship or such like, and Weaver has principally agreed to give enough money for that. (CPP [T])

8.D.3. Special Problems for Female Immigrants[79]

Being chairman of the Department of Mathematics at Wheaton College, a women's college in Norton, Massachusetts, the female refugee and noted applied mathematician Hilda Geiringer wrote on June 4, 1947 to the British SPSL:

> One very obvious remark is that it is much harder for women than for men to attain academic positions of distinction. I am speaking particularly of the East. The great Universities, Columbia, Harvard, Yale, Princeton, Cornell, etc. are practically closed to them. A typical example is the women's College of Harvard, Radcliffe College, where the instructors are the Harvard Professors while women are used as tutors, supervisors, etc. There are, of course, a few exceptions (very few indeed) but, ceteris paribus, the possibilities for women are near to zero. The distinguished women colleges are almost exclusively undergraduate colleges (with the exception of Bryn Mawr College which has about 500 undergraduates and about 120 graduate students).[80]

Between 1939 and 1944 Geiringer, who had officially reduced her age by two years in order to improve her chances for employment, was at Bryn Mawr College in Philadelphia in temporary positions, the same college that had employed Emmy Noether until her death in 1935.

8.D.4. Political Mistrust on the American Side[81]

When Weyl had proposed to Alexander Weinstein, his former student in Zurich, that he immigrated on the Russian quota since this was not fully used, Weinstein's sister-in-law commented on that in the following letter to Weyl in August 1940:

> Among the American consuls the opinion seems to prevail that a human being born in Russia (even in the tsarist one) is inevitably a communist and that one has accordingly to put additional obstacles in his way.[82]

---

[79]See also the case study on Emmy Noether below.

[80]SPSL, box 279, f. 3 (Pollaczek-Geiringer, H.), folio 61v.

[81]See on this point also remarks in chapter 9.

[82]OVP, cont. 33, f. Weinstein, Alexander, 1939–46. Susan Engreen to Weyl, Los Angeles, August 25, 1940 (T).

Weyl hastened to discard this idea, writing in September that same year to the American consul in Lisbon about Weinstein:

> I am sure that he has no Communist leanings at all. His father was a well-to-do doctor, and both his parents were killed or starved to death during the Bolshevik revolution. Dr. Weinstein never considered going back to Russia, even though he went through some years of extreme hardship as an exile.[83]

8.D.5. The Priority of Private Foundations and Pure Research Institutions in Helping the Immigrants

Norbert Wiener from MIT in Cambridge, Massachusetts, supported the integration of refugees strongly. However, in tune with his personality, Wiener acted rather individually and not within the bounds of an organization such as the AMS. In December 1934, he wrote an article in the *Jewish Advocate* titled "Aid for German Refugee Scholars Must Come from Non-Academic Sources." Here Wiener stated that immigrants did not generate revenues for American universities, being as a general rule of little use in teaching freshmen or undergraduates. Therefore financing for them had to stem from nonacademic sources, foundations for the most part, in order to prevent xenophobia and anti-Semitism.[84]

In January 1936, Richard Courant reported to his friend James Franck about xenophobic remarks by the former Hilbert student and current president of the Rockefeller Foundation, Max Mason, who had addressed the New School for Social Research in New York City:

> Mason welcomed the New School because it is best suited to solve the problem of refugees outside the domain of American universities. On humanitarian grounds his speech did not leave me with a favorable impression. It simply reflects the well-known fact that there is resistance against immigrants everywhere and people in responsible positions feel compelled to comply with that sentiment.[85]

The policy of Harvard University and in particular of its president James Conant was wavering with respect to the refugee problem. However, it was strictly against using the endowment of the university for that purpose. A leading Harvard mathematician also stated this in 1939: "I see no

---

[83]Ibid. Weyl to the American Consul in Lisbon, September 3, 1940.

[84]Wiener Papers, MIT, box 11, f. 537. Newspaper clipping, *Jewish Advocate*, Boston, December 14, 1934, pp. 1 and 4. See also Wiener's later article in the same paper, dated February 5, 1935 and quoted in chapter 6.

[85]Courant to Franck, January 16, 1936. CPP Miscellaneous (T). Obviously Mason alluded to the University in Exile within the New School, which finally was supported by the Foundation after the outbreak of the war. See Johnson (1952), pp. 366–67.

chance that Harvard is going to have funds to help subsidize an asylum fellow in mathematics."[86]

8.D.6. The Restricted Scope and Possibilities Available to the German Mathematicians' Relief Fund

This Fund was indeed almost a lifesaver for some immigrants, such as Wolfgang Sternberg,[87] who never found an adequate position in the United States. In some cases the money was cabled abroad, for instance for Ernst Jacobsthal and Paul Kuhn as refugees in Sweden (coming from occupied Norway), the former being "without clothes and money."[88] The very small sums provided by the Relief Fund, which, at one point, led to the total depletion of its resources after two one-time payments of $200 to Fritz John and Max Zorn, is documented by Courant's correspondence with Hermann Weyl (CPP). This correspondence includes a round-robin letter by Weyl dated January 18, 1936, which says that in 1934 he and Emmy Noether had proposed reducing by 1 to 4 percent the incomes of immigrants having obtained positions in order to support those still without. The American mathematician Norbert Wiener, in a letter to the Emergency Committee in February 1935, commented on the Fund with the following words:

> It seems rather pitiful to solicit such funds from poorer of such scholars, although Weyl can very well afford it.[89]

Given the precarious financial situation the majority of immigrants found themselves in it is no wonder that Weyl stated in 1936 that his proposal had largely been ignored, and he therefore asked to deduct at least 1 percent from incomes generally for that purpose. In a letter from March 1935, Courant pointed out to Weyl the wide variety of initiatives existing to support immigrants making Weyl's proposal to combine all forces in the Relief Fund unrealistic:

> To my horror I notice that I have not reacted so far to your and Emmy Noether's circular. I am very sorry that my financial situation which is presently more than problematic does not allow me to send more than $20. . . . Anyway I am concerned

[86]Mathematics chairman J. L. Walsh to H. Weyl, May 26, 1939, Shapley Papers, HUA, box 6C, file: S. See chapter 9 for the partial success Shapley's Asylum Fellowship Plan had nonetheless.

[87]The repeated payments to Sternberg by Weyl, even as late as March 22, 1945, are documented in the Oswald-Veblen Papers, OVP, cont. 33, two folders Sternberg. One gets the impression that Weyl may have paid part of it from his own pocket.

[88]OVP, cont. 31, f. Jacobsthal, E., 1938–44. Weyl to R. Courant and R. Baer (March 17, 1943).

[89]N. Wiener to Edward R. Murrow, February 25, 1935, Copy, EC, box 2, f. Bernstein, Felix, 1933–35.

that the people close to the Notgemeinschaft [he meant probably Duggan's Emergency Committee; R. S.] fail to show appreciation for your efforts. In any case I feel obliged to support—as far as I can—the Notgemeinschaft directly.[90]

In 1936 Weyl wrote in a letter to Heinz Hopf in Zurich (Hopf had just tried to raise support for Heinrich Grell):

The enclosed circular is originally destined for the German mathematicians in America, however, Swiss Francs [Fränkli] are as welcome as dollars.[91]

8.D.7. Further Motives for Xenophobia: Mental Borders, Anti-Semitism, Differences in the Science Systems, Professional Jealousy

In 1927 when the appointment of one English statistician Wilson (?) was discussed at Brown University in Providence, R. G. D. Richardson, still officially a Canadian citizen,[92] wrote to Birkhoff at Harvard:

With one foreigner Tamarkin in the department, we feel that it might be a considerable risk to take another one such as Wilson. Englishmen do not adapt themselves very quickly to American ways, and generally they do not wish to do so.[93]

When Richardson was interested, in 1942, in hiring the French physicist Leon N. Brillouin (1889–1969), he had still qualms about national idiosyncrasies:

His wife is Polish. Some people say that she is Jewish or part Jewish; I do not know. I have not met her, but Brillouin himself makes an excellent appearance and does not seem to have that air of superiority and provinciality we sometimes associate with French scientists.[94]

Brillouin would, indeed, fit into the American environment. He subsequently made a career at Columbia University in New York City.

In a letter to the EC on October 19, 1941, Charles G. Laird (University of Idaho, Pocatello) talked about problems experienced by the immigrant Max Dehn, resulting from professional jealousy:

Some faculty members have suggested professional jealousy, and the circumstances make such a suggestion plausible. Dr. Dehn has an enviable interna-

---

[90]Courant to Weyl, March 21, 1935. Weyl Papers, ETH Zurich, Hs 91:66 (T).

[91]Weyl to H. Hopf, January 20, 1936 (T). Weyl Papers, ETH Zurich, Hs 91:281. Hopf did contribute, as it becomes clear from another letter by Weyl, dated March 17, 1936., ibid.

[92]This is an example for the fact that Canada and the United States were and are much less foreign countries to each other than to other countries. See Archibald (1950).

[93]Richardson Papers, BUA, Correspondence, May 17, 1927, f. 23 B.

[94]BUA, Appl. Math. Div. I.90, R. to H. Wriston, February 2, 1942.

Figure 41 *Max Dehn (1878–1952). The cofounder of modern topology was also deeply involved in work on philosophical and historical problems of mathematics. After his expulsion from Frankfurt he finally found a very subordinate position in the small Black Mountain College in North Carolina. After the war the authorities in Frankfurt missed the opportunity of giving Dehn a dignified rehabilitation.*

tional reputation, the head of the department has a master's degree from a small institution, and has given little evidence of scholarly interests since obtaining his degree some years ago. . . . It is possible, also, that the head of the department objects to Dr. Dehn on racial or nationalistic grounds.[95]

James Franck warned his friend Courant in 1935 against too active a "passion for organization" ["Organisationslust"] in view of the existing xenophobia:

I have to tell you that I have heard from unobjectionable and well-meaning sources that the mathematicians in New Haven and in Princeton and maybe

[95]EC, box 6, f. M. Dehn, 1941–44.

elsewhere are somewhat angry at you because you allegedly promote people to positions, in short behave like an administrator and do not show restraint as a foreigner should. . . . Xenophobia here seems on the rise and we have to reckon with it. . . . We must not forget that we once deceived ourselves about the safety of the ground we were living on. So let us stick together and show extreme caution.[96]

A letter by Richardson to Harald Bohr, dated June 29, 1933, reads:

Courant is an able administrator as well as mathematician and would doubtless find a subordinate position in America not to his liking. On the other hand, on account of his lack of knowledge of American conditions, I think that no American university is likely to put him in a position of responsibility.[97]

In a personal letter from 1933 to Hans F. Blichfeldt (1873–1945) in California, Richardson, a mere administrator with fewer achievements in mathematical research than Courant, called the latter an "arch schemer."[98] Richardson admitted that Veblen and the Bohr brothers disagreed with this judgment, but nevertheless warned Blichfeldt against Courant's appointment in California:

He would want to organize the whole of the Pacific Coast and would be interested in that type of thing rather than in creating some mathematics.[99]

The following quotation by the American applied mathematician Warren Weaver, who had befriended Courant during his emigration, gives a clear indication to what extent anti-Semitism figured virulently in a large number of judgments. Weaver remarked as late as 1941 in his diary with regard to Courant, a frequent guest of his at the RF headquarters in New York City: "WW says C. is an applied mathematician of the first rank and an exceedingly energetic organizer. The dangers in the situation are that C. is rather too energetic an organizer and that he would not necessarily use good racial judgment about his appointments."[100]

In the aftermath of World War II, at the beginning of a second wave of emigration from Europe, new problems arose alongside existing ones. This

[96]Franck to Courant, November 1, 1935. CPP Miscellaneous (T). The last sentence has been used as an epigraph in this chapter. See above.

[97]R. G. D. Richardson to Harald Bohr, June 29,1933, Richardson Papers, BUA, Correspondence, box: German-Jewish Situation, f. H. Bohr.

[98]There was both actual jealousy on the part of Richardson, which would even increase in the competition for an institute on applied mathematics in the USA in the late 1930s, and an old conflict between Richardson and Courant from their time in Göttingen. In letters to others Richardson accused Courant of having been negligent in quoting Richardson's old papers on difference equations and differential equations. See Reid (1976), pp. 227–29.

[99]Richardson to Blichfeldt August 8, 1933, Richardson Papers, BUA, Correspondence, box: German-Jewish Situation, f. H. Blichfeldt.

[100]Quoted from Siegmund-Schultze (2001), p. 201.

is indicated in a letter by Harvard mathematician Joseph L. Walsh (1895–1973) to the theoretical physicist John H. Van Vleck (1889–1980) of May 1945:

> An American is preferable to a European (an Asiatic, or African), for the latter frequently is unable to understand our undergraduate methods . . . A man who has remained in Europe during the war is a . . . risk unless he is known to have continued active research during the war period.[101]

8.D.8. Decline of Xenophobia in Connection with Political Events on the Eve of World War II

When supporting the immigration of Otto Neugebauer, AMS secretary Richardson acknowledged that a new situation had arisen. In a letter to the Emergency Committee he wrote on December 11, 1938:

> I was called on to act as consultant to the Emergency Committee on Displaced German Scholars a few years back and was in large part instrumental in establishing several mathematicians in this country (Hans Lewy, Gabriel Szegö etc.). The time came, however, when it seemed to me that we had taken in to the country all the mathematicians (15 or more) that America could absorb without arousing opposition from those on the ground and creating unfortunate antiforeign sentiment. Believing that I voiced the opinions of the majority of mathematicians, I advised Dr. Murrow at that time that we should discourage further importations. . . . But a new situation has arisen in Europe.[102]

This "new situation" is explained in a letter dated March 13, 1939 to Stephen Duggan of the EC written by Oswald Veblen, who worked closely with Richardson:

> My impression is that we are not far from the saturation point in the more prominent universities; but that there are still many less well known academic institutions in which refugees could be placed with substantial advantages both to the individual and to the institution. . . . There is a great deal of evidence that the events subsequent to Munich, particularly the pogrom in Germany and the artificial stimulation of racism in Italy, have had a profound psychological effect in this country, which makes opportune a new effort to salvage cultural values.[103]

---

[101]Walsh to Van Vleck, May 15, 1945. HUA, Math. Dept. UAV 561, Correspondence and Papers 1911–62, UAV 561.8, box 1939–42, f. Ahlfors.

[102]EC, New York City, box 26, f. O. Neugebauer, 1938–44.

[103]EC, box 151, f. O. Veblen.

8.S. Case Studies

8.S.1 The Case of the Female Emigrant Emmy Noether

Emmy Noether was without doubt the most talented and creative female mathematician in the first part of the twentieth century, and maybe even of all times. After her dismissal in Göttingen in 1933, she lived for one and a half years in American emigration, until her premature death after routine surgery in April 1935. Albert Einstein wrote a moving obituary in the *New York Times*. This has been translated into English by Abraham Flexner who changed the original somewhat and called "the last years the happiest and perhaps most fruitful of her entire career."[104] While Noether's influence on O. Zariski and R. Brauer in Princeton in 1934–35 shall not be disputed,[105] the free translation by Flexner seems somewhat exaggerated. Above all the question must be raised, in the context of the present book, whether or not the Americans managed to provide a position for Emmy Noether commensurate with her abilities and fame, after the Germans in the 1920s clearly failed to do so due to anti-Semitic, antifeminist, and other political prejudices.[106]

The topologist at Princeton University, Solomon Lefschetz, while setting great store by Noether's algebraic research, failed to submit an application urging her attachment to a prestigious university boasting a strong mathematics faculty. After only a brief referral to the various ways in which she was persecuted in Germany he contended himself with writing the following to the Emergency Committee in June 1933:

> It occurred to me that it would be a fine thing to have her attached to Bryn Mawr in a position which would compete with no one and would be created ad hoc; the most distinguished feminine mathematician connected with the most distinguished feminine university.[107]

This was what finally happened, although there were constant problems connected with financing the position, as Noether was not ideally suited to elementary teaching.[108] Recognizing these problems, Oswald Veblen,

---

[104]A. Einstein: "The late Emmy Noether"; *New York Times*, May 4, 1935, p. 12. A comparison between the German original and Flexner's translation is in Siegmund-Schultze (2007).

[105]Some remarks on that are in chapter 10.

[106]Extensive literature on Emmy Noether has been published meanwhile, including Aleksandrov (1935/81), Brewer/Smith (1981), Dick (1981), Kimberling (1981), Lemmermeyer and Roquette (2006), Tobies (1997/2008), Tollmien (1991), van der Waerden (1935), and Weyl (1935). However, the question of the adequateness of American support has not been discussed systematically in any of these papers.

[107]S. Lefschetz to Emergency Committee, June 12, 1933, EC, box 84, f. Noether, E.

[108]See the quotation on Noether's "eccentricity" by the president of Bryn Mawr College in chapter 9.D.4 below.

THE LATE EMMY NOETHER.

Professor Einstein Writes in Appreciation of a Fellow-Mathematician.

To the Editor of The New York Times:

The efforts of most human beings are consumed in the struggle for their daily bread, but most of those who are, either through fortune or some special gift, relieved of this struggle are largely absorbed in further improving their worldly lot. Beneath the effort directed toward the accumulation of worldly goods lies all too frequently the illusion that this is the most substantial and desirable end to be achieved; but there is, fortunately, a minority composed of those who recognize early in their lives that the most beautiful and satisfying experiences open to humankind are not derived from the outside, but are bound up with the development of the individual's own feeling, thinking and acting. The genuine artists, investigators and thinkers have always been persons of this kind. However inconspicuously the life of these individuals runs its course, none the less the fruits of their endeavors are the most valuable contributions which one generation can make to its successors.

Within the past few days a distinguished mathematician, Professor Emmy Noether, formerly connected with the University of Goettingen and for the past two years at Bryn Mawr College, died in her fifty-third year. In the judgment of the most competent living mathematicians, Fraeulein Noether was the most significant creative mathematical genius thus far produced since the higher education of women began. In the realm of algebra, in which the most gifted mathematicians have been busy for centuries, she discovered methods which have proved of enormous importance in the development of the present-day younger generation of mathematicians. Pure mathematics is, in its way, the poetry of logical ideas. One seeks the most general ideas of operation which will bring together in simple, logical and unified form the largest possible circle of formal relationships. In this effort toward logical beauty spiritual formulae are discovered necessary for the deeper penetration into the laws of nature.

Born in a Jewish family distinguished for the love of learning, Emmy Noether, who, in spite of the efforts of the great Goettingen mathematician, Hilbert, never reached the academic standing due her in her own country, none the less surrounded herself with a group of students and investigators at Goettingen, who have already become distinguished as teachers and investigators. Her unselfish, significant work over a period of many years was rewarded by the new rulers of Germany with a dismissal, which cost her the means of maintaining her simple life and the opportunity to carry on her mathematical studies. Farsighted friends of science in this country were fortunately able to make such arrangements at Bryn Mawr College and at Princeton that she found in America up to the day of her death not only colleagues who esteemed her friendship but grateful pupils whose enthusiasm made her last years the happiest and perhaps the most fruitful of her entire career.

ALBERT EINSTEIN.

Princeton University, May 1, 1935.

Figure 42 *Einstein on Noether in the* New York Times. *Einstein's obituary of Emmy Noether in the* New York Times, *May 4, 1935.*

from the nearby Institute for Advanced Study in Princeton, proposed the following in a letter to the director, Abraham Flexner, in February 1935, shortly before Noether's unexpected demise:

> The actual action agreed upon by the professors of the Institute with regard to a grant is to set aside $1500 to be used as a grant for Miss Noether in the year 1935–36 in case other means of support should fail. This proposal was agreed to (1) in recognition of the fact that Miss Noether has been conducting a seminar last year and this, without compensation, and (2) in view of our appreciation of her intrinsic worth. . . .
>
> I am inclined to think that the view of our group towards further commitments would be something like this: that we should be glad to see further grants made during a period in which an effort was being made to place her permanently at Bryn Mawr or elsewhere; moreover, that in saying this we should be conscious of the possibility that this might become a permanent commitment on the part of the Institute. There is no doubt that, apart from the uniqueness of her position as a woman mathematician, she is quite obviously one of the most important scientists who have been displaced by the events in Germany. Therefore even a permanent commitment could be nothing but creditable to the Institute.[109]

So it seems that even under American conditions a permanent appointment—rather than a mere "commitment"—for a female mathematician,[110] regardless of her stature in research, raised difficulties at even such a prestigious institution as the IAS. Liberal men such as Veblen had to expressly insist on the point that a "permanent commitment could be nothing but creditable to the Institute." One gets, however, the impression that Noether's subsistence—if not a prestigious position—in the United States would have been secured under any circumstances, given her obvious importance as a mathematician. At the same time one feels that the shock of Emmy Noether's premature death in 1935 aroused a bad conscience among American mathematicians for not having offered her a proper position at the IAS—one to which she would have been ideally suited due to her abilities in research and her lack of proficiency in the English language. In fact, in December 1935 a Memorial Fund in honor of Emmy Noether was established under the leadership of the IAS, in particular its director Abraham Flexner and Oswald Veblen. The following letter written by the mathematician G. A. Bliss to Veblen says something about the extent of that fund, the establishing of which (except for Einstein) no immigrant was directly involved in, thus making it a purely American matter:

---

[109]O. Veblen to A. Flexner, February 28, 1935, IAS Archives, Faculty Files, Oswald Veblen, box 32, f. Veblen, 1934–35.

[110]See the respective remarks by Hilda Geiringer quoted above in this chapter.

The amount desired to be raised took my breath away a little in comparison with the sums which we have been able to raise hitherto in honor of distinguished mathematicians in this country.[111]

8.S.2. A Case of the Exploitation of Immigrants by an Engineer at Cornell (M. G. Malti)

On July 6, 1939, Hermann Weyl wrote to Artur Rosenthal (then still in Europe) after he had tried and failed to find a suitable position for him:

As I have already told you the arrangement with Kansas City has unfortunately failed. I therefore strongly advise you to accept the offer from Malti. I realise it is hardly a tempting position. Malti knows next to nothing about mathematics, he has written a book on the Heaviside Calculus. You will have regular office hours, eight hours a day, during which you have to do mathematics. Warschawski, who was with Malti before and who is a modest person, has complained bitterly about him. But there is the connection to Cornell, and if Snyder gives his support, you have a good chance of obtaining a position at the Mathematics Department of Cornell.[112]

Rosenthal, a scholar of great range, known internationally for his articles on point sets and real functions in the German *Encyclopädie der Mathematischen Wissenschaften*, at that time fifty-two years of age, had decided to accept the position no matter what.[113] The position was, however, no longer available in September 1939, and Rosenthal found something at Ann Arbor (Michigan) soon after. However, less prominent men such as Stefan Warschawski, Fritz Herzog, Michael Golomb,[114] and Wolfgang Sternberg had to, one after the other, accept the position offered by Malti. The latter happened to be of Armenian, not American origin, a fact that may have made it more difficult for Sternberg to adapt to the situation. In fact, Sternberg (in several letters to Weyl) hinted at the differences in mentality between him and Malti.[115] As a mathematical assistant to Malti he received only $100 per month and continued to be dependent on the occasional handout from Weyl's Relief Fund.

---

[111]Bliss to Veblen, January 2, 1936, OVP, cont. 9, f. Noether, Emmy.

[112]OVP, cont. 32, f. Rosenthal, Artur, 1938–42 (T).

[113]Ibid., Rosenthal to Weyl, September 10, 1939.

[114]See Golomb's respective quotation in chapter 10 (D) in connection with applied mathematics.

[115]On June 4, 1944 Sternberg wrote to Weyl: "Malti is Armenian by birth and belongs to the Arab race. He has a certain oriental slyness and has this way reached his position merely by sitting on his post." OVP, cont. 33, f. Sternberg, Wolfgang, 1939–44. The two extensive folders on Sternberg in the OVP also contain much information on Sternberg's problems, on his difficult character and on repeated payments to him in the range of $100 from the Relief Fund.

8.S.3. *Five Case Studies about Academic Anti-Semitism in the USA*

8.S.3.1. CONSIDERATION OF ANTI-SEMITISM IN THE POLICIES
OF THE RELIEF ORGANIZATIONS

On January 9, 1939 the Emergency Committee wrote to Eureka College in Eureka, Illinois, alluding to previous mail including information regarding four dismissed German mathematicians "whom we believed to be Aryan and Protestant."[116] However, the EC now had second thoughts about whether one of the four, Gustav Mesmer,[117] as the committee put it, "fits the [college's] racial and religious conditions." As to the three others, Max Dehn, Ludwig Hopf, and Hans Schwerdtfeger, the EC assumed (wrongly in the first two cases) that they would fulfill the conditions of Eureka College.

Many refugees were aware of the racial and religious policies of some Americans. In a letter to Weyl, Hans Fried from Vienna wrote on July 11, 1940:

> I am of Jewish confession, while my wife and daughter are Lutherans. Since it is of utmost importance to me to find a position I would be willing—if this is sufficient—to be baptized.[118]

Weyl rephrased this a little bit in his letter to the American Friends Service Committee of July 16, 1940, apparently in order to avoid giving too unfavorable an impression of opportunism:

> He writes to me that he has considered for some time being baptized, and asks whether he would be acceptable if he took this step now.[119]

One German mathematician, whose field of mathematics was closely linked to the Americans (topology), also held a certain appeal to them on a social level. This was Kurt Reidemeister, who, as it turned out, was not really dismissed by the Nazis but "only" transferred for disciplinary reasons (from Königsberg to Marburg) as discussed in chapter 4. Veblen first wrote to Richardson about him on June 11, 1933:

> Reidemeister has been "beurlaubt" though not a Jew nor particularly inclined to the left.[120]

---

[116]EC, New York City, box 6, f. M. Dehn, 1933–1940.

[117]Gustav Mesmer (1905–1981), assistant at the Kaiser Wilhelm Institute for Fluid Dynamics at Göttingen, was a mechanic and engineer rather than a mathematician. See "List of Displaced Scholars" LDS. According to Mesmer's own communication in the Poggendorff biographical dictionary of scientists (vol. VII) he was in still in Germany in 1940 and held a position in the United States since 1950. This suggests that he was not racially persecuted.

[118]OVP, cont. 30, f. Fried, Hans, 1940–41 (T).

[119]Ibid.

[120]Richardson Papers, BUA, Correspondence, box: German-Jewish Situation, f. O. Veblen. The letter has been quoted in more detail in chapter 4.

Veblen added in another letter to Richardson, on August 22, 1933:

> At present I think Reidemeister is the most attractive of the German possibilities. He is "Aryan" . . . and extremely able.[121]

8.S.3.2. EXAMPLES OF AMERICAN NATIONALIST AND RACIST PROPAGANDA AIMED AT IMMIGRANTS

Norbert Wiener's article in the *Jewish Advocate* of December 1934, in which he pleaded for measures to prevent xenophobia, was criticized in the *Boston Evening Transcript* in an extremely xenophobic tone, articulated by one Morrison I. Swift.[122] This position bore similarities to Bieberbach's racism in Germany, and Wiener felt compelled to write a letter to the editor of the *Boston Evening Transcript,* in which he says:

> I must emphatically dissent from the thesis that American students must have only American teachers. The habits of thought of the American may differ from those of the foreigner, but so do those of the Northerner from those of the Southerner, those of the New Englander from those of the Westerner. . . . The scope of science is world wide. In this age of dissolution and decay it is one great thing that is resisting the disintegration of our culture into provincialism and barbarism. These exiled colleagues of ours have something of value to say to us and to our children. We should pay every attention to conflicting claims to our charity, and should avoid, as far as possible, exciting a quarrel between our foreign friends and our unemployed American colleagues.[123]

8.S.3.3. PROBLEMS IN RELATIONSHIPS BETWEEN ASSIMILATED (IN PARTICULAR BAPTIZED) AND ORTHODOX JEWS IN AMERICA

Wolfgang Wasow wrote about his experiences with tradition-conscious American Jews immediately after his arrival in 1939:

> The morning after my first night in New York Lilo asked me to buy her . . . some food at a nearby grocery store. That store . . . had two meat counters. One of them was much more expensive than the other. I asked the salesman for the reason. He answered that the meat that cost more was kosher. I bought some from the cheaper side and heard him comment angrily: "This is why God has brought all that misfortune on the German Jews." . . .
>
> This was my first contact with the ambivalent attitude of most American Jews to us: We were Jews in the eyes of God—and of Hitler—to them and therefore

[121]Ibid.

[122]M. I. Swift, "Our Own Teachers First," *Boston Evening Transcript*, December 19, 1934, part 2, p. 2.

[123]Wiener to editor of *Boston Evening Transcript*, December 26, 1934, part 2, p. 2. Also in Wiener Papers, MIT, Correspondence, box 1, f. 40.

Figure 43 *Norbert Wiener (1894–1964). This important American mathematician from the Massachusetts Institute of Technology supported the creation of pensions for foreigners. He warned against xenophobia and helped individual immigrants such as Otto Szász.*

Wolfgang R. Wasow
(1909–1993)

Figure 44 *Wolfgang Wasow (1909–1993). This student of Richard Courant in Göttingen worked on the foundations of Ludwig Prandtl's boundary layer theory in aerodynamics. Wasow writes in his unpublished memoirs (1986) about his adventurous way of emigration (finally ending up with a professorship in Madison, Wisconsin).*

deserved assistance, although we were detestable, godless, arrogant outsiders. The bulk of American Jews descended from Jews in Eastern Europe, who had always been legally discriminated against and forced to take their ethnic and religious traditions seriously.[124]

[124]Wasow (1986), pp. 209–10. On Wasow see also O'Malley (1995).

Felix Bernstein, the statistician from Göttingen, had similar experiences, but nevertheless viewed the western European tendency for assimilation as advisable in America. In May 1933 he wrote from New York City to Albert Einstein, who was then in Europe:

> There is a problem in the national concentration of those parts of Jewry that are not willing to assimilate very much so far and are therefore sooner or later in danger, in all countries, of experiencing the fate of the German Jews. I see this danger very clearly here in New York. It is unthinkable in the long run that in the factual [erklärten] capital of a country two million Jews live in more or less total separation from the rest of the population and gain increasing political influence without facing the danger of the formation of resistance, particularly in case of a sudden turn of the political situation. Had not the leader of the Ku Klux Klan proved to be a swindler several years ago, something similar like in Germany could have happened here, maybe slightly moderated in an American way, but causing similar feelings of injustice.[125]

Six years later, in 1939, Bernstein tried to convince Einstein to write a conciliatory letter to Henry Ford (Detroit), known for his anti-Semitic positions in the 1920s but who had recently turned table and criticized the oppression of Jews in Germany. In his letter to Einstein, Bernstein said that there was a need for haste due to Ford having been rejected by Zionist circles in the United States. Accordingly there was concern that he would reestablish contact with Germany. Albert Einstein, who at times shocked American Jews by his support of Zionist ideas,[126] replied rather sharply, using the reserved "Sehr geehrter Herr" in addressing his long-term acquaintance Bernstein:

Dear Mr. Bernstein,
Your proposal is a good example for that lack of dignity which I have painfully experienced in many German Jews. You might say that political acting has to do with deliberate reasoning, not with dignity. But I am not of this opinion. Acting out of a healthy feeling is superior to any cleverness [Schlauheit], be it only because the other is also clever. What I despise ethically, I do not do. Such is the case here.

With friendly greetings Yours.[127]

---

[125]F. Bernstein to A. Einstein, May (undated), 1933, Einstein Papers, Jerusalem, 49 250-3 (T). Philip Roth's fictitious "Office of American Absorption (OAA) . . . encouraging America's religious and national minorities to become further incorporated into the larger society" (Roth 2004, p. 85), is a satirical reaction to concerns similar to the ones articulated by Bernstein.

[126]Porter (1988), p. 412.

[127]A. Einstein to F. Bernstein, January 5, 1939, Einstein Papers, Jerusalem, 52 564 (T).

8.S.3.4. THE ANTI-SEMITISM OF GEORGE DAVID BIRKHOFF

Throughout the 1930s, George David Birkhoff was the undisputed doyen of the American mathematicians. His fame was based on his work in topological dynamics, beginning with his fixed point theorem (1913) confirming a conjecture by Henri Poincaré, connected to the three-body problem of astronomy. Birkhoff's opinion was of particular importance to many, also in political respects. In the time before the forced emigrations, Eberhard Hopf from Germany, who was not Jewish and who—as a fellow of the International Education Board in 1930—had contact with Birkhoff, wrote about reservations on Birkhoff's part toward him, which had only recently been mellowed:

> Birkhoff is indeed so much nicer toward me since he got information about my confession.[128]

In his article of 1981, which provides valuable source material, Nathan Reingold quotes Birkhoff several times making anti-Semitic remarks, for instance from a letter to Richardson of 1934. In this letter Birkhoff expresses his suspicions about the Jewish mathematician S. Lefschetz of Princeton, who in his opinion could very well misuse his position as editor of *Annals of Mathematics*:

> He will get very cocky, very racial and use the Annals as a good deal of racial perquisite. The racial interests will get deeper as Einstein's and all of them do.[129]

A functionary of the Rockefeller Foundation, Warren Weaver, noted in his diary the following from a conversation with Birkhoff on October 13, 1934:

> B. speaks long and earnestly concerning the "Jewish question" and the importation of Jewish scholars. He has no theoretical prejudice against the race and, on the contrary, every wish to be absolutely fair and sympathetic. He does, however, think that we must be more realistic than we are at present concerning the dangers in the situation, and he is privately (and entirely confidentially) more or less sympathetic with the difficulties of Germany. He does not approve of their methods, but he is inclined to agree that the results were necessary. . . . He feels that the outstandingly able Jewish scholars have positions, and that there is a serious danger that a second flight, forced upon our attention by means of a highly organized propaganda, will compete unfairly with our own men who are equal or superior to them in ability.[130]

---

[128]Hopf to Tamarkin, December 14, 1931, Tamarkin Papers, BUA, box: Correspondence (A–H), f. E. Hopf.

[129]Reingold (1981), p. 321, quoting from Birkhoff's letter to Richardson, May 18, 1934.

[130]Siegmund-Schultze (2001), pp. 200–201. As was seen above, even Weaver, Courant's friend, was occasionally given to anti-Semitic remarks.

Figure 45 *George David Birkhoff (1884–1944). The professor at Harvard University, who after his proof of H. Poincaré's "last geometrical theorem" (1913), gradually became the undisputed leader of the American mathematicians had serious doubts about immigration due to the academic unemployment in the United States. He also revealed anti-Semitic sentiment.*

Jewish mathematicians such as Norbert Wiener, R. Courant, and Oscar Zariski,[131] among them immigrants after 1933, repeatedly accused Birkhoff of xenophobia and anti-Semitism. Reingold quotes from a 1936 letter written by Norman L. Levinson (1912–1975) in Princeton to his teacher Norbert Wiener, who was on bad terms with Birkhoff, in which one reads:

> P.S.: Einstein has been saying around here that Birkhoff is one of the world's greatest academic anti-Semites.[132]

In a letter to Harald Bohr in 1936 the immigrant Richard Courant discussed the question of whether or not Birkhoff should become coeditor of the *Grundlehren* series of the Springer publishing house:

> My principal reservation against Birkhoff is the following: There may occur over the years the situation in which my ultimate withdrawal from the Yellow Collection becomes inevitable. In this event I do not want to be testator to a Nazi like Birkhoff. I would find it much more reassuring if a friend like Veblen or Hardy or Newman, possibly also Stone, then automatically showed solidarity with me.[133]

Further documents from Birkhoff's papers at the Harvard University Archives leave no doubt about his anti-Semitism.[134] Birkhoff's address to the fiftieth anniversary of the American Mathematical Society in 1938 provoked irritation among American mathematicians and immigrants alike. The following is the much-quoted passage from this speech:

> With this eminent group among us, there inevitably arises a sense of increased duty toward our own promising younger American mathematicians. In fact, most of the newcomers hold research positions, sometimes with modest stipend, but nevertheless with ample opportunity for their own investigations, and not burdened with the usual heavy round of teaching duties. In this way the number of similar positions available for young American mathematicians is certain to be lessened, with the attendant probability that some of them will be forced to become "hewers of wood and drawers of water." I believe we have reached a point of saturation, where we must definitely avoid this danger.[135]

[131]Zariski's recollections have been called contradictory though, even by his biographer, Parikh (1991), p. 41. That his ambiguous description of Birkhoff was based rather on "contradictions" in Birkhoff's behavior will be discussed below.

[132]Reingold (1981), p. 322.

[133]Courant to H. Bohr, October 8, 1936, CPP (T).

[134]Birkhoff Papers, HUA, 4213.4.5. box 1, file: personal (1937–38), letters to G. D. Evans, February 17, 1938 and to A. S. Gale, dated November 4, 1937. In the last letter one has also the stereotypic allusion to a connection between "Jewish blood" and "superficiality," applied to the noted topologist and immigrant from Austria, K. Menger. Menger had apparently a different, better impression about Birkhoff's opinion of him. See Menger (1994), p. 162.

[135]Birkhoff (1938), p. 277. See also the epigraph above.

Birkhoff's position was from the outset very controversial. AMS secretary Richardson, himself occasionally given to anti-Semitic remarks, wrote to Birkhoff on September 20, 1938, not long after the anniversary, about the incriminating passage:

> p. 277, first half: A number of persons who had bought the volume of Addresses (not all Jews) expressed marked disapproval of the sentiments written here.[136]

A particularly strong reaction came from Abraham Flexner (1866–1959), director of the Institute for Advanced Study in Princeton, who felt that the standard of American mathematics could only be raised and strengthened by the influx of competent foreigners and that the need for their services would increase in the coming years. He wrote to Birkhoff on September 30, 1938:

> If we would place fifty Einsteins in America, we would probably within the next few years create a demand from other institutions for several hundred, and this is also true of Birkhoffs and Veblens and Moores. Let us keep firmly in front of our eyes our real goal, namely the development of mathematics, not American mathematics or any other specific brand of mathematics, just simply mathematics. . . . Hitler has played into our hands and is still doing it, like the mad man that he is. I am sorry for Germany. I am glad for the United States.[137]

Birkhoff's colleague and friend Veblen, who cooperated closely with Flexner at the IAS, did not concur either with formulations such as those in Birkhoff's speech, even though he did at one time share his and Richardson's concerns about a super-saturation of the American job market. There also exists proof that Birkhoff's anti-Semitic remarks found approval among some American mathematicians.[138] In 1936, Einstein confirmed the existence of similar resentment in Princeton in a letter to Hans Reichenbach, who wished to go to the United States from his first host country, Turkey:

> Carnap told me recently that they had told him expressly in Princeton they did not want to appoint Jews. So all that glitters is not gold and nobody knows what will happen tomorrow. Maybe savages are the better humans after all.[139]

---

[136]Birkhoff Papers, HUA 4213.2, box 12, file: R (1938).

[137]Birkhoff Papers, HUA 4213.2, box 12, file: D–F (1938).

[138]See examples given above in 8.D.7 and remarks by the statistician Edwin B. Wilson (1879–1964), as quoted in Siegmund-Schultze (1994), p. 311.

[139]Einstein to H. Reichenbach, May 2, 1936, Einstein Papers, Jerusalem, 20 118 (T).

What separates, however, Birkhoff's speech from similar remarks by others is the publicity it received and the influence of its originator. Not surprisingly, Birkhoff's son, Garrett, has, in retrospect, tried to downplay his father's anti-Semitism and to stress instead G. D. Birkhoff's legitimate concerns for the American educational system. Allegedly, with remarks like those on the anniversary of the AMS, his father was trying to save the American system of science, which he considered to be "more human and personal" than the European systems.[140]

Finally, given George David Birkhoff's unabashed anti-Semitic remarks, how is one supposed to interpret the fact that he was friendly to and helped Jewish mathematicians such as S. Ulam, Tullio Levi-Civita (1873–1941), Jacques Hadamard (1865–1963), and Otto Szász? In his autobiography of 1976, Ulam, for one, gratefully acknowledges help received from Birkhoff. The latter's friendship with the Italian Levi-Civita and the Frenchman Hadamard is confirmed by Marston Morse (1892–1977).[141] Although Hadamard's immigration to the United States was supported by several Americans and by German immigrants (Rockefeller Foundation, John R. Kline (1891–1955) of the AMS, H. Weyl, E. J. Gumbel), AMS secretary Richardson (Birkhoff's friend) first tried to advise him against extending an invitation to Hadamard, due mainly to his advanced age.[142] However, Birkhoff did extend a helping hand to Hadamard, as documented in a grateful letter by Hadamard to Birkhoff, dated October 16, 1941.[143] Regarding the Hungarian-Jewish mathematician O. Szász, expelled from Frankfurt in 1933, Birkhoff wrote to Richardson on November 15, 1935:

> I feel that I can speak with authority concerning not only his scientific ability but also his character as a man. . . . I admire him very much.[144]

However, this preferential treatment of individual immigrants on Birkhoff's part cannot invalidate the charge of anti-Semitism, especially if one interprets this as prejudice effectively implemented without looking at any person in particular, thus actually discriminating against that person from the outset. In other words, the fact that a prejudice could, at times, be overcome in connection with specific individuals does not change the principally prejudiced treatment directed toward the same individual.

---

[140]Birkhoff (1976b), p. 66.

[141]Ulam (1976); Morse (1946), p. 358.

[142]Richardson to Birkhoff, December 16, 1940, Birkhoff Papers, HUA, 4213.2, box 14, file: P–R (1940). See also below in chapter 9 a letter by Richardson to Gumbel about Hadamard, dated April 16, 1940.

[143]Birkhoff Papers, HUA, 4213.2, box 15, file: H–K (1941). Birkhoff seems to have cooperated with H. Shapley in this matter. See the Shapley Refugee Files in the Harvard University Archives.

[144]Birkhoff Papers, HUA, 4213.4.5, box 1, file: personal 1936.

8.S.3.5. DECLINING ACADEMIC ANTI-SEMITISM IN THE USA AFTER 1945

Aerodynamicist Kurt Hohenemser experienced remnants of academic anti-Semitism even after the war. As a victim of National Socialism he had been preferentially treated in his application to immigrate after 1945. He wrote to Richard von Mises in 1948:

> What I have experienced here in St. Louis of Nazism and anti-Semitism in private contacts with former Germans even exceeds what I got to know in Germany. Unfortunately this attitude is not restricted to former Germans.[145]

Predominant, however, was a decline in academic anti-Semitism, explained in the following words by the historian and differential geometer Dirk J. Struik in an interview I had with him on December 18, 1991: "One of the advantages of the foundation of Israel was that academic anti-Semitism disappeared."

The emigrant Michael Golomb goes even further back in time with his explanation, to when the Americans entered the war in 1941. In his letter to me, Golomb summarizes several reasons for existing academic anti-Semitism in the United States and for its eventual decline:

> Have I experienced academic antisemitism in the USA? I certainly have. When I arrived here (in early 1939) there were very few Jewish mathematicians (or other scientists) employed by American colleges and universities, and those few were limited to schools in the Northeast. It was common knowledge and I was able to confirm it from personal experience that there existed a numerus clausus for Jewish students in most academic institutions, but especially in engineering schools. When I came to Purdue (in 1942) there was one Jew among the appr. seven hundred faculty members, a physicist, who was hired by the Head of the department, who was a converted Jew. Before my appointment was finalized, I had to be interviewed by the Dean of the School of Science. His first question was, whether I was Jewish. When I confirmed this, he asked me whether I was aware that Jews were not liked in that part of the country. I do not believe that this universal antisemitism in academic circles was motivated by concerns about economic competition, it was prevalent among the older tenured faculty, not much among the younger people, who had not yet found a position or obtained tenure. Very few of the immigrant Jewish scholars found in those years academic positions commensurate with their qualifications, but this was not due to the existing antisemitism, the non-Jewish immigrants did not fare any better. All this changed soon after the entrance of America into the war. There was a big demand for every kind of scientifically schooled personnel, men and women, Jewish and non-Jewish, domestic and

---

[145]Hohenemser to von Mises, September 25, 1948. Richard von Mises Papers, HUA, 4574.5, box 9, f. July 48–Jan. 49.

foreign. This change, brought about by the necessities of the war, became a permanent feature of the American cultural scene. Antisemitism and xenophobia have disappeared from academia and research institutions. I never felt resentment among my non-Jewish colleagues when I got meritorious preferential treatment.[146]

[146]Letter by M. Golomb to the author, July 19, 1993.

NINE

Acculturation, Political Adaptation, and the American Entrance into the War

> The intimacy of the coffee house had to give way to the distance and strangeness of the American lifestyle, and so they were for the most part happy but not glücklich.
>
> —L. Coser 1988[1]

> Another bar between the foreign professor and his students was the difference in attitude which characterized the European as distinguished from the American professor. The former had developed to a fine art the technique of social distance from his students.
>
> —M. R. Davie 1947[2]

> I think it should be accepted as a norm that only in rare cases can we expect the European scholar to be placed in an institution in academic life in America corresponding to what he had in Europe before the disaster.
>
> —L. Wirth 1940[3]

> I think German Fascism will get a really good hiding in the end, as they deserve. Moreover, there will be great social changes during the war and even more after the defeat of the German government, which will mean the biggest social progress that mankind has seen so far.
>
> —Wolfgang Sternberg 1941[4]

[1]Coser (1988), p. 100, on the immigrants in the United States (T), with "glücklich" being the German word for "happy." The image of the "coffee house" as a place of human and scientific communication in Europe appears quite frequently in the literature on emigration, often in a rather nostalgic vein. See also Ulam (1976), pp. 33–34.

[2]Davie (1947), p. 306.

[3]Stated in a discussion between the Rockefeller Foundation and the EC in 1940: EC, box 148, f. Rockefeller Foundation, 1940–45, L. Wirth to T. B. Kittredge.

[4]Emigrant Wolfgang Sternberg to the friend of his youth, Richard Courant, after the entrance of the United States into the war, on December 14, 1941, CPP (T).

WHILE "governmental apathy"[5] existed about refugees' entering the United States, and although immigration policies were tighter compared with around 1900, conditions for acculturation, once the immigrants were in the country, were relatively favorable, particularly with respect to opportunities for employment: "Compared to the action of other countries offering a haven, the situation in the United States was favorable in this respect. There was no federal legislation restricting employment opportunities except that forbidding employment to 'enemy aliens' in certain war industries."[6]

9.1. General Problems of Acculturation

Nevertheless, the existence of general social problems of acculturation for the European scientific immigrants could not be ignored. These included insufficient language skills,[7] the surprising informality of social contact between students and lecturers in the United States, and the unusually high teaching loads due to the "service function" of the American university. The German mode of scientific discussion was often quite different from that of the Americans. The difference between the rather individualistic European working style and the more cooperative manner of the Americans is highlighted in many surviving documents (D).

Americans sometimes criticized immigrants for their lack of participation in administration,[8] while at the same time the "exaggerated" organizational activity of people like Courant evoked fear, as shown by Richardson's reaction, mentioned above. Thus the immigrants had to tactfully strike a balance.

The refugees were expected to familiarize themselves gradually with American publishing outlets and to publish in English.[9] In this respect immigrants met additional obstacles. The American Mathematical Society charged a fee for publication in their outlets,[10] the payment of which was not always guaranteed by the host universities or by the emergency organizations.[11]

The professional success of the immigrant depended on the age of the

---

[5]Krohn (1987), p. 32.

[6]Davie (1947), p. 394.

[7]For the language problem, see ibid., p. 306.

[8]Wilder (1989), p. 201.

[9]Even Einstein, who was usually not given to "political correctness," chided C. Lánczos at one point for continuing to publish in Germany (Stachel 1994, p. 218).

[10]This fee is mentioned as "page charges" in Archibald (1938), p. 35.

[11]The Rockefeller Foundation, for one, refused to take over these additional costs for the immigrants it supported. See W. Weaver to M. H. Ingraham, November 15, 1934, RAC, R.F. 1.1, 200 D, box 125, f. 1543.

individual, the date of immigration, and on the particular discipline and the demand for it. Although most of the Austrian immigrants arrived rather late in the United States (around 1938), they seem to have succeeded rather well in emigration, particularly when compared to their previous, often precarious professional situations (often as schoolteachers and actuaries).[12] This seems to be connected to the "modernity" of some of the mathematical fields they represented (logic, topology, applications in economics, statistics). With regard to the loss of reputation often felt by immigrants, most of the examples cited will be of individuals previously occupying full professorships in Germany. Most of them were in their prime with regard to age. For the older ones among them (Korn, Bernstein, Hellinger, Sternberg) there were additional hurdles to overcome. Of the prominent German mathematicians, the one who probably found it most difficult to adapt to the United States was the former director of the institute for mathematical statistics in Göttingen, Felix Bernstein.[13] The younger immigrants, most of whom had also achieved conspicuous success in Germany, as in the case of Richard Brauer,[14] often experienced problems due to the principle of seniority prevalent at American universities.

In addition, acculturation was impeded by more general problems of "mentality." The immigrants were generally more politically aware than their American hosts. Many of the Europeans had different standards of sexual morality than the Americans. The loss of reputation due to emigration; lower salaries, in particular lack of pensions; and lack of home help—all of these factors played a role during acculturation, as did the effect of diminished scientific expertise on the part of the immigrants, due to longer periods of emigration spent away from their true vocation.[15]

Several of these factors impeding acculturation were of a general nature and did not just concern scientists or mathematicians. They have been discussed repeatedly in the literature.[16] In a sense, the dividing line as to mental attitudes, politics, and morality existed not so much between American and European scientists as between immigrants and the broader American public. Most notorious is the so-called Bertrand Russell Case in which the English mathematician and free-thinking philosopher Bertrand

[12]This is described in Dahms (1987), p. 101, particularly with respect to the careers of K. Menger and K. Gödel, but one might add the names of F. Alt, E. Lukacs, A. Wald, and others as well.

[13]More on Bernstein's fate below in the case study S.2. There were others who were less prominent, such as Wolfgang Sternberg, and had a similarly harsh fate. In the case of E. J. Gumbel, political factors played a decisive role as well (case study S.1 below).

[14]Brauer's biographer, Feit (1979), p. 20, complains that Brauer was appointed in Toronto in 1935 only as assistant professor. Brauer had been Privatdozent in Königsberg between 1925 and 1933, which was at least equivalent to an assistant professorship in America.

[15]Wasow, who had to teach nonmathematical subjects for several years in Italy in order to survive, mentions this latter point particularly (Wasow 1986, p. 229).

[16]For example, Davie (1947), pp. 306ff.

Russell (1872–1970) was hindered by a public smear campaign in 1940 from assuming a position secured for him by his academic colleagues at the City College of New York for 1941–42.[17]

In the following parts of this chapter and in the Documents part several of these more general problems of acculturation intellectuals experienced will be presented with respect to mathematicians. Emphasis will be placed on *four problems*: (1) the tension between politicization and the need for political adaptation many immigrants experienced, (2) the assignment to research institutes of outstanding immigrants "unable to teach," (3) the problem of providing adequate pensions for elder immigrants, and (4) the involvement of foreigners in war research. With respect to the latter two points, teaching and war research, problems more specific to mathematics will also be taken up. However, this will be done in more detail in the following chapter on the "impact of mathematical immigration."

9.2. Political Adaptation

The grave upheavals of emigration and of the world war could not fail to influence political experience and opinion of the previously quite often apolitical emigrants. Similar to the existential experience of the First World War two decades before, emigration led to an at least temporary politicization of many of the mathematicians affected. This politicization could be expressed in social optimism for the future, as in the remarks by Wolfgang Sternberg quoted above. It often ended, however, in pessimism, when emigrants realized the contradiction between scientific progress and the apparent immutability of social circumstances, in particular during rearmament and failed "denazification" ("Entnazifizierung") in Germany and Austria after the war. Throughout the Cold War reference to Nazi rule and the war were focal points in the arguments of the former emigrants when reflecting on science and society.[18]

In the United States of the 1930s the majority of the immigrants shared the views of the Democrats around FDR, hoping as they did for a more liberal policy of immigration and for more help for their countrymen back in Germany (D). Some refugees contributed to a broadening of the political horizon of some of the American mathematicians. At times they came into conflict with more conservative American mathematicians.[19]

---

[17]See Kallen and Dewey (1941).

[18]In 1971, Kurt Mahler wrote in his memories: "With the extremists on both sides of the spectrum trying to destroy basic research and replacing it by their dogmas of intolerance and power madness, we seem to be on the way to new dark ages." Quoted from Poorten (1991), p. 379. See also chapter 11.

[19]Lipman Bers, for one, a stateless Jew and a socialist, had particular experiences that were not shared by many in the United States.

However, the politicization of the immigrants was mostly a *relative* one in the sense that it was remarkable only against the backdrop of previous and expressly self-declared apolitical positions of many German-speaking mathematicians and of their American hosts. Moreover, there were changes of opinion in the opposite direction, too. While many apolitical German mathematicians became relatively politicized during emigration, the representatives of the rather marginal field of logical positivism of the Vienna Circle, who, prior to their emigration, had often supported radical political positions, critical of the state of the European society and of the distribution of social goods, became relatively de-politicized as a result of their uprooting, at least in the long run.[20]

On the whole, when uttering political opinions all immigrants had to exercise extreme care. The same is true for their communication with the mathematicians remaining in Germany and also for their official statements about Germany. Some of the immigrants' remaining economic relations with Germany, such as pension entitlements, played a role in this context as well.[21]

They had to exercise caution in the American environment. This environment remained predominantly apolitical at the universities, in spite of the temporary "rise of the American left"[22] in the 1930s. Immigrants were expected to behave accordingly and appropriately. Mathematicians who had been expelled in Germany for specific political reasons[23] often tried to trivialize their political activities in the Weimar Republic (D). This strategy had already been practiced by the political emigrants in their various, if mostly futile, efforts to prevent their dismissal in Germany.[24] It was now used in their efforts to obtain American citizenship. Applicants for positions in the United States were often expected to refrain from any political utterances even against Hitler's Germany.[25] If they did not comply they were suspected of having extreme leftist or rightist

---

[20]See the recent book, Reisch (2005). Dahms (1987), pp. 104–5, points in this connection to the lack of a tradition of a labor movement in the United States. In Feferman and Feferman (2004), p. 248, the political pressure on R. Carnap during the McCarthy red-scare period is made responsible for his increasing political restraint.

[21]See chapter 7. A. Berliner did not wish public protest against his dismissal by Springer as editor of *Die Naturwissenschaften* by A. Flexner, the director of the IAS (Courant to Berliner, September 17, 1935, CPP). Also Courant wrote in the same vein to van der Waerden (October 15, 1935), because he was concerned about the sales prospects of Springer's in the United States.

[22]Diggins (1973).

[23]Instead of the racial pretext, which was also political.

[24]See above chapter 4.

[25]Duggan and Drury (1948), p. 190, report on a case of a refugee who was dismissed in the United States "for one reason only—his outspoken and determined criticism of Nazi Germany." The authors surmise that anti-Semitism was also involved in this case.

political positions. Prominent refugees who did not keep silent, such as Einstein, were viewed with mistrust and disapproval by conservative Americans, regardless of whether they were Jewish or not.[26] In order to support their relations to Jewish emergency organizations some emigrants stressed their Jewishness more than they would have done in Germany, even before the Nazis had come to power. Political refugees such as E. J. Gumbel remained outsiders even during their emigration (S). With the attainment of American citizenship and the beginning of the war some immigrants dropped their attitude of political reserve and became involved in afterwar plans for Germany, only to be frustrated by the actual developments in Germany and to be forced into political restraint and caution again under McCarthy.

9.3. Problems of Adaptation in Teaching and Research

The emphasis placed on freshmen and undergraduates at American colleges was puzzling to European immigrants, more so as this brought out the problems with language. However, it also became clear very soon that the pressure felt by most immigrants was so big that they were willing to put up with not just the considerable loss of professional reputation but also much higher teaching loads. Even one of the most creative German mathematicians of those years, the number theorist Carl Ludwig Siegel, was expected at first to take part in regular academic teaching. However, due to the expansion of the Institute for Advanced Study, opportunities evolved to protect exceptional talents such as Siegel from normal teaching routines (D). But even the IAS was no empty social space, and the people there had difficulty tolerating eccentrics such as Siegel.[27] In a few cases, problems adapting to American conditions were instrumental in refugees' returning to Germany after the war.[28]

---

[26]Porter (1988), p. 330, mentions R. A. Millikan's criticism of Einstein. At the same time, Einstein was heavily attacked by Jews remaining in Germany who did not have his prescience as to the future fate of the Jewish people. See Grundmann (2004), pp. 279–80. On A. Flexner's problems with Einstein see more below in the Documents section.

[27]Siegel, from his arrival at the IAS on April 15, 1940, until 1944, received a yearly stipend of $3,000, which was, however, for the most part provided by the Rockefeller Foundation. See Archives IAS, Faculty, Veblen, box 33, f. School of Mathematics. Misc. According to the same source, the Hungarian Paul Erdös (1913–1996), when he was at the IAS, received only once a rather modest stipend, $750, 1939–40. On the problems of Erdös, who was as socially unconventional as Siegel, and his relations to the IAS, see Pach (1997), pp. 46–47.

[28]Among them Reinhold Baer and Siegel; the latter had even returned to Germany once before the war, as outlined in chapter 5.

9.4. Age-Related Problems and Pensions

Age was an important factor preventing some mathematicians from emigrating at all.[29] It caused others to delay their emigration. Others who did emigrate faced problems in the host country of earning enough to be able to provide them with a decent pension at a later date (D). The problem of pensions for immigrants was made worse by the very different American retirement system. As seen previously, this was recognized early on by Norbert Wiener (1934). It was later emphasized by Weyl and Courant in their discussions with the Emergency Committee in the early 1940s. Even in the mid-1940s[30] and after the war no general solution was found for it. As late as 1949 a round-robin letter was sent out (probably by Courant) reminding the addressees of the old Noether-Weyl Relief Fund and asking for a continuation of that fund on behalf of the older immigrants (D).

9.5. The Influence of War Conditions

From the start of the war the situation for immigrants changed in several respects. Academic unemployment was on the decline but political suspicion was on the increase, at least in some quarters.[31] To counter this, even before the war many emigrants had begun to write letters to each other in English.[32] A crucial point was whether or not one had obtained American citizenship; the position of noncitizens and foreigners[33] became more precarious during the war. In the United States there was and still is a general five years' waiting period before naturalization. Noncitizens were divided into "enemy aliens," basically those from Germany and Japan, and ordinary noncitizens. There was frequent ambiguity about the classification of immigrants from Austria, which had officially become a German province in 1938. Although the German-speaking enemy aliens—unlike the Japanese[34]—were not, as in England, interned, restrictions were imposed

[29]The examples of Blumenthal, Hausdorff, and others have been mentioned in chapter 5.

[30]This is clear from a letter by H. Weyl to the EC, dated December 14, 1944, in which he expresses concern about the future of Max Dehn: "His age seems an almost unsurmountable obstacle." EC, box 6, f. M. Dehn, 1941–44.

[31]Apparently there were different opinions about the precautions to be taken. Veblen in his testimony for Pollaczek quoted above (chapter 6) found that the "antifascists who come here from Europe are extremely pro-American and ready to help."

[32]For instance Weyl's letter to Gumbel in 1936 as quoted in chapter 7.

[33]For the latter see the problems of the aerodynamicist, Heinrich Peters, below in D.

[34]The internment of the Japanese noncitizens resulted in the accusation of racism against the American government, particularly in historical retrospective.

on residence permits. Participation in war research was practically ruled out (D). Initiatives to ease these restrictions, by the EC and for a time by a special "Committee Utilization of Talents of Refugee Scholars" were apparently to no avail.[35] Classification as an enemy alien could also lead to discrimination regarding to copyright.[36] However, the exigencies of war could also lead to early naturalization, as apparently in the case of Abraham Wald, whose statistical sequential analysis was of particular importance to the American war effort.[37]

Once the immigrants had become citizens, no further restrictions were applied to their participation in war research. This was particularly apparent in the case of Richard Courant (D), who became a member of the Applied Mathematics Panel under Warren Weaver, within the Office of Scientific Research and Development (OSRD).

The material presented in the Documents part of this chapter discusses the question of to what extent "acculturation" was "successful" for the individual mathematicians and to what degree "failure" occurred among immigrants fortunate enough to have survived and eventually reached America.

9.D. Documents

9.D.1. The General Requirement of "Adaptability"

Problems adapting to American society on the part of immigrants and visitors were not new in 1933. The foreign Rockefeller fellows of the 1920s had had their specific problems. On October 21, 1927, the topologist Heinz Hopf wrote in a rather jocular tone from Princeton to H. Kneser in Germany:

> Otherwise we try to acclimatize in America, that means to learn to live with it. For instance we have learned that the aim of civilization is a maximum of centralization of all human activities, so that it is difficult to have breakfast at home instead of in an ice-cream hall, and that one has to go to the shoemaker in order to polish one's boots.[38]

---

[35]See in this connection, below in D,the offer by Gustav Bergmann, who had come to the country in 1938, and the reaction to that offer.

[36]Siegmund-Schultze (1997).

[37]Wallis (1980), p. 330.

[38]Heinz Hopf Papers, ETH Zurich, Hs 621:854 (T). By "polishing the boots" Hopf alluded to the lack of a maid at his furnished room, according to a letter to R. Brauer, March 10, 1933, ibid., Hs 621:306a. Many emigrated European housewives complained about the unavailability or expense of home help in the United States.

After 1933, however, emigrants really had no choice in the matter. They had to be adaptable. Courant remarked in a letter of 1935 to his long-time friend Wolfgang Sternberg:

> It is simply a fact that one has to be more elastic and adaptable in these times.[39]

In the same year Courant again emphasized the importance of adaptability when writing to Max Born about the chances of Fritz John in Lexington, Kentucky:

> The position itself is very promising, if John adapts and proves his worth.[40]

In a letter of June 1937 to Harald Bohr, Courant bemoaned the critical attitudes of some immigrants, even those of second rank, which could endanger the prospects of others during future immigration:

> Unfortunately, Siegel is not the only critic who has made the people here suspicious. (After all they would find something for Siegel and for Artin here at any time.) But unfortunately other Europeans such as Baer, John, even Busemann have reflected on America in a less than positive and often disapproving way. Nothing, however, is more damaging to the relations between the Americans and the immigrants than the Americans' assumption that the immigrants consider them with European condescension and deem them not equally civilized.[41]

Carl Ludwig Siegel was not only known for his individualism and eccentricity but also for his sensitivity for the suffering of other human beings. After Siegel in a state of delusion made a temporary return to Europe, he wrote to Courant on March 22, 1939 from the French town of Nice, describing the cruelties of the November 1938 pogrom in Germany and the indifference many German professors showed with respect to these atrocities. However, Siegel then added:

> I no longer have the hope, which led me to America four years ago, of finding a tolerable position abroad. My character is too clearly developed, and I can no longer suppress my asocial instincts and individualistic tendencies. I can no longer adapt, I am too much of a Prussian.[42]

---

[39]Courant to Sternberg, May 30, 1935, CPP (T). Earlier (on March 30, 1935) Courant had advised Sternberg: "Perhaps school teaching or teaching at a technical school."

[40]Courant to Born, on July 26, 1935, CPP (T).

[41]Courant to Harald Bohr, June 11, 1937 (T). CIP, Bobst Library, New York City, box 3.

[42]Copy in Veblen Papers, Library of Congress, cont. 12, f. Siegel, 1935–60. Original (hand) in Courant Papers, Bobst Library, New York City, box 3 (T).

Figure 46 *Carl Ludwig Siegel (1896–1981). The versatile mathematician (number theory, function theory, celestial mechanics) was one of the few "voluntary emigrants" in the sense that he was neither "racially" nor "politically" persecuted. He went to the United States "only" for the sake of unimpeded scientific communication and finally—as a pacifist—due to the war. Siegel's wavering between Germany and America reflects the inner conflicts of many emigrants.*

9.D.2. Problems Arising from the Loss of Status Due to Emigration and from the Widespread Principle of Seniority in Academic Promotions[43]

Richard Courant, an able applied mathematician, wrote on May 2, 1933 in a realistic vein to Abraham Flexner of the IAS in Princeton:

> It goes without saying that I do not aim at one of the splendid positions at your Institute but would be content with a modest position. (CPP [T])

In Courant's case it nevertheless required some persuasion to convince him of the restricted opportunities in the American academic job market and to get him to put up with a lower social position than the one he had previously held in Germany. In December 1933 his friend, the physicist James Franck, informed him about a recent offer from New York University for him, which had been obtained by Warren Weaver of the Rockefeller Foundation: "You simply say yes and come in the fall. Courant, I am so relieved. Please do not carp, saying Columbia is more exquisite or such."[44] In 1934 Gabor Szegö wrote from Königsberg to J. D. Tamarkin in the United States:

> I have no idea what salary is needed preventing a professor in the U.S. and his family from dying of hunger. But I am far from particular in my demands.[45]

In 1935 Hermann Weyl from the IAS in Princeton wrote with a tinge of racism to Pólya (Zurich) about Siegel:

> Siegel has been with us since mid-January and is giving an outstanding lecture course. He too does not want to return to Germany, if this can be avoided; he would prefer the most obscure colored college in America to that. But it now seems very difficult to find a position even for such an outstanding mathematician.[46]

Arthur Korn wrote to von Kármán in Pasadena on July 9, 1936:

> I am now willing to accept any position where I can apply my knowledge in mathematics, physics, and electrical engineering.[47]

---

[43]Again the conditions in the United States will be stressed here. However, there were comparable conditions also elsewhere, particularly in Great Britain, where K. Mahler was still assistant lecturer in Manchester in 1943, ten years after his immigration. See Poorten (1991), pp. 374–75.

[44]J. Franck to Courant, December 6, 1933, CPP (T).

[45]Askey (1982), p. 5 (T).

[46]Weyl Papers, ETH, Zurich, Hs 91:421, Weyl to G. Pólya, February 16, 1935 (T). Weyl's quotation recalls another one by Feller, given above, and it reflects the only marginal representation of Afro-Americans in American academia at that time.

[47]Kármán Papers, Caltech, Pasadena, 79.17 (T).

On Hans Rademacher's career at the University of Pennsylvania, the editor of his Collected Papers remarked in 1974: "In those years, the length of faithful service to the institution and not professional excellence, was the main criterion for promotions—a fact that was forcefully explained to the somewhat surprised assistant professor by a most self-assured dean."[48] Similarly, Richard von Mises, the former director of the Institute for Applied Mathematics in Berlin, did not enter the United States without problems. Dean Harald M. Westergaard of the Harvard University Graduate School of Engineering wrote to him on June 7, 1939:

> I was pleased to receive your cablegram according to which you will accept an appointment as Lecturer of Applied Mechanics for the year 1939–40, without salary but with an obligation to present some Lectures bearing on mechanics. I regret that our funds do not permit us to offer a salary, but we shall welcome you here.[49]

Thus von Mises had to "buy," at least at the beginning, his minor position as a lecturer at Harvard University (i.e., he had to provide the salary himself), in order to acquire a lifesaving visa to the United States. The same had to be done for his assistant and future wife, Hilda Geiringer.[50] Von Mises became a full professor at Harvard in 1945.

Another immigrant-mathematician who had to cope with a considerable loss of status and with an enormous teaching load was function theorist Karl Löwner from Prague. The Americans discovered his "genius" rather late, in connection with war research:

> Carl and his family came to Louisville, Kentucky, in 1939 under rather modest circumstances. Loewner was not the self-advertising type and very few people realized what values were hidden in Louisville. During the Second World War Tamarkin brought Loewner to Brown, where his genius was recognized.[51]

Aurel Wintner, who had come to the United States prior to 1933, wrote in 1950, shortly after the death of Ernst Hellinger:

> He undoubtedly was one of the leaders of his generation, and I believe that this country has lost much by not affording him such a position for which he would have been entitled and in which he could have exerted his influence on American mathematics to a fuller extent.[52]

[48]Grosswald (1974), p. xvi.

[49]Richard von Mises Papers, HUA, HUG 4574.5, box 3 (folders 1935–44), f. 1939.

[50]In her case at least part of the money for her first year at Bryn Mawr College seems to have come in 1939 from her brother in England, the musicologist Karl Geiringer (1899–1989), who would later on have a position at Boston University from 1941. See EC, box 10, f. Geiringer, Hilda, 1933–40. See also Appendix 4.3.

[51]Memorial address on Löwner by M. Schiffer and G. Szegö. Szegö Papers, f. 22, Miscellaneous, undated, 1968. See also below D.

[52]In a letter to the American W. T. Reid, as quoted in Rovnyak (1990), p. 27.

9.D.3. *Different Traditions in Teaching and Unfamiliar Teaching Loads*

One of the epigraphs of this chapter stresses the "technique of social distance" between teacher and student, cultivated in Europe and different from American traditions. It seems, however, that the immigrants were by and large willing to adapt to American standards in this respect.

In May 1935, R. G. D. Richardson wrote the following in a letter of recommendation to the University of Kentucky for Hans Lewy from Göttingen:

> Dr. Lewy prefers to teach graduate students. . . . This is partly because he feels that he is less likely to displace an American. But Lewy is intelligent enough to know that professors cannot choose in these particulars and that he must himself adapt to American ways.[53]

In January 1936, Richardson still felt the need to advise Lewy:

> The undergraduate teaching is the main work of nearly everybody in academic work in mathematics in America, and this is universally true of the younger men. We have to stand or fall by our success in this particular.[54]

Lewy indeed showed the necessary ability to adapt, maybe because, being in his early thirties, he was still relatively young. Other immigrants about the same age as Lewy also had little problem adapting to the different traditions in teaching, as for example, Karl Menger:

> Unlike many refugee intellectuals, Menger liked the United States, and soon felt completely at home in his new country. Other ex-European mathematicians were offended at being required to teach such elementary subjects as trigonometry; Menger enjoyed teaching undergraduate courses.[55]

Unfortunately older mathematicians such as Otto Szász had more difficulty adapting. The mathematics department of MIT had the following to say about Szász in 1934:

> His English is far from perfect and his training was of course that of a professor in a German university which is no guarantee of success with subclassmen.[56]

Indeed, the sometimes low level of mathematics at which immigrants were expected to teach was unusual to them. When in 1936 Emil Artin

---

[53]Richardson to P. P. Boyd, May 4, 1935. Richardson Papers, Correspondence, box: German-Jewish Situation, f. Hans Lewy,

[54]R. to Lewy, January 17, 1936. Richardson Papers, Correspondence, box: German-Jewish Situation, f. H. Lewy.

[55]Menger (1994), p. xii. From the introduction by the editors.

[56]March 19, 1934. Richardson Papers, Correspondence, box: German-Jewish Situation, f. O. Szász.

was preparing to emigrate from Hamburg due to the danger his Jewish wife was in and implications concerning his job, Courant invited him to give guest lectures at New York University on the condition that "you are willing to adapt to the level of students with little or even no mathematical background, who are, however, motivated and interested."[57]

The considerable teaching load at some American universities even dissuaded some immigrants from accepting job offers, especially if the offers were only temporary ones. This applied to Gumbel, Geiringer, and Fried, who were considered candidates in the efforts to find a successor to Karl Löwner, who had been teaching eighteen hours a week at the University of Louisville, Kentucky. Richardson wrote to Gumbel about that in 1943:

> In this way it might be a stepping stone to some position in an American college.[58]

In 1941, Alfred Brauer (older brother of the better-known mathematician Richard Brauer), who had a good reputation as an academic teacher in Berlin, was considered for a job at an American high school. He refused the post, due partly to language problems and partly to insufficient understanding of the American social environment:

> Although I was teaching for a time at a German High-School, I shall not be able to teach at an American school. A school-teacher must not only have the ability to teach but must have also full understanding for the children. . . . I think that I am too short a time in this country to meet these requirements.[59]

For some mathematicians the teaching requirements were simply too demanding, as with Reinhold Baer, who had taught at Urbana, Illinois, since 1938. He was one of the few emigrants who went back to Germany after the war: "In the European context, many of these courses would be taught at school rather than university. Baer really hated this elementary teaching. The thought of escaping from it played a major part in his decision, many years later, to return to Germany."[60]

9.D.4. Extraordinary Solutions for Outstanding Immigrants

After the wave of expulsions had started in 1933, the Rockefeller Foundation in New York City tried to maintain traditional policies of restricting

[57]Courant to Artin, July 27, 1936 (T). Courant Institute Papers, file Artin, 1935–58.

[58]Richardson to Gumbel, November 16, 1943, Richardson Papers, Correspondence, f. 112 G.

[59]A. Brauer to EC, July 2, 1941. EC, box 4, f. A. Brauer. In 1939 Brauer had found a temporary appointment at the IAS in Princeton where he was officially Hermann Weyl's assistant but mainly used his time to build the Institute's library. In 1942 Brauer received a permanent appointment at the University of North Carolina in Chapel Hill.

[60]Gruenberg (1981), p. 342.

support to outstanding mathematicians with proven records as researchers. In September 1933, one officer of the RF wrote in a letter to Louis J. Mordell in Cambridge, concerning the proposed mathematicians Mahler, Baer, Heilbronn, B. Neumann and F. Behrend, who had no place where to go:

> The requirement of a post for a fellow to return at the expiration of his fellow-ship must be maintained. . . . The young mathematicians whom you mention are young scientists of promise to be sure, but not mature enough to fall within the group of eminent deposed scholars, whom the officers in Paris are authorized to assist through the Special Research Aid Fund appropriated from New York.[61]

Originally even Carl Ludwig Siegel was expected to earn a living as an academic teacher at a lower level. From New York City, Richard Courant wrote to Hermann Weyl in early 1935:

> Siegel was here last weekend and appeared to be rather depressed about the idea that he may have to teach, as an "unskilled worker," theoretical physics in Cincinnati.[62]

At about the same time, Emmy Noether had problems finding a permanent position and only occasionally received invitations to give guest lectures at the IAS. On March 21, 1935, three weeks before her death, the EC noted the following about Noether's temporary appointment at a women's college near Princeton:

> President Park of Bryn Mawr College came in yesterday to discuss plans for Emmy Noether. She said Emmy Noether was too eccentric and unadaptable to be taken on permanently at Bryn Mawr but that she would like to keep her two years more.[63]

After a while exceptional solutions were found for the most prominent immigrants such as Siegel and Gödel, as indicated by the following extract from a note dated March 10, 1935 from the files of the Rockefeller Foundation: "Both Gödel and Siegel queer—could not teach classes in Univ. Inst. Adv. Study can capitalize their genius."[64] The two mathematicians were eventually given positions at the IAS, where they were exempt from regular teaching obligations. However, they did not obtain full professorships there either; this kind of position was awarded to Gödel only after the war, while Siegel had returned to Germany by then.

---

[61]L. W. Jones, Paris, to L. J. Mordell, September 9, 1933, Mordell Papers 23.1.
[62]Courant to Weyl, February 15, 1935 (T). Weyl Papers, ETH Zurich, Hs 91:65.
[63]EC, box 26, f. E. Noether, 1935.
[64]RAC, R.F. 1.1. 200 D, box 143, f. 1770.

Another exceptional mathematician with problems in adapting socially was Stefan Bergman(n).[65] Although colleagues at Brown University in Providence granted him a "touch of genius" he did not receive preferential treatment, at least not until the end of the war. In a letter from Richardson to the president of Brown University, H. M. Wriston, in late December 1945, it is also intimated that Bergman's brand of theoretically very sophisticated applied mathematics was over their heads and the needs of the then emerging group of applied mathematicians[66] at Brown:

> His work is very recondite and his method of presentation so obscure that very few people can or will read it. Bergman is impossible as an undergraduate teacher and only with very advanced students is he really inspiring and useful. From the standpoint of research accomplished, he . . . indeed has a touch of genius. . . . With all his desire to accommodate himself to his environment and his colleagues, he is alien and will never achieve any popular favor. . . . He is not an applied mathematician though some of his ideas have very important connections with foremost problems in hydrodynamics. . . . I am not able to handle him. . . . He will add scientific prestige to any group.[67]

Wriston in his answer dated December 28, 1945, categorically refused to continue Bergman's employment. It was only in 1952, at the age of fifty-seven, and after a rather subordinate position at MIT in Boston, that Bergman found a professorship at Stanford University in California.[68]

9.D.5. Individualistic European versus Cooperative American Working Style[69]

In 1936 C. Carathéodory, who then held the Carl Schurz Guest Professorship at Madison, Wisconsin, tried to convince AMS President S. Lefschetz to give support to a lecture tour on historical topics to be undertaken by Otto Blumenthal; Lefschetz replied rather rudely and sarcastically:

> I am only the President of the American Mathematical Society (until the end of this year) and not its Führer! . . . There exists a specific Committee of the American Mathematical Society in charge of the visiting lectureships.[70]

[65]He was the creator of the kernel function in the theory of orthogonal analytic functions.
[66]See Prager (1972).
[67]R. G. D. Richardson to H. Wriston, undated, late December 1945. BUA, Appl. Math. Div. II 154 (H. Wriston).
[68]Schiffer and Samelson (1979).
[69]Analogous examples for physics are given in Schweber (1986).
[70]Lefschetz, in his letter dated December 7, 1936, suggested that von Kármán, who had asked Carathéodory to help Blumenthal, should write to Birkhoff, who was chairman of that committee. Kármán Papers, Caltech, Pasadena 5.4. Nothing came out of it or of the plans for Blumenthal's trip.

Wolfgang Wasow, who came to the United States in 1939, made the following observations about the different working styles of Americans and Germans (Europeans):

> The differences between American and German ways came out in many minor daily occurrences. In Germany, whenever a small group of people were doing something the tendency was that one person would soon establish him (or her)self as the leader and, at least in the short run, this leadership would be informally accepted. In America the tendency was to settle even minor issues by a majority vote, and if special tasks were to be assigned it was done the same way.[71]

Apparently, Richard von Mises was one of those "Germans" (though Austrian by birth) mentioned by Wasow used to deciding matters by themselves and to expecting acceptance of their decisions from others. Some subliminal conflicts between von Mises and R. G. D. Richardson are palpable in their correspondence. They indicate problems that the relatively late immigrant von Mises had in coping with the mentality of his new environment. In von Mises's case, a certain elitist attitude came into play, typical for him,[72] but which also expressed the traditional social rank of the German professor. For example, Richardson reacted in a letter, dated September 14, 1942, to a previous letter by von Mises in which the latter had apparently complained about the enlarged board of editors for the planned journal *Quarterly of Applied Mathematics*, fearing lack of opportunities for individual editors in influencing decisions. In order to convince von Mises, Richardson stressed the national and patriotic importance of the planned journal and pointed to differences in attitudes between Europeans and Americans:

> I believe that there is a considerable difference between the situation here and in Europe and that policies which would be successful in one country might easily be failures in others. Here in America we manage things by cooperation . . . general policies are laid down by a group and not by an individual. . . . My own desire is to make it possible to use all the talent we have in America in this terrible struggle for the existence of civilization.[73]

---

[71]Wasow (1986), p. 221.

[72]Widman (1973), p. 94, reports on von Mises's time of emigration in Turkey that his "distanced manner did not facilitate his contacts with Turkish students and colleagues" (T).

[73]Mises Papers, HUA, HUG 4574.5, box 3, f. 1942. Von Mises remained reserved and withdrew from the foundation of the *Quarterly*.

9.D.6. Problems of Moral Prudishness in the United States: The Extreme Case of Carl Ludwig Siegel

For Siegel, a bachelor sharing accommodation with female friends[74] in Princeton, one of several reasons for his short-term return to Germany in the 1930s and for his final return after the war was (what he considered) the sexual prudishness of the American society. With his strong individualism Siegel was given to sweeping generalizations and often to absurd comparisons.[75] He wrote to Courant in September 1935:

> It would be meaningless to escape the sadism of Göring's only to get under the yoke of Mrs. Eisenhart's notion of morality. . . . Please do not be offended that I do not like your America.[76]

It was probably remarks of this kind that led Harald Bohr in 1936 to wish that Siegel "could express himself more responsibly and coherently, as it is certainly not his intention to make propaganda for the Third Reich."[77] In fact and not surprisingly, Siegel, who held strongly pacifist convictions, found it even more difficult to adapt to militaristic Nazi Germany than to "prudish" America. At the outbreak of the war, he immigrated once again to the United States, this time via Norway. Not unexpectedly he again ran into problems in the United States. After the war Siegel complained in a very emotional letter to Oswald Veblen about the "Gestapo-minded government" of the United States refusing to give his female friend Hel Braun an entry permit.[78] Hermann Weyl found this letter unacceptable, since at that time the U.S. government had to be critical in its evaluation of entry permits for Germans, and Braun had neither been dismissed nor otherwise persecuted during the Nazi years. She could therefore not claim preferential treatment.[79] In this case Siegel realized that he had gone too far. He apologized in a letter to Veblen one week later:

> I feel ashamed of my violent reaction whose cause is in no proportion to the sufferings of other victims of modern history.[80]

---

[74]A Betty Backe and the mathematician Hel Braun (1914–1986). Braun's memories (1990) of the 1930s give anecdotal information on Siegel's emigration.

[75]See also below in chapter 10 Siegel's "comparison" between the Nazis and Bourbaki.

[76]C. L. Siegel to R. Courant, September 18, 1935, CPP (T). I thank Herbert Mehrtens (Braunschweig) for drawing my attention to this letter.

[77]H. Bohr to Courant, August 24, 1936, CPP (T).

[78]Siegel to O. Veblen, June 30, 1946. OVP, cont. 12, f. Siegel, 1935–60.

[79]Ibid., Weyl to Veblen, July 9, 1946.

[80]Ibid., Siegel to Veblen, July 8, 1946.

9.D.7. Language Problems[81]

Kurt Mahler and Richard Brauer both admitted that at the time of their emigration they had only "reading knowledge of English."[82]

Hermann Weyl's 1946 memories of his education at a classical high school (Gymnasium) in Germany around 1900 indicate that he knew no English at all, and as a result he only belatedly took notice of Bertrand Russell's logicism:

> I grew up a stern Cantorian dogmatist. Of Russell I had hardly heard when I broke away from Cantor's paradise; trained in a classical Gymnasium, I could read Greek but not English.[83]

The same is reported by Fritz Herzog, who had fled from Berlin in 1933 and who in 1980 in response to the question "When did you learn the language?" from the American H. Strauss, said:

> I started two months before leaving Germany. In Gymnasium, English was not among the languages we learned.[84]

The immigrant Lipman Bers talked about the great level of tolerance shown by American students to the language problems of their teachers, although it remains unclear whether Bers's experience is representative:

> I am convinced that in no European country would students tolerate teachers whose language they could hardly understand.[85]

9.D.8. The Need for Publications in the Language of the Host Country

Richard Courant wrote to Carl Ludwig Siegel in 1934 giving him advice on how to increase his chances of an invitation to Princeton:

> According to Veblen the submission of your work in the Annals will have a positive effect.[86]

---

[81]See also Weyl's letter of resignation written to the Nazi Prussian Ministry of Culture October 9, 1933, printed in Schappacher (1993), pp. 81–83, where Weyl reflects on the importance of the German language for him. See also in chapter 10 Halmos commenting on the simple German in van der Waerden's *Moderne Algebra*.

[82]Poorten (1991), p. 372, and Feit (1979), p. 4.

[83]Weyl Papers, ETH Zurich, Hs 91a:17, talk at Bicentennial Conference Princeton, first version, December 1946, p. 12.

[84]IBD microfilm, reel 26.

[85]Bers (1988), p. 242.

[86]Courant to Siegel, October 8, 1934, CPP (T). Note also Courant's conflict of loyalty because of his connection to Springer, who edited *Mathematische Annalen*, the German counterpart to *Annals of Mathematics*. See also below in chapter 10 Courant's discussion with Richard Brauer on the publication of the latter's *Algebra*.

In 1936 Gabor Szegö wrote to Courant from St. Louis that he had been granted a subsidy from Washington University for the publication of his book on orthogonal polynomials:

> Although there is no obligation connected to it, I believe that, under the given circumstances, it is more appropriate to publish the book here in America.[87]

9.D.9. Support by Immigrants for Economic and Social Reform, in Particular for New Deal Positions

The following judgment by Coser might refer more to humanists and social scientists, but it could possibly apply also to the more politically active and leftist among the refugee-mathematicians, such as Gumbel, Rademacher, Courant, and Carnap:

> When in the 1930s the German socialist immigrants came to the U.S.A., New Deal was in its prime. Thus many of them saw New Deal as a phenomenon that largely corresponded to their reformist political and social ideal of a planned, reform-oriented state economy.[88]

In 1934 Courant wrote to the physicist Max Born, comparing the economic problems of Europe with those of the United States:

> The situation here is certainly filled with uncertainties and stress, too. But for one thing, one does not understand and feel them so clearly, and for another, they cannot be compared in seriousness and danger to those in Europe. However, I believe that in America a fundamental change of the entire social organism is under way as well and that the main dispute between the two hostile camps is whether recovery can be attained by a return to the old business methods or whether new forms of economic and social life have to be found.[89]

Harald Bohr in Copenhagen wrote to his friend Courant in November 1936:

> Our mutual hope that Roosevelt would become re-elected has splendidly and luckily come true.[90]

---

[87]Szegö to R. Courant, February 5, 1936, CPP (T). See in this connection also Courant's request as quoted above in chapter 7 (D).

[88]Coser (1988), p. 96 (T).

[89]R. Courant to M. Born, October 18, 1934, CPP (T).

[90]H. Bohr to R. Courant, November 18, 1936, CPP (T). Courant, on his part, had written from Copenhagen to D. Flanders in New York on July 10, 1936, signing with "Heil Roosevelt."

9.D.10. Pressure to Adapt Politically and Political Mistrust against Immigrants on the Part of the Americans

The following is quoted from a letter, written by the Jewish director of the Institute for Advanced Study in Princeton, Abraham Flexner, on September 28, 1933 to another Jew, Felix M. Warburg (New York City). Flexner condemns Albert Einstein's explicit stance against the Nazi government and his "everlasting publicity," while Flexner pleads for appeasement:

> Last night Professor Lefschetz, who holds the highest professorship in mathematics in Princeton University and is himself a Russian Jew, came to see me and asked me if I could not in some way shut Einstein up, that he was doing the Jewish cause in Germany nothing but harm and that he is also seriously damaging his own reputation as a scientist and doing the Jewish situation in America no good.
>
> I may add for your private information that I am seriously concerned as to whether it is going to be possible to keep him and his wife in this country. I have been pleading with them all summer to show the elements of common sense, and their replies have been vain and foolish beyond belief. You have doubtless noticed in the morning paper that the German Government has retracted in part its attitude towards Jewish merchants. Einstein is simply making it as hard as possible for the German Government to climb down. . . . Though he is of course not a Communist, he is now only partially a Pacifist. . . . His presence on the platform will do no good to anybody. . . .
>
> He and his wife are better taken care of today than they have ever been in their life if they will only behave themselves. Other German Jewish scholars like Franck and Haber, both Nobel Prize medalists, have actually given up their posts either voluntarily or through suppression and allowed the world to judge, with the result that they are more highly esteemed than ever and their dignity has hurt the German Government a good deal more than Einstein's everlasting publicity.[91]

R. G. D. Richardson, the secretary of the AMS, took a principal stance in 1933 against "propaganda" by foreigners and—clearly being suspicious of Emmy Noether in particular—advised in this respect the dean of the graduate school of Bryn Mawr College near Philadelphia:

> It should be appropriate for you . . . to inform Professor Noether that no propaganda—racial, political, or religious—should be contemplated."[92]

In August 1933 Richardson had written to Hans Lewy, who was being considered for a position at Brown University:

[91]IAS, Archives, General, f. Warburg, Felix M.

[92]Richardson Papers, BUA, Correspondence, box: German-Jewish Situation, f. Miscellaneous Letters, Richardson to E. M. Schenck, November 3, 1933.

It is to be taken for granted that no propaganda of a political or racial character is to be carried on during your stay at Brown University. Brown does not pass judgment on the questions involved in the recent revolution in Germany. We wish only to be helpful to individuals and to promote the cause of science.[93]

Hans Rademacher, one of the few leftist scholars of the Weimar Republic who had consequently been dismissed by the Nazis for political reasons, tried to dispel concerns abroad that he had been too active politically. He wrote to Harald Bohr in Copenhagen in 1934:

I have not ingratiated myself [herangedient] with the new rulers but rather kept an objective reserve as it is my nature. But I have fulfilled my official duties in full loyalty and dignity.[94]

Wolfgang Wasow wrote the following on the immigration of his mother and brother to the United States in 1939. The two had fought on the republican side in the Spanish civil war against the rebels under Franco who were supported by Hitler's Germany:

To the American officials my mother and Holger had to emphasize that they were refugees from Nazi persecution but not otherwise politically involved. In particular, it was necessary to play down the political aspects of their activities in Spain.[95]

One acting dean at Idaho did not dare to propose Max Dehn for a position at his school. He wrote the following about Dehn to the Emergency Committee, which had been willing to subsidize the position:

When he first came we had protests from the American Legion and from individuals, charging that he was or might be (1) a Nazi spy or (2) a communist, since he traveled across Russia and got here via Pacific.[96]

The daughter of the statistician Felix Bernstein, the statistician Marianne Bernstein-Wiener (1917–2005), wrote to me in 1998:

We got the information . . . that Bernstein would be permanently appointed at the Biology Department at Columbia. . . . Nothing came out of it, also in the case of the Math. Dept. of Columbia. . . . The reasons were given to my father by E. Kasner [chairman for mathematics at Columbia; R. S.]. . . . The Trustees

---

[93]Richardson to Lewy, August 12, 1933. Richardson Papers, Correspondence, box: German-Jewish Situation, f. H. Lewy.

[94]EC, box 29, undated [1933], (T), f. H. Rademacher, 1934–44.

[95]Wasow (1986), 266.

[96]J. R. Nichols (Idaho) to EC, March 3, 1942, EC, box 6, f. M. Dehn, 1941–44. See also Dawson (2002) for Dehn's flight with the Trans-Siberian Railway.

(Wallstreet) claimed that we German emigrants were all communists and they were not prepared to finance us.[97]

A rare example of an immigrant daring to criticize, if only slightly, American politics, at least those of the past, was Hermann Weyl. In his obituary of Emmy Noether, Weyl took issue with the behavior of the Americans and their allies after the First World War, which in his opinion had promoted German nationalism. Weyl, who was firmly established at the IAS and had an outstanding scientific reputation, was probably the only one who could safely talk about "breaking by the Allies of the promise of Wilson's Fourteen Points,"[98] thereby hinting to the contradictions in the treaty of Versailles of 1919 and its inherent potential for conflict. Even Weyl found it prudent to omit mentioning that the original (American) proposals by U.S. president and university professor Woodrow Wilson (1856–1924) had been rejected by the American Senate.

9.D.11. Waning Political Restraint on Immigrants after Obtaining American Citizenship and the Impact of the American Entrance into the War

Long-time immigrants to the United States felt comparatively more at liberty to utter political opinions, such as John von Neumann in a rather prescient and realistic letter from Budapest on October 4, 1938, to Marshall Harvey Stone (1903–1989), shortly after the Dictate of Munich:

> How does the present European settlement impress you? I think, that there is some good in it, since it gives Tchekoslovakia [sic] the frontiers which it should have had in the first place, after 1918—or at least approximately so. But I don't think that the next general war is more than postponed by it, and not by much, either. England's weakness and indecision has only become more obvious, and this, after all, is the main motive for any European war. In any case, it seems, that the armament race is going on at top speed, and that is a more reliable criterium [sic] than anything the "statesmen" say.[99]

The head of the Applied Mathematics Panel of the American war research organization OSRD (Office of Scientific Research and Development under V. Bush), Warren Weaver, wrote an obituary of Richard Courant in

---

[97]M. Bernstein-Wiener to me, undated, received February 11, 1998 (T). Without taking this anecdotal evidence too seriously, one has to admit that there was a connection between xenophobia, anti-Semitism, and anti-Communism in certain American circles that later came to the fore in the McCarran Act of 1952 and in Senator J. McCarthy's political witch hunt.

[98]Weyl (1935), p. 208.

[99]M. H. Stone Papers in the AMS Archives, Providence, RI, box 37, f. 17.

1972. In it, Weaver included the following passage describing an event immediately after Courant had obtained his American citizenship in 1940:

> When the small group of United States mathematicians responsible for organizing the Applied Mathematics Panel of the Office of Scientific Research and Development during World War Two was considering, in view of the classified nature of many of their projects, whether it was prudent to include in the top governing committee, a man—however distinguished and able—who had been a member of the Imperial German Army during World War One, it was, however, the conviction of all those involved in the decision that Courant was deserving of complete trust. When he was about to attend the first meeting of the Panel, a particularly wise member, Thornton C. Fry, said, "We must, from the very outset, make Courant realize that we view him, with no conceivable reservation, to be one of us." Accordingly when the meeting was ready to convene, and there was a somewhat timid rap on the door, Fry sprang up, opened the door, extended his hand, and said with enthusiasm, "Dick, come in."
>
> In view of the almost reverential respect with which titles were viewed in the great German universities, it may well be the case that this most distinguished professor and Herr Direktor had never before in his adult life been so informally saluted. But Fry's idea was an inspiration. For from that moment on Courant was, quite without question, one of us.[100]

In August 1941, J. C. Hunsaker (1886–1984) of the Massachusetts Institute of Technology wrote a letter to Heinrich Peters (1901–1982), who had come to MIT in 1935 but now found himself in Germany, having been caught there during a vacation when war broke out in 1939. Hunsaker advised Peters not to come back to the United States because the situation for citizens from Germany and its allies had become complicated, although the United States had not yet entered the war:

> There is no doubt that this country has gone a long way in both its material and moral preparation for war. Relations with the Axis Powers are definitely strained, and the position of their citizens here is not a happy one. M.I.T., as you would imagine, is more and more concerned with national defense projects. Many of the faculty have left to serve in the Army or Navy, or in the munitions industry. Our Wind Tunnel is working two shifts for the industry and has a "no admittance" sign on it. Even our own students can't go in. Under such conditions, you would not wish to be here.[101]

---

[100]Weaver (1972), pp. 148–49.

[101]Hunsaker to Peters, August 25, 1941. MITA, AC4, Office of the President, box 169. In a letter to me dated March 18, 1994, Eric Reissner confirmed that Peters could not return to Boston and that he later became a professor at a Technical University in São Paulo, Brazil. Peters, has not been included in the list of emigrants in Appendix 1 (1.1), because he was oriented more toward engineering than toward mathematics. See Poggendorff VIIa and VIII.

There was sometimes ambiguity about the classification of Austrian citizens due to the occupation of Austria by Nazi Germany in 1938:

> The Department of Justice issued a regulation on February 5, 1942, according to which Austrian emigrants in the U.S. generally had to be considered as non-enemy aliens. Nevertheless Austrians were occasionally treated very differently, and the declarations of American authorities whether they acknowledged the "Anschluss" or not, remained ambiguous until the Moscow declaration of November 1, 1943.[102]

In 1942 Hermann Weyl alluded to these differences when recommending Max Dehn and Hilda Geiringer for the mathematical part of the program of military education for students:

> Dehn, being German, is technically speaking an "enemy alien," whereas Mrs. Geiringer, a native of Austria, is not. But there is no doubt whatsoever of the political reliability of either of them.[103]

Shortly after the United States entered the war, in December 1941, R. G. D. Richardson reported to Brown University president H. Wriston about Oswald Veblen's problems as a "guard" for foreigners at the Institute for Advanced Study. Richardson called the IAS rather inappropriately a "mild sort of concentration camp" and proposed similar measures in Providence:

> <u>Refugees</u>: Veblen reports that the Institute for Advanced Study with 20 enemy aliens has a real problem and that he has to stay on the job instead of going off for scientific meetings or a holiday. He has collected complete information regarding the status of each alien and this is being tabulated and given to the police for their information. Princeton University will not allow these people on the campus (because of defense work in the physics laboratory); they are also advised not to leave town. Veblen says that there will be soon a national order restricting the travel of aliens; so his brood will not attend scientific meetings for fear of not being able to get back. The Institute seems to be a mild sort of concentration camp. Some of these ideas may be worth Brown's consideration.[104]

Hans Samelson, who in 1941–42 had a stipend (externally funded) at the IAS, indicated to me in 1994 that the restrictions were not too severe:

> There were restrictions on "enemy aliens," but they were not severe. In Princeton I was not allowed to enter several university buildings where war work was

---

[102]Eppel (1988), p. 986 (T).

[103]Weyl to E. J. Moulton, February 28, 1942. Northwestern University Archives, Evanston, Department of Mathematics, 11/3/19/2, box 11, file 19: Hilda Geiringer.

[104]Richardson to Wriston, December 19, 1941, BUA, Appl. Math. Div. I 90 (H. Wriston).

going on. For travel beyond 18 miles from my domicile I had to seek permission from the Attorney General; I asked several times and it was always granted.[105]

Richard von Mises of Harvard University mentioned travel restrictions for enemy aliens when writing on May 8, 1942 to Richardson at Brown University in Providence about a Summer School for Applied Mechanics at Brown (eighty miles from Boston), to which he had been invited:

> I am grateful to you for offering me your help concerning my traveling permit. Since I shall be busy at Harvard during the first part of your session and somewhat longer, I shall have to commute between Cambridge and Providence. Later on I hope to be able to stay in Providence. Thus I would need a permit for both commuting once or twice a week and for changing my residence. It would be a great help if you could send me a letter, addressed to the U.S. District Attorney McCarthy, in Boston, in which my situation is explained. I shall get another letter from the Harvard Graduate School.[106]

In October 1941, prior to Pearl Harbor and the United States entering the war, the head of the Emergency Committee in Aid of Displaced Foreign Scholars, Stephen Duggan, took a critical stand in his report on the restrictions already imposed upon immigrants. Referring to the Summer School at Brown, organized by Richardson, he wrote to the latter, apparently expressing admiration for Richardson's ability in coping with these problems:

> What was surprising to me . . . was to note that out of the six visiting professors listed on the faculty, four—Dr. von Mises, Dr. Friedrichs, Dr. Prager and Dr. Bergmann—were refugee scientists. Certain refugee scholars are finding extreme difficulty in getting within many miles of defense projects. Every now and then we will get a young refugee worker in the field of aerodynamics who complains bitterly because he cannot apply his training experience to vital defense problems.[107]

Even after the United States ended the war, Richardson's group at Brown University apparently offered opportunities to involve noncitizens into defense projects, if limited to general tasks only. With regard to the transfer of Karl Löwner (having changed his name to Charles Loewner) from the University of Louisville, Kentucky, to Brown University in Providence, Richardson wrote in 1943:

> We are eager to have Loewner come here because he can be of very considerable use in the war effort. I suppose he is the most distinguished mathematician

[105]H. Samelson to the author, May 2, 1994 (T).

[106]BUA, Appl. Math. Div. I 48.

[107]Richardson Correspondence, f. 87 D (Duggan, 1940–42), S. Duggan to Richardson, October 15, 1941.

in the state of Kentucky, and while he has the handicap of being a foreigner, we think that he would be a valuable addition to our group.[108]

Regarding the possible involvement of foreigners in mathematical cryptography (an area of much stricter conditions of secrecy), the War Department wrote to Brown's President Wriston on January 7, 1942:

It is necessary that the candidates be . . . American citizens, preferably native born, who have no close relatives or ties in foreign countries.[109]

Gustav Bergmann, who had come to the country in 1938 and who at the time was at the department of philosophy at the state-run University of Iowa, offered his services to the American war effort, when he wrote to the EC on January 5, 1942:

Up to the outbreak of this war I thought that our place was in the laboratory and the study and not in the public and political life of the nation, the less so because we could so easily have appeared to speak in our own cause. Now however the die is cast and we may very well ask ourselves: What can we contribute to the national effort in the present emergency?[110]

While Stephen Duggan of the EC showed interest in this letter, other Americans, having discussed the matter in a Committee for the Utilization of Talents of Refugee Scholars, were more diffident. One of them actually wrote: "In most cases, government employment of such scholars on research or in work of information is impossible, for the greater part of them, since they are aliens."[111]

In 1944, with the imminent defeat of Nazi Germany, Paul Tillich (1886–1965, theologian and refugee to the United States), issued "A Declaration of the Council for a Democratic Germany," which he also sent to refugee-mathematicians such as Richard von Mises, who up until then had kept his political opinions to himself.[112] The German declaration reads:

---

[108]R. to J. H. Simister in Louisville, November 22, 1943. BUA, Appl. Math. Div. II 95 (Charles Loewner).

[109]BUA, Appl. Math. Div., I 90 (H. Wriston).

[110]EC, box 171, f. Committee Utilization of Talents of Refugee Scholars. The head of the EC, Duggan, wrote on the letter: "This is a splendid idea!" On the reaction to Bergmann's offer see below.

[111]Both Bergmann's letter and the commentary are in EC, box 171, f. Committee for the Utilization of Talents of Refugee Scholars, January 1942.

[112]When still in Istanbul, von Mises refused participation in peace initiatives by his Dutch colleague J. M. Burgers (1895–1981). In a letter dated May 17, 1938, von Mises said he did not believe in the effect of such initiatives anymore after Hitler had invaded his country, Austria. Von Mises Papers, HUA, HUG 4574.5, box 3, f. 1938. For Burgers see Alberts (1994). Von Mises apparently did not sign Tillich's declaration either.

Many of those who left Germany as opponents of Nazism and who are now living in the United States have been reluctant, so far, to participate in political discussions. They believed this restraint was a matter of tact, which they owed to the country that has become a second home country for some, a generous asylum for others.[113]

9.D.12. Personal Failure of Immigrants in the United States, Due to Age- and Pension-Related Problems

In 1935, Courant, in a comforting letter to his long-time friend Wolfgang Sternberg in Jerusalem, commented on the difficult and precarious situations several immigrants or would-be immigrants to the United States found themselves to be in:

> It is not true that all other people have found jobs. Of the mathematicians who lost their jobs without having found something I can mention: Rothe, Rogosinski, Bernays, F. Levy, Blumenthal, Hopf in Aix-la-Chapel, Lüneburg, Paul Hertz, most of them with family.[114]

Courant refrained from mentioning Felix Bernstein, who, despite his fame in set theory and biological statistics, fared particularly badly and was only able to get temporary jobs.[115]

On the U.S. career of the noted topologist Max Dehn, Siegel wrote in 1965:

> The prestigious universities would have found it inappropriate to offer Dehn a low-paid job. So they preferred to ignore his presence.[116]

In 1940 Emil Julius Gumbel, himself without a permanent position, tried to do something for Jacques Hadamard when Hadamard had to flee German-occupied France. Gumbel was grateful to Hadamard for having previously come to his aid upon his arrival in France, the first country in his emigration. Accordingly, he wrote to Richardson of Brown University. Richardson, however, finding Hadamard too old and no longer useful, replied:

> While it is true that Brown University would be glad to welcome Professor Hadamard and pay him a small stipend if he were passing through, we recommend that he not be invited to come to this country, and I have said this to

---

[113]Von Mises Papers, HUA, 4574.5, box 3, f. 1944 (T). For Tillich's emigration see Greenberg (1996).

[114]Courant to Sternberg, July 19, 1935, CPP (T). Courant's allusion to the families of immigrants was apparently meant to show Sternberg, who did not have a family, that his situation could be worse. By "F. Levy," Courant probably meant F. Levi, who finally went to India.

[115]See case study 2 below and Appendix 4.4.

[116]Siegel (1965), p. 20 (T).

other persons. In his day, Hadamard was a great figure in mathematics and he has visited this country and received a warm welcome, but he is now old and has not done anything significant in mathematics for a decade. There are persons in Europe whom the mathematicians of this country would much prefer to have come and there are persons in this country who might make a tour under the auspices of some mathematical group with more success than Hadamard.[117]

H. Weyl, G. D. Birkhoff, and the astronomer Harlow Shapley of Harvard University were among several colleagues who eventually gave their support to Hadamard. As there was no tangible benefit in employing older immigrants, Americans like Shapley used auxiliary arguments in order to support them, such as the need to increase the national prestige of their universities. But "Charity was never his [Shapley's] argument for support."[118]

One key problem in the case of older immigrants such as Bernstein, Dehn, and Hadamard, was securing adequate pensions for them, since they would not have enough time themselves to save for their retirement. This problem had been discussed very early by Norbert Wiener in his December 1934 article in the *Jewish Advocate*. Wiener remarked that in Germany professorial pensions were of a similar magnitude to professorial incomes and added: "This they cannot have here, but, we do not wish to see them threatened with a pauper's old age."[119] Toward the end of the war, when rumors were circulating about the Emergency Committee being suspended, Hermann Weyl again took up the problems, as yet unresolved, relating to pensions. He wrote to Stephen Duggan in May 1944:

In general it has been impossible to make adequate provision for older refugees who are about to reach the age limit, for retirement. I will mention a few names in my own field—Jacques Hadamard, Erich Marx, Felix Bernstein, Max Dehn, Fritz Reiche—for whom this question is acute right now. For many more it will become acute in a few years. What arrangements can be made for them? I am afraid that unless some financially powerful American organization offers a general and radical solution, this problem will become interwoven with the appalling problem of settling the account with the future German Government

[117]Richardson to Gumbel, April 16, 1940, copy. Birkhoff Papers 4213.2, box 14, file P–R. Richardson announced in his letter that he sent a copy to G. D. Birkhoff, expecting the latter would support his negative position. However, in the case of the refugee Hadamard, Birkhoff was willing to help. See chapter 8 for Birkhoff's otherwise rather negative response to the refugees.

[118]Jones (1984), p. 236.

[119]Wiener's article in the *Jewish Advocate* in Boston, dated December 14, 1934, has been mentioned in chapter 8. It was titled "Aid for German Refugee Scholars Must Come from Non-Academic Sources."

for all the violations of property and property rights perpetrated by the Nazis. Indeed most of the refugee scholars who came from Germany have legal claims for salaries or pensions which they did not give up when they left the old country, and which even the Nazis have recognized to a certain extent. Whether this is not perfectly illusory I do not know.[120]

Five years later, in 1949, a round-robin letter, probably written by Richard Courant, made an appeal for help with the pensions for immigrants:

In 1934, a "Mathematicians' Relief Fund" was organized for assistance to scientists who had left Germany and who had not yet established themselves in this country or else-where. Some years ago this fund was discontinued since the time of pressing need seemed to have passed. However, tragic cases have developed recently; among the mathematicians who have come to this country, there are some who are destitute, old, sick, and without right of pension or annuity. It appears that a revival of the old fund on a somewhat broader basis is an urgent necessity. We appeal to you for a contribution to make help in such cases possible. Payments should be made to: Mathematicians' Relief Fund, Institute for Advanced Study.[121]

9.S. Case Studies

9.S.1. The Tragic Fate of a Political Emigrant: Emil Julius Gumbel[122]

Emil Julius Gumbel was politically an outsider in the community of American mathematicians, and, to a smaller extent, an outsider in his research subject of mathematical statistics. He was Jewish by Nazi definition, but, as previously mentioned, had already fled from Germany to France in 1932. Even before that, on October 2, 1931, the head of the German Academic Exchange Service (Deutscher Akademischer Austauschdienst) in Berlin, A. Morsbach, had warned Stephen Duggan of the Institute of International

---

[120]Weyl to Duggan, May 2, 1944. EC, box 103, f. H. Weyl. Weyl discussed the same problem in his correspondence with H. Shapley. The "recognition" of pensions by the Nazis, which Weyl mentions with some hesitation, was, however, indeed the exception. Permission for emigration was often given only on the condition of relinquishing all demands. The latter also resulted in legal problems after the war, when the immigrants claimed compensation.

[121]Undated [1949]. AAMS, MS 75.5. J. R. Kline Correspondence, box 34, f. Courant 1949. I thank L. Beaulieu for this information. In the same file there is a letter by Courant to Kline (June 6, 1949), mentioning the particularly precarious situation of Sternberg and Bernstein and citing Emmy Noether as the one who had proposed the original Relief Fund.

[122]See also chapter 7, particularly the case study on the fate of Gumbel's book *Freie Wissenschaft* (1938).

Education (New York City) against the advisability of allowing Gumbel to hold "guest" lectures in the United States due to his "political standing." Duggan and Dunn, who both knew Gumbel personally and who would both at a later stage sit on the Emergency Committee, shrugged these warnings off as being politically motivated.[123] Shortly before the Nazis came to power, on January 6, 1933, another American, Nathanael Peffer, wrote a letter to President Max Mason of the Rockefeller Foundation, mentioning Gumbel's contacts with Abraham Flexner of the IAS (Princeton) and with Stephen Duggan. Peffer wrote:

> This seems to me a man who not only deserves helping out but could add something to American scholarship. . . . (Gumbel speaks fluent and scholarly English, of course.)[124]

As mentioned in Peffer's letter, although at least three other individuals wrote to Mason on Gumbel's behalf, help from the RF was not forthcoming for him until much later. According to Peffer's letter, even the liberal Stephen Duggan was "fearful of the political complications" for Americans, giving refuge to a political activist such as Gumbel. Moreover, in the beginning there were problems as to how to justify including Gumbel among persons to be given support by the EC, as one letter by E. R. Murrow of the EC to the "American Civil Liberties Union," of July 21, 1933 shows:

> Unfortunately the fact that Gumbel was dismissed before the outbreak of the current madness has prevented his name from being included on many lists that have been submitted to us.[125]

Eventually, Gumbel received temporary support from the Rockefeller funds. However, after the German occupation of France in June 1940, his situation in this country became unbearable. Apparently in a concerted action, Albert Einstein of the Institute for Advanced Study and the two leading American statisticians, Harold Hotelling (1895–1973) and Samuel S. Wilks (1906–1964), succeeded in persuading Alvin Johnson of the New School for Social Research in finding a position for Gumbel in August 1940. Einstein, in his letter to Johnson, turned Gumbel's political achievements into a positive feature, stressing how Gumbel had published material about right-wing political assassinations in the early Weimar Republic:

[123]See the letter by Dunn to Duggan, April 27, 1932. Dunn Papers, American Philosophical Society.

[124]RAC, R.F. 1.1, 500 S, box 23, f. 234.

[125]E. R. Murrow to Roger N. Baldwin, July 21, 1933. EC, box 12, f. Gumbel, E. J., 1934–40. Baldwin replied, however, on July 25, that help in a case like Gumbel's was particularly justified and important.

His most valuable achievements, however, are publications of outstanding political interest. . . . He is, therefore, very well known in Germany and would, of course, in the case of his extradition, face not only the death penalty but the most cruel torture as well. To save the life of this extraordinary person, is, in my opinion, a high moral duty.[126]

Hotelling added:

Since the collapse of France it is reported that the Nazis are seeking to put Dr. Gumbel in confinement.[127]

While Einstein may have been well intentioned in stressing Gumbel's political achievements knowing Alvin Johnson's liberal mind,[128] the two statisticians placed great emphasis on Gumbel's statistical work being "of excellent quality" (Hotelling), "ranking with the best done in Germany before Hitler came to power" (Wilks).[129] Help from the New School arrived just in time for Gumbel. The latter had written from Marseille on July 18, 1940, to his American acquaintance, the geneticist Leslie C. Dunn, asking for help "to save this naked life."[130] The Rockefeller people, who did not generally fully share the aims of Johnson's school,[131] apparently continued to be reserved about Gumbel's politics. A Rockefeller functionary, who interviewed Gumbel in 1942, found to his astonishment that Gumbel was "slightly less adaptable than the rest of the family, and even a little critical of US academic and scientific circles."[132]

In a speech given in 1964 in New York, Gumbel had the following to say about the tragic fate of many political emigrants and on the relation between professional work and political convictions:

Emigration is the most cruel thing that can happen to a politically interested person. Suddenly reality stops. The emigrant lives exclusively in the past, world history stops the day he leaves the country. . . . And each emigrant who comes

---

[126]Einstein to A. Johnson, August 7, 1940. RAC, R.F. 1.1, series 200, box 50, f. 583.

[127]Ibid., H. Hotelling to A. Johnson, August 7, 1940.

[128]Although Johnson was "anxious to keep Nazi and Communist propagandists out of the country" as he declared in his autobiography which, however, was written under the conditions of McCarthyism and red hunt in the United States (Johnson 1952, p. 370).

[129]Ibid., Wilks "To whom it may concern," August 1, 1940. There might be implied in this evaluation a slight note of condescension with respect to German statistics at large. On Gumbel's statistical work see Hertz (1997).

[130]Gumbel to Dunn, July 18, 1940. Dunn Papers, American Philosophical Society. Dunn responded with a cable for which Gumbel thanked him on August 6, 1940.

[131]The director of the school, A. Johnson, said of the time before 1939: "The Foundation had never been very cordial to the New School or to me" (Johnson 1952, p. 366).

[132]RAC, R.F. 1.1, series 200, box 50, f. 584. Interview with Prof. Emil Gumbel, Mrs. Gumbel, and Mrs. Page (Mrs. Gumbel's mother), January 21, 1942. Signed "A.M."

later is the enemy of the earlier one: You haven't come out in time, you tried to collaborate with the enemy. Emigration is totally unreal. All previous fights are fought once again in one's imagination. . . . And I have tried to distance myself from these fights.[133] After all, besides my convictions, I had also a profession. And I always performed my profession independently of my convictions. I have never tried to make my convictions a profession but always considered them as luxury.[134]

9.S.2. A Case of Failed Accommodation by an Older Immigrant: Felix Bernstein

In 1896, at the age of eighteen, as a first-semester university student at his place of birth, Halle, Felix Bernstein achieved fame as a mathematician by proving what was later named the Cantor-Bernstein equivalence theorem in set theory. Later on he became director of the Institute for Mathematical Statistics at the University of Göttingen. As an immigrant to the United States he had the advantage of being in the country since 1933 and of having connections there from previous stays. Unfortunately, his age (fifty-five by 1933) was against him. In spite of frequent praise by colleagues for his accomplishments and the importance of his field in applied mathematical statistics[135] his career in the United States stagnated.

As early as 1934 there was a note in the files of the Rockefeller Foundation (from which Bernstein had received help on a temporary basis):

> B. was the one definite misfit among the displaced scholars aided by the committee.[136]

With his display of dissatisfaction and insistence on his merits Bernstein did not endear himself to the functionaries of the New York Emergency Committee either, as a drastic note in the files of the SPSL in London reveals, based on discussion in July 1935 with E. Murrow of the EC:

---

[133]This remark contradicts somewhat Gumbel's actual behavior and experiences during emigration. His fight against emigrant Arnold Bergsträsser (1896–1964), whom he suspected to be a Nazi spy, was very real and no "imagination." See Gumbel's Papers at the Regenstein Library in Chicago.

[134]The manuscript of the talk exists only in German, which is translated here. Regenstein Library, Chicago, Gumbel Papers, box 4, f. 6: "Erinnerungen eines Aussenseiters," 23 pp., p. 9. On Gumbel's problems to find a job in the United States even after the war, see chapter 11.

[135]On this, several statements by Americans will be mentioned in chapter 10.

[136]RAC, R.F. 1.1, 200 D., box 130, f. 1603. The note was from May 21, 1934.

Figure 47 *Felix Bernstein (1878–1956). The mathematician had become well known for his work in Cantor's set theory (ca. 1900) and as head of the Göttingen Institute for mathematical statistics and insurance mathematics in the 1920s. In exile in America from 1933 onward, Bernstein never found an adequate position and was at times under financial pressure.*

Mr. Murrow: No assurance. No chances. Could not make necessary adjustment. "First rate pest." Emergency committee carrying him on next year, but that is final.[137]

In 1943, Ross G. Harrison, the biologist at Yale and then chairman of the National Research Council, wrote to Duggan from the EC:

> Professor Felix Bernstein was in to see me this morning. His situation is truly desperate . . . most recently New York University. Now having reached the age of 65, he is compelled to retire without having accumulated any reserves or pension rights.[138]

In the same year, Hermann Weyl wrote sketchily about his years with Bernstein in Göttingen, and on Bernstein's problems in emigration coping with the obvious loss in social status. In this letter, which was directed to Harlow Shapley (Harvard), Weyl writes, among other things, about Bernstein:

> His personality not too pleasant, mainly because he seems to feel the necessity of convincing himself at every moment of his own superiority.[139]

Two years later, in 1945, Duggan wrote to Chancellor H. W. Chase from New York University in reply to Chase's request for a contribution by the EC to a pension for Bernstein:

> The new committee is definitely not undertaking pensions because they involve too large a sum of money to be effective. The Emergency Committee, as you know, has closed its program.[140]

In 1946, Bernstein wrote to his friend Albert Einstein in a depressed mood:

> I am seriously worn out by the hardships of the past 10 years. I purposely asked for at least a temporary position in Princeton. I would have regained my strength there more easily. . . . As an immigrant I have hardly been welcomed here for my teaching. But scientifically I am still capable of considerable achievements. . . . I cannot see that the difference in quality between my results and those of other immigrants is so large as to give me no right to demand equal consideration. A modest place for me has to be found, where I will be left in peace.[141]

---

[137]SPSL, box 277, folio 390.

[138]Harrison to S. Duggan, Washington, DC, October 7, 1943. EC, box 2, f. Bernstein, 1936–45.

[139]Weyl to H. Shapley, June 5, 1943. EC, box 2, f. F. Bernstein, 1936–45. The passage concerning Bernstein is reproduced in its entirety as Appendix 4.4.

[140]Ibid. Duggan to Chase, February 9, 1945.

[141]Einstein Papers, Jerusalem, 56 510-2, undated 1946 (T).

After the war, in March 1949, the now sixty-eight-year-old Bernstein wrote a desperate and rather pretentious letter to Einstein:

> I have been occupying professorships here for 16 years with a total income of 53,000 dollars, of which I had to spend 2,000 for my work. The pension amounts to 23 dollars a month. Of course I could not save anything from this income. . . . I do not know of any case which is as horrible as mine. . . . Many American mathematicians feel that grave injustice has been done to me. . . .
>
> I have great respect for mathematicians such as Goedl [*sic*] and Morse, who have created a direction of research of their own. I also admire Wedderburn, but I cannot admit that there is any justification for giving average people [Mittel-größen] such as Veblen, von Neumann, Bochner and others such preferential and overpaid positions. . . . and allow, at the same time, a real talent go into decline.[142]

Judging by the amount Bernstein mentioned in his letter, it appears that his income was not exceptionally low compared with other immigrants. However, he had apparently suffered extraordinary expenses (including medical bills) for his family (wife and two children). Einstein replied to Bernstein on March 5, 1949, saying that his letter was exaggerated and had filled him with deep concern. Einstein found the scientific plans, described in some detail by Bernstein in his letter, unrealistic, almost grotesque, in fact, given the "formidable difficulties in modern physics" (T).

Three years later and four years before his death, even Bernstein's American citizenship was placed in jeopardy due to his no longer having sufficient income in the United States On June 24, 1952, L. C. Dunn, the geneticist at Columbia, wrote to the vice president of that university:

> Much as I sympathize with Professor Bernstein's desire to retain his American citizenship and to have a dollar income to help him to do this, I do not believe that there was every [*sic*] any obligation on the part of the University to provide the means for this.[143]

In fact, 1952 was the very same year that the U.S. Congress passed the infamous McCarran-Walter Act. This led, in several cases, to the exclusion of prominent individuals from immigration, and for some it even implied loss of citizenship after a stay abroad. It was apparently in this context that Felix Bernstein's widow Edith wrote a letter to then Massachusetts

---

[142]Bernstein to A. Einstein, undated, March 1949. Einstein Papers, Jerusalem, 57 648-1/2/3 (T). Einstein's reply is dated March 5, 1949 and has the archival number 57 649.

[143]American Philosophical Society, Philadelphia, Library, Dunn Papers. The letter went in copy also to Felix Bernstein.

senator, John F. Kennedy, who replied on January 23, 1959 with the promise: "I myself intent [*sic*] to introduce or co-sponsor a measure to alleviate some of the worst injustices of the present legislation."[144] The Immigration and Nationality Services Act of 1965 eventually replaced the McCarran-Walter Act.

[144]Marianne Bernstein Papers, with Ron Wiener, Sarasota, FL.

TEN

The Impact of Immigration on American Mathematics

The enrichment of American scholarship as a result of this emigration can scarcely be overstated. . . .

To use this particular discipline [mathematics] as an illustration, one may cite the brilliant faculty in mathematical research which had been developed at the University of Göttingen, ironically enough with funds supplied by the International Education Board in 1926, and which was dispersed in the thirties almost en masse. If Hitler had set out with benevolent intent to build up America as the world's great mathematical center, he could hardly have achieved more successfully the result which his ruthlessness accomplished. Welcomed in the universities of the United States, a substantial portion of the members of this faculty, together with other refugee mathematicians, have made a contribution in mathematical theory and applied mathematics which can scarcely be measured, but which had high practical usefulness in the development of war effort in the Second World War.

—R. B. Fosdick 1952[1]

THE institutional and cognitive dimensions of mathematics within the process of emigration have been left largely untouched so far in this book. Such a discussion, however, is essential when inquiring about the impact of emigration and its consequences for mathematics. Impact itself has to be defined more precisely first. The impact emigration had on mathematics both in the countries of origin and the host countries varied greatly. It is not easy weighing the losses due to emigration in the originating countries up against the gains made by the host countries due to immigration. The problem is particularly difficult as emigration was a dynamic historical process that opened up qualitatively new developments and cannot be compared to a zero-sum game. Part of the "qualitatively new" in this development was the impact on "world mathematics" as a whole, and it requires a

---

[1] R. B. Fosdick, president of the Rockefeller Foundation for many years, in Fosdick (1952), pp. 277–78.

broader perspective to appropriately discuss the impact of emigration and the interconnection between its social and cognitive dimensions.

Examining the global development of the discipline of mathematics, one has to take into account two further factors:

1. that development was also determined by more general and long-term trends, such as the changing relationship of mathematics and technology,[2] and greatly improved means of scientific communication,[3] factors not necessarily connected to emigration

2. the events concerning the United States and Germany were only one part—although arguably the most important part—in the overall process of mathematical migration and of changing international relations in mathematics at that time.

Undisputedly, the impact of mathematical immigration (not only restricted to German-speaking immigrants) to the United States was of great importance, arguably even more than in other sciences—something that perhaps was most visible in applied mathematics.[4] After the war, an unnamed prominent refugee-mathematician (presumably French: J. Hadamard?) noted the exceptional role played by mathematics, thereby also alluding to a then still existing divide between Europe and America with respect to the educational systems:

I believe that in my field the influx of European scholars had a deep and remaining effect on American institutions. Not only did it improve and intensify the graduate work in mathematics in many institutions but also it influenced the general attitude of mathematicians concerning theoretical and applied knowledge. Mathematics in the United States before 1930 was rather theoretical without much interest for applications. In the last ten years, especially during the war, this has changed.

In spite of close collaboration of American scholars of American and European background, there is still considerable confusion about the aims, common features, and diversities of American and European universities and secondary schools . . .

Exchange of young university teachers (perhaps also secondary school teachers) between the United States and various European schools should be resumed and if possible intensified as soon as normal traveling and living conditions

---

[2]In this respect the advent of digital computers in the 1940s and the rise of industrial mathematics were of preeminent importance.

[3]Here one has to consider, among other things, that there was almost no civil air traffic in the 1920s and 1930s in Europe and none between Europe and America. Also mathematics departments had almost no technical assistance or assigned travel money at that time.

[4]Siegmund-Schultze (2003a) and below. According to Davie (1947), p. 312, even more than in mathematics and the sciences immigration to the United States had the most impact in musicology and cultural history.

return. This is a vital affair for international understanding and for the improvement of our teaching standards.[5]

This statement also makes it clear that scientific emigration had consequences for the whole of Europe and that postwar developments were to a considerable degree not just a German-American but also a European-American affair.

As intimated in the quotation by the anonymous French mathematician, the differences between American and European secondary and university educational systems remained apparent even after the war, despite all the institutional changes, in particular those connected to research-level mathematics.

The quotation by the French mathematician also indicates principally two major effects at a cognitive level by which mathematics in the United States benefited from immigration—and not limited to German immigrants. These effects were increased competitiveness in research,[6] and a rise in academic applied mathematics. The latter influence of immigration was interwoven with the conditions of war and with institutional changes as described above. It was also connected to inner-mathematical developments and changes in attitude and mentality that will be more closely described in 10.4 and 10.5 through the examples of purer mathematical domains such as abstract algebra.

In this chapter two aspects of the impact of mathematical emigration will be discussed, a more general and a more specific one. On a more general level the "impact" is viewed from several global, national, biographical, or nonmathematical perspectives. In this context the relative effects of gain and loss due to emigration will be discussed to some extent, pointing out the differing mathematical traditions in the two main countries of origin, Germany and Austria. On a more specific level against the background of those more general tendencies, the concrete impact of emigration on American mathematics in institutional, cognitive, and stylistic respects will be discussed. On the cognitive side the examples of applied mathematics and abstract algebra will be emphasized.

As in previous chapters, the discussion is complemented and illustrated by documents and case studies. Since this chapter can only rely on the few previous historical studies available, part of the discussion will be necessarily sketchy[7] and is meant to stimulate further historical research.

[5]Quoted from Duggan and Drury (1948), p. 131.

[6]AMS secretary Richardson (1936), p. 210, showed with statistical tables that even in the period 1930–33 fewer than half of the American PhDs in mathematics ever published a paper.

[7]This is particularly true for the case study on the impact of Emmy Noether's algebra, where the presentation of the historical material—interesting in itself, at least in the opinion of this author—gets preference over its final evaluation.

10.1. The "Impact of Immigration" Viewed from Various Global, Biographical, National, or Nonmathematical Perspectives

Evaluating the impact of emigration, its losses and gains, is strongly dependent on the perspectives and interests of the observer,[8] and even the critical historian can but achieve an approximate success in his/her effort to give an objective picture. There are, in particular, certain "intangibles." The following quotation describes these nicely for mathematics:

> There may be important influences which are due to migration and are of a more intangible nature than the transfer of interest and research activity in a special field from one country to another but which cannot easily be documented. The style of teaching and writing are important in every field; in mathematics, the emphasis on abstraction and axiomatics, or on motivation and historical context, or the establishing of connections between different fields are examples of phenomena which are subject to the influence of the migration of scholars. The same is true for the standards of training and education, including the planning of a curriculum.[9]

Several of these "intangibles," insomuch as they do not merely concern discrete scientific "styles" and "tacit knowledge,"[10] can, in fact, be made more explicit. Differences and changes in forms of scientific communication can be analyzed insofar as they manifested themselves in changes in the infrastructure of science.

The multilayered connections among changing communication habits, mathematical migration, and the development of reviewing journals, in particular the breaking of the German monopoly on mathematical reviewing, have all been investigated in considerable detail by this author.[11] It has been shown that the founding of the American *Mathematical Reviews* in 1940 was the result of an overlapping of long-term developments in the system of mathematical communication and short-term political changes, specifically, emigration. It has also been argued that changing patterns in mathematical research, above all the rise of "structural mathematics" in the 1930s, had an essential influence on the methods of publication adopted by mathematicians. This concerns in particular the mathematicians' use of language and their reflection on historical sources. The manner of reporting

---

[8]See the remarks in chapter 1 on gains and losses, internationalization and "Americanization."

[9]Chandler and Magnus (1984), p. 199.

[10]This is a notion often used in the theory of science to describe those parts of science that cannot be explicitly formalized and that require direct, personal contact between teachers and students, i.e., oral communication. See Polanyi (1966).

[11]Siegmund-Schultze (1993a) and (1994).

on mathematics in mathematical reviewing and encyclopedias changed correspondingly.[12]

More strongly oriented toward the strictly cognitive dimension of mathematics than the general work on mathematical communication referred to above is the *The History of Combinatorial Group Theory* (1982) by Bruce Chandler (born 1931) and Wilhelm Magnus (1907–1990), the latter a postwar immigrant to the United States. The book, written by mathematical insiders, impressively establishes the multilayered connections to the social dimension of mathematics and to the changes in mathematical communication, due to emigration.[13] Among other things, the authors investigate the breaking of the monopoly of German literature in combinatorial group theory and the rise of the English language.[14] Chandler and Magnus also give due consideration to a more recent tendency in literature on the history of emigration that looks at the work of the emigrants compared to the work of their nonemigrated colleagues and the work of mathematicians elsewhere. One of these more recent publications on emigration expresses this in the following words: "Whoever investigates emigration must not restrict his/her perspective on the group of emigrants. He/she has to take entire sub-disciplines as the object of investigation."[15] In combinatorial group theory, where literature until 1933 was almost exclusively German, the immigration of K. A. Hirsch and B. H. Neumann to Great Britain, and of R. Baer to the United States, as well as the emigration of fifteen other mathematicians working in the discipline, is among the reasons that today's research in this field is distributed almost evenly over the entire mathematical world.[16] Although the authors admit that combinatorial group theory is too narrow a subject to be able to give the full picture for the trends in all algebra, the study is representative in the sense that it is treating a clearly distinguishable subdiscipline of traditional "pure" mathematics. Chandler and Magnus's book therefore stands for the investigation of the continuous (as opposed to the merely disruptive) tendencies within the process of emigration, which were determined by inner-mathematical research: "The fact that combinatorial group theory has, so far, not been influenced by the practical needs of science and technology makes it possible for us to use combinatorial group theory to exhibit the role of the intellectual aspects of the development of mathematics in a

---

[12]This already applied on the different styles of the two German reviewing journals, the old *Jahrbuch über die Fortschritte der Mathematik*, and the modern *Zentralblatt für Mathematik*, founded in 1931 by the future emigrant Otto Neugebauer.

[13]See Chandler and Magnus (1982), in particular the chapters "Modes of Communication" and "Geographical Distribution of Research and Effects of Migration," pp. 187–200.

[14]See on the latter point even more systematically the German paper by Göbel (1988–92).

[15]Fischer (1991), p. 47 (T).

[16]Chandler and Magnus (1982), pp. 193–97.

clear-cut manner."[17] While the authors seem to reserve the "intellectual" for the inner-mathematical domain, it was exactly the deep changes in the relationship between mathematics and its neighboring disciplines and fields of application during emigration and war research that led to a new concept of mathematics altogether and that determined the connection between the social and the cognitive dimensions.

Another important publication aiming at combining the two dimensions, but arguing—unlike Chandler and Magnus—more on the level of applied mathematics, is the book by Hanle, *Bringing Aerodynamics to America* (1982). The "early emigrant" Theodor von Kármán, who is at the center of Hanle's book, was, together with Courant, a key figure in the transfer of academic applied mathematics from Germany to the United States,[18] with von Kármán being closer to engineering mathematics than Courant. Hanle's book is therefore an important contribution to the broader understanding of the history of emigration and its impact. The rise of several sub- and neighboring disciplines of mathematics would hardly have happened— at least not in this form—without the oral communication, and the meeting and mutual influence of various mathematical, scientific, philosophical, and social traditions, enforced by emigration. One has to think for instance of mathematical statistics and the analytical philosophy of science, both fields being established at that time at American universities under the influence of immigration, although that influence was also exerted by non-German-speaking immigrants.

Up until this point the impact of emigration has been discussed mainly with respect to its effects on global mathematics and in relation to general changes in mathematical research practice. From this purely scientific (cognitive) perspective the impact must be considered to be positive, as emigration accelerated changes in the communication structure of mathematics that had been underway before, and counteracted a mutual national isolation of mathematical cultures, at least in the short term.

The perspective on the impact of emigration varies yet again when looking at it from the standpoint of national mathematical cultures. Chandler and Magnus emphasize that for the transfer of mathematical ideas to be successful a certain maturity is required within the mathematics of the receiving culture. The authors confirm in their book what the immigrant Lipman Bers said in 1980, that "research in a country will profit from the immigration of scholars only if it is already on a level comparable to that of the immigrants."[19] However, an indisputable fact that also concurs with

[17]Ibid., p. v.

[18]See below and Siegmund-Schultze (2003a). The recent book (Eckert 2006) stresses in addition the importance of an even earlier immigrant, Max Munk.

[19]Chandler and Magnus (1982), p. 193. Similar observations have been made for the conditions of the acculturation of European physics in the United States via emigration. See Schweber (1986).

the observations of contemporaries is that emigration did, indeed, change the course of the history of mathematics and that the considerable gains within American mathematics due to immigration have to be balanced against the losses on the European, especially the Germans', side. From the position of historical methodology considerable problems are involved in evaluating both the European losses and the American gains, which include the post hoc fallacy and the problem of counterfactual arguments, as well as the lack of a *tertium comparationis* of the undisturbed system.[20] Moreover, developments in the United States and in Europe after World War II led in part to further emigrations and in part to a reincorporation of scientific disciplines (but not necessarily of scientific personnel), making it difficult to treat the problem of the impact of forced emigration from the Nazis. Postwar developments also showed that short-term losses in Germany due to emigration often had a positive outcome later, and that in many cases, if emigration had not taken place, the losses would have been even greater due to isolation and the partial stagnation of research in Germany. Even the war-related seizure of German copyright by the American Alien Property Custodian in 1942 had a favorable side effect for German mathematics—the unclear outcome for the authors as to their publication rights notwithstanding—as it kept the American public aware of, and familiar with much of the work done in Germany, often even in its original German language.[21] This became important for the resumption of scientific communication after the war, something even representatives of the German publishing house Springer (temporarily deprived of its publication rights during the war) admitted.[22]

In order to evaluate the impact of emigration one has to examine shortcomings already apparent in several mathematical disciplines in the German-speaking realm prior to emigration, just as one also has to consider the differing traditions in Germany and Austria (D).[23] The biometric tradition in mathematical statistics, begun in England, was known in Germany by the 1920s but was largely restricted to Felix Bernstein's institute in Göttingen. In economically based statistics the influence of the German-speaking emigration was stronger from the "periphery," namely Austria, than from Germany. The late immigration to the United States of the

---

[20]Thiel (1984), p. 228; Fischer (1991), p. 43. Coser, following Max Weber, recommends the use of the method of "Gedankenexperiment" ("thought experiment"), and he sees an essential effect of emigration in the "deprovincialization" of the American intellectual life. See Coser (1988), p. 100. See also remarks in chapter 1.

[21]See Richards (1994). The problems even for victims of National Socialism, such as Max Born, to regain their publication rights after the war is described in Holl (1996), pp. 178–81.

[22]Siegmund-Schultze (1997), p. 158.

[23]See the remarks on the differences on the German and Austrian mathematical cultures in chapter 1.

Austrians E. Helly, A. Wald, and E. Lukacs contributed to the development of mathematical statistics in this country, although progress in this area had already been made due to the influence of British, Scandinavian, and Polish research.[24] Furthermore, original American stimuli, in particular industrial statistics (W. A. Shewhart and Th. Fry), were probably equally important in the development of mathematical statistics in the USA.[25]

In mathematical logic (K. Gödel) and topology (K. Menger), the more influential immigrants were again from Austria; in topology they encountered a strong American tradition (O. Veblen, J. W. Alexander, R. L. Moore), that benefited greatly from the early immigration of non-German-speaking countries (S. Lefschetz, E. R. van Kampen). The rise of analytical philosophy of science in the United States, previously mentioned, occurred at the intersection of the theory of probability, foundations of mathematics, and philosophical research. It was considerably influenced by two Germans, who had, however, been passed by before in Germany and in one case had moved to the Austrian, later the Czech "periphery": Hans Reichenbach (Berlin) and Rudolf Carnap (Vienna/Prague). This tradition was later continued in the United States by Carl G. Hempel (a student of Reichenbach), who in 1933 was a young emigrant from Berlin.[26] Gödel and Menger too had close contacts to the Vienna Circle. This was important for their research in logic and topology around 1930, even though they always stressed their originality and independence.

As a result of emigration reencounters or reconsiderations of former connections among the differing streams of German-speaking mathematicians frequently occurred, influenced by new developments in international mathematics. The probabilist Willy Feller, who had been influenced by the Swede Harald Cramér at the first stop of his emigration, joined the mainstream of stochastics, which was strongly determined by A. N. Kolmogorov's 1933 book. Feller did this together with Americans (e.g., J. L. Doob) and Austrians (e.g., E. Lukacs). Philosophers of mathematics such as Reichenbach and Carnap, who had been allies prior to emigration, became even closer in the United States. In Iowa, the Vienna philosopher of mathematics, Gustav Bergmann, started cooperating with the psycholo-

---

[24]Hunter (1996). The immigration of R. von Mises and H. Geiringer to the United States in 1939–40, who were also closely related to the Austrian scientific culture, has to be considered as a special case here, because von Mises maintained his own and peculiar concept for the foundation of probability and statistics and became rather isolated in this respect, particularly during emigration.

[25]Bayart and Crépel (1994).

[26]Much has recently been published on the Vienna Circle of Logical Positivism and the somewhat related Reichenbach group in Berlin, particularly by F. Stadler. See also Danneberg et al., eds. (1994).

gist Kurt Lewin (1890–1947), who prior to emigration had been a member of Reichenbach's Berlin Society for Scientific Philosophy.[27]

For mathematics in Germany after World War II the reinclusion of stochastical, topological, and functional-analytical traditions, which had been in danger of being hindered in their development in the 1920s and which were even less cultivated after 1933, was of eminent importance.[28] The reinclusion of disciplines did not always work, as is clear in the case of the analytical philosophy of science in the tradition of the Vienna Circle, reincorporated in Austria too late to make up for the losses due to emigration.[29]

On an individual level, emigration often brought about a change in the scientific orientation of the persons involved. More often than not emigration triggered decisions essentially made a while back. For Hans Hamburger, who had been expelled from Köln, his earlier disappointment with the lukewarm reception of his work in differential geometry was probably a trigger for a total change of his field of research during emigration (henceforth operator theory).[30] With Richard Brauer emigration led to the end of his work on the theory of algebras, for which he had received international acclaim due to the joint proof with Helmut Hasse and Emmy Noether of the "Hauptsatz" (1932). Brauer's biographer Feit wrote that "the theory of simple algebras became dormant for over a generation."[31] For Brauer global disciplinary developments went hand in hand with a change in circumstances—lack of personal success being less of an issue here.[32]

Seen from the perspective of nonmathematicians the impact of mathematical emigration may have looked quite different. It seems to be evident that contemporary American university and college administrators did not at that time, as a rule, consider mathematics as the most important discipline to be supported. This is, for example, documented in the files of the Emergency Committee in Aid of Displaced Foreign Scholars,[33] which had commissioned a Gallup Survey for the years 1939–41. Question number 3 in this questionnaire for all American universities and colleges asks: "Are there disciplines in which it is felt that a unique contribution can be made by refugee scholars, not to be matched by Americans?" A

[27]Lewin's application of topology on Gestalt psychology, however, was observed rather skeptically by mathematicians. S. MacLane wrote on December 5, 1950 to G. A. Austin on Lewin's attempts: "This is . . . a colossal failure of Mr. Lewin, whose so-called topological psychology was very clearly not topology" (M. H. Stone Papers, Archives AMS 38.17).

[28]See Fischer et al., eds. (1990).

[29]Stadler (1988), p. 121.

[30]This according to an assumption in Hoheisel (1966).

[31]Feit (1979), p. 6.

[32]This is not to deny the personal disappointments Brauer had to experience due to the failure of his project of a textbook on algebra. See the case study below.

[33]EC, Box 153, f. Gallup Survey, 1939–41.

total of 376 American universities and colleges replied. Subjects classed as being of "unique" importance were first of all "language and literature"—mentioned sixty-one times. Mathematics came further down the list, being classed only twice as "unique," and coming in nineteenth place out of a total of twenty-eight fields of study. Not surprisingly, mathematics was not given priority in the minds of college administrators. Thus the eventual establishment of mathematics was accomplished—to a large extent—by the efforts of insiders. The need for mathematics had to be "constructed" by the interest groups involved, although the exigencies of war did finally prove the need for mathematics and physics, especially in the training of military personnel. Obviously, the high degree of internationalization in mathematics, while making the process of acculturation and transplanting easier in several respects,[34] did not automatically guarantee a strong demand for the field in the host countries. Local and social conditions could and did work against it, and its lack of "uniqueness" (as understood by some American authorities) carried a danger of competition with young Americans.

Taking into account these methodological problems of gains and losses and of the chosen historical perspective, several specific impacts on American mathematics need to be discussed, on institutional, organizational, and cognitive lines.

10.2. The Institutional and Organizational Impact

The German-speaking emigration was crucial to the creation of new mathematical centers in the USA.[35] Developments on the West Coast (California),[36] the expansion of research institutes such as the IAS, and specialized institutes for applied mathematics stand out in this respect. Also institutions hitherto less known in mathematics such as the University of Notre Dame in Indiana, near Chicago, or New York University (Graduate School for Mathematics under R. Courant) were "put on the mathematical map"[37] by German-speaking immigrants. In Courant's graduate

---

[34]D. P. Kent (1953), p. 242, says: "Those occupations having a body of knowledge internationally known and applicable, like medicine, engineering, or mathematics, or those arts having a medium of expression universally accepted, like music or painting, fare best in the transplanting."

[35]Similar evaluations are probably true for other host countries too, such as the UK. Poorten (1981) reports on the role of K. Mahler in the rise of mathematics at Manchester.

[36]With respect to Berkeley, however, other factors like the appointment of the American G. C. Evans (1887–1973) and the impact of the Polish statistician, J. Neyman (1894–1981), should not be underestimated.

[37]This formulation is by Menger (1994), p. 216, on his efforts in Notre Dame from 1937. On the modest beginnings of Courant's Institute see the quotation by W. Wasow below in (D).

school and in the mathematical department at Brown University in Providence, Rhode Island, under R. G. D. Richardson, immigrants gained preeminent influence due to their expertise, especially in applied fields of mathematics during the war.[38] As mentioned in chapter 2 the impact of foreigners was mainly (with the exception of the IAS, which had just been founded in 1932 before the "forced immigration" started) restricted to new mathematical centers and was of lesser importance in centers already in existence. In this respect there was a certain parallelism with developments in physics as outlined by K. Hoch. The synthesis between theoretical and experimental physics then happening in the United States was largely caused by European theoretical physicists being given appointments in the hinterland, at smaller colleges.[39]

Further areas where immigration made an impact were mathematical publishing and the structure of the school and university systems. In both areas, however, the impact seems to have been scattered and based on individual initiatives rather than being systematic and thoroughgoing. The publication of influential textbooks by immigrants (including some first published in German during the 1930s) such as the ones by Courant and Friedrichs (1948), Courant and Hilbert (1953/62), and Feller (1950/66) bear mentioning.

There existed, however, a degree of conflict between immigrants and Americans concerning publication routines and in their opinions on educational systems. Some immigrants aimed at a fundamental change in the American system of secondary and university education, but their proposals met with relatively little success.[40] Some of these proposals were a response to criticism from Americans, alleging that immigrants were just interested in research and not willing to take a broader responsibility for their new country. For the immigrants, however, it was always a psychological tightrope to what extent they should show an interest in and a willingness to take responsibility, at the same time not appearing overly intrusive. In any case there was a readiness among the more liberal and internationally minded Americans to learn from the immigrants' experiences, as is shown in an effort made in 1935 by a member of the Emergency Committee, E.

[38]Similarly, at the University in Exile of the New School for Social Research in New York City, foreigners dominated the whole structure due to the European traditions in social research. See Krohn (1987), p. 11. At this school also German-speaking mathematical immigrants such as F. Alt and E. J. Gumbel, and French mathematicians such as J. Hadamard and A. Weil, were temporarily employed.

[39]Hoch (1983), p. 237. See also Schweber (1986). The difference in mathematics, compared to physics, was, however, that the new centers emerged close to established institutions such as Brown, or at least in central places such as New York City.

[40]One of these initiatives, undertaken by Courant, Weyl, and the American and earlier Dutch immigrant A. Dresden, is documented below in this chapter.

R. Murrow, a future American leading journalist, to publish a collection of foreign opinions and to ask immigrants such as Emmy Noether for contributions.[41]

10.3. The Impact of German-Speaking Immigration in Applied Mathematics

Before going into detail about the immigration of applied mathematics to the United States, historical background about German and American traditions in applied mathematics is necessary.[42]

The establishment of applied mathematics as systematic research and training in academic surroundings came rather late to the United States—late compared to Europe—dating from around the late 1930s, with the imminence of World War II acting as a decisive stimulus. After the war the United States took the undisputed lead in the field.[43] The development of applied mathematics in the United States in the period mentioned came as no surprise, but that it happened so late, did. This, in fact, has often been described by contemporaries and historians as a "paradox": they found it astonishing and in need of explanation that the American culture, so well known for its technical achievements and its pragmatic philosophical spirit, excelled in fields of pure mathematics such as topology and mathematical logic early in the twentieth century, but lagged behind in the application of mathematics.

The dominant role model within the establishment of research universities in the United States at the end of the nineteenth century was the ideal of German science and its university system. German mathematics was the mentor of American pure academic mathematics.[44] German applied mathematics—a rather new field dating from the German reform movement in mathematics and engineering around 1900, led by Felix Klein in Göttingen—would, thirty years later, once again, become the single most important foreign stimulus for the establishment of applied mathematical research in academic America. Unlike the earlier assimilation of the role model of German (pure) mathematics, which functioned mainly through a direct study by Americans in Germany, the influence of German applied

---

[41]Apparently, this initiative came to nothing. Emmy Noether declined to participate due to alleged lack of experience in the United States, proposing Richard Courant instead. See EC, NYC, box 84, f. Emmy Noether, and Appendix 4.1.

[42]The following remarks are basically a short version of my argument, in Siegmund-Schultze (2003a), which in itself is an abstract of an extensive (408 pages) unpublished study (1998) by me that was funded by the Deutsche Forschungsgemeinschaft (Bonn).

[43]Lax (1989).

[44]Parshall and Rowe (1994).

mathematics was less direct and less monopolistic or dominant.[45] There was, however, a threefold German influence exerted by way of *ideas*, *persons*, and *ideals*: *ideas*, such as David Hilbert's direct methods in the calculus of variations, as revealed in Walter Ritz's (1878–1909) variational method in differential equations of 1908, or Ludwig Prandtl's boundary layer paradigm in fluid mechanics; *persons* (emigrants such as Theodor von Kármán and Richard Courant); and *ideals*, such as the principle of "combined research and training" (Humboldt) also in applied mathematics, as Courant would propagandize it after his arrival in the United States in 1934.[46] These three lines of influence were intimately connected to one another; for instance, Hilbert's spirit in analysis—or what his students made of it[47]—manifested itself in definite mathematical ideas, conveyed by persons such as Courant and were, at the same time, an expression of the social ideal of mathematics in Göttingen around 1900 and in its later development.[48]

By the time Courant, in Göttingen in the 1920s, declared the Kleinian reforms as already completed and warned against further institutionalization of separate institutes of applied mathematics at the universities,[49] work in these fields had not even begun in the United States. No specialized research journal or a society comparable to the German Society for Applied Mathematics and Mechanics (GAMM) was to be found. Engineers were more or less shunned by mathematicians,[50] with some notable exceptions such as MIT, where electrical engineers worked alongside mathematicians such as Norbert Wiener. Another exception was the Guggenheim-funded GALCIT in Pasadena (California), where, from 1929 on, the "early immigrant" from Germany/Hungary, von Kármán, developed an academic school of engineering mathematics, with the emphasis on aerodynamics (Hanle 1982). In comparison, the well-equipped state laboratories of the National Advisory Committee on Aeronautics (NACA) were less capable adjusting to or had problems adapting to theoretical research. This was probably due

---

[45]In chapter 3 the influence in applied mathematics of the Russians Sokolnikov and Timoshenko, the Swede Grönwall, and the Irishman Synge is mentioned as well.

[46]Reid (1976).

[47]It is well known that Hilbert did not contribute a single line to the book known as "Courant/Hilbert": *Methoden der mathematischen Physik* (1924/37), which was finally translated into English and thoroughly modernized after the war (1953/62).

[48]See Pyenson (1979), Rowe (1989), and Tobies (1991).

[49]There was a dispute in 1927, between Richard Courant in Göttingen and Richard von Mises in Berlin, on the pages of *Die Naturwissenschaften* where von Mises stressed the continued necessity of a separate field "applied mathematics." See Mises (1927).

[50]Mathematical physicist Warren Weaver, who later in World War II would head the Applied Mathematics Panel, was surprised, in 1930, "at the emphasis given, in the discussion [on a journal for applied mathematics; R. S.], to the field between Mathematics and Engineering" (Butler [1997], p. 79).

to their lack of connection to university-based formal education and training programs. For example, Max Munk (1890–1986), who immigrated in 1920 and was a former student of Ludwig Prandtl's in Göttingen, was at NACA but lacked students.[51]

The American publishing industry still lagged behind the German, Springer-dominated industry even in the 1930s, at least in regard to monographs (Siegmund-Schultze 1997) and in respect to applied mathematics at large. This can clearly be seen from the history of the six-volume monograph *Aerodynamic Theory: A General Review of Progress under a Grant of the Guggenheim Fund for the Promotion of Aeronautics*, edited by W. F. Durand between 1934 and 1936. Durand needed European authors, among them Germans, to contribute to his volumes. He failed to find an American publisher and had to resort to Springer; the latter jumped on this occasion to expand overseas.

Instrumental in spreading Hilbert's spirit in mathematical physics[52] among Americans were the Courant/Hilbert book *Methoden der mathematischen Physik*, volume 1 (1924) and the seminal paper "On the Partial Difference Equations of Mathematical Physics" (henceforth CFL) written by Courant and his students Kurt Friedrichs and Hans Lewy in 1928.[53] CFL was reprinted in the *IBM Journal of Research and Development* as late as 1967 on the grounds that it was "one of the most prophetically stimulating developments in numerical analysis . . . before the appearance of electronic digital computers. . . . The ideas exposed still prevail."[54] CFL gained special importance in the history of numerical analysis because it includes the rudiments of numerical stability. Seen from the point of view of modern (abstract) analysis the methods used in Courant/Hilbert and in CFL were rather traditional and simple. For instance, the integrals occurring were seldom in the modern sense of Henri Lebesgue.[55] The so-called Finite Element Method in boundary value and eigenvalue problems, which Courant also brought to American soil (even though appreciated rather late)[56] dates back to Hilbert's use of minimizing sequences in rescuing the

---

[51]Eckert (2006).

[52]This spirit is not identical with Hilbert's general approach toward the foundations of physics, based on mathematical axiomatics (Corry 2004). This approach was, however, equally palpable especially among differential geometers in Princeton (Butler 1992/97).

[53]The influence of CFL on applied mathematics in the United States during the war and after has been discussed recently by Amy Dahan (1996). See also Goldstine (1972).

[54]Courant, Friedrichs, and Lewy (1928/67), foreword to English translation, p. 213.

[55]Weyl (1938), p. 602.

[56]According to Zienkiewicz (2000), p. 10, Courant's work of 1943 on the finite element method "had to lie in obscurity" for many years because Courant was no engineer and did not realize applications in that field.

Dirichlet principle of the calculus of variations around 1900.[57] So, Courant/Hilbert, CFL, and (Courant 1943) brought in classical analysis and Hilbert's tradition and built bridges between mathematicians, physicists, and engineers.

Even before von Kármán and Courant came to the United States in 1929 and 1934 respectively, German results in Courant and Hilbert and CFL (the "Hilbert spirit") as well as in aerodynamics were recognized on the other side of the Atlantic. Since many Americans (O. Kellogg, M. Mason, R. G. D. Richardson) had studied in Göttingen in the first decade of the twentieth century, Hilbert's spirit—this time also in applied mathematics—did not fall on unplowed ground. However, the Americans who had been in Göttingen at that time were (with the possible exception of Kellogg) not among those who would have the most bearing on research.[58] This may account for why it took great personal effort on the part of Courant and his students to rekindle interest in the spirit of Hilbert in America as far as applied mathematics was concerned.

This leads the discussion finally to immigration. In addition to Americans starting to get a grasp of European ideas and to implement them on American soil, both human resources (immigrants) and a change in the infrastructure of research were required to facilitate the transplantation of European, German, traditions in applied mathematics.[59] There is no doubt about the decisive role played by mathematicians and engineers from Europe and particularly Germany, such as Courant, von Neumann, von Kármán, and von Mises, in helping to introduce a new style into American applied mathematics. There was a great need for original and creative thinking on an individual basis in applied mathematics just as in pure research. This is what many Americans with applied interests (Bateman, Wilson) lacked and what European individualistic immigrants had in excess.[60]

In many cases the immigration and employment of German-speaking mathematicians such as A. Basch, G. Bergmann, S. Bergmann, F. Bernstein, W. Feller, K. Friedrichs, H. Geiringer, M. Herzberger, G. Kürti, A. Korn, H. Lewy, W. Prager, and A. Weinstein was expressly justified and facilitated by a shortage of qualified Americans in applications (D). For some younger mathematicians the pressure toward applications on the eve of

---

[57]See Williamson (1984), Zienkiewicz (2000), and Courant (1943).

[58]But the Americans were very often influential in the organization of science and mathematics, for example the president of the Rockefeller Foundation (Mason) and the secretary of the American Mathematical Society (Richardson).

[59]Reingold (1981).

[60]Henry Bateman (1882–1946) and Edwin Bidwell Wilson (1879–1964) were, in fact, important bridge builders to the European traditions, but they were more of the receptive, collecting type and in their mathematical tastes were rather conservative.

war brought about a reorientation of their mathematical work.[61] Among these younger immigrants were M. Golomb, F. Herzog, F. John, R. Lüneburg, F. Theilheimer, W. Wasow, and others (D). Established applied mathematicians, such F. Bernstein, who had a difficult time during emigration, sometimes perceived such changes in the subject areas of the younger mathematicians as pure opportunism, or as a personal threat to their own careers (D). A similar demand for applied mathematics in several other host countries influenced the immigration of Prager, Geiringer, and von Mises to Turkey, the success of W. Romberg in Norway, and the reorientation toward hemodynamics (dynamics of blood circulation) by H. Müntz.[62] Accordingly, the stopovers of immigrants in other countries, before they arrived in the United States, also have to be considered. This is clear from the example of W. Feller (D), who was influenced by the applied stochastics of Harald Cramér in Stockholm. Occasionally, previous interests in applications shown by immigrants were given new life in the United States. For instance, F. Alt had been interested in mathematical economics prior to his arrival in the United States from Austria (D). Immigrants such as Alt, Feller, Helmer, Wald, and Tintner, contributed, with refugees from other European countries, to the mathematization of subjects such as statistics, economics, optimization, and computing, which partly in the United States, but above all in Germany (somewhat in contrast to Austria) had a slow development prior to emigration.[63] In his Graduate School for Mathematics at New York University, Courant, from the beginning, around about 1935, stressed the unity of pure and applied mathematics, as well as the combination of research and teaching. This was in rather direct contrast to the Institute for Advanced Study in Princeton,[64] which focused

[61]The same applied also to the careers of young mathematicians in other countries, among them the mathematicians W. Magnus and R. Moufang, who remained in Germany. For Moufang see Chandler and Magnus (1982), p. 123. Discrimination against women-mathematicians in pure mathematical domains was involved here as well.

[62]See chapter 6. Müntz is not considered a German-speaking refugee from Nazi Germany here, as he was expelled by Soviet Russia.

[63]Celebrating the impact by G. Tintner, Fox reports on C. F. Roos's "frustrations as a mathematical economist in the United States in the late 1920s" (Fox et al. 1969, p. 16). Franz Alt, however, in a letter to Patti W. Hunter, dated February 16, 1996, stressed the "well developed" field of economics in the United States, especially in descriptive statistics, when he arrived in 1938: "We found a great body of economic statistics collected and published by the Federal Government the likes of which did not exist in Europe."

[64]Several applied mathematicians such as Gumbel and Bernstein were turned down in their efforts to get stipends from the IAS. Its first director, Abraham Flexner, who was on good terms with Courant, wrote to the latter on September 12, 1933: "I have . . . been making every effort to find a permanent post for you in America. The difficulty at the Institute is that, in Veblen's judgment, your field lies outside that which the School of Mathematics is undertaking to cover" (Courant Papers, Bobst Library, New York City, box 4).

on traditional pure research areas—only to find itself, ironically, caught up during the war in the new developments in computing through its leading member John von Neumann (see below). However, the real upswing for Courant's school came during the war. Courant obtained his American citizenship just in time to be able to play an important role in the war effort. He induced immigrants such as K. Friedrichs and F. John to direct their careers more toward applied mathematics.[65] In his efforts at Brown University to build a center for applied mathematics and mechanics, the former secretary of the AMS, Richardson, relied strongly on the talent and experience of immigrants.[66] In this respect Richardson cooperated closely with Stephen Duggan from the Emergency Committee in New York City. He wrote to Duggan as late as October 1941: "America has woefully neglected its applied mathematics and we must get to the point where we can hold our own with Germany."[67] Duggan called a Richardson memorandum, discussing the consequences of this neglect, an "eye opener."[68] Although the systematic utilization of immigrants within war research was limited due to the problem of citizenship, Richardson succeeded in involving mathematicians such as K. Löwner and W. Prager in the more theoretical parts of the work.

It is difficult to evaluate to what extent and how quickly the influence of German ideas and people together with the overall requirements of industry, military and other fields of application led to a change in American institutions and ideals of research in applied mathematics. Thornton Fry's report (1941) and Roland G. D. Richardson's article (1943) seem to indicate that—except for a few institutions[69]—there was still no systematic endeavor made in the United States in the first years of the war to train applied mathematicians within an academic environment, at least not at institutions outside Brown and New York University.

World War II both spurred European immigrants on to adapting to American ideals and made Americans reconsider their relation to applications and to intensify collaboration with government bureaucracies. In the beginning, American pure mathematicians such as M. Morse had

---

[65]On the effect of Courant's citizenship see the quotation by Warren Weaver in chapter 9. For Friedrichs see Reid (1983).

[66]As to the involvement of H. Lewy and W. Prager see the following remark by Richardson from 1942, quoted in Rider (1984), pp. 167–68: "In large part because the field we are cultivating here at Brown has been neglected in America, the faculty is almost exclusively composed of Europe-born mathematicians." For the history of the center for applied mathematics at Brown see Prager (1972).

[67]Richardson to Duggan, October 22, 1941. Richardson Papers, Correspondence, BUA, f. 87 D (1940–42).

[68]Ibid. Duggan to Richardson, October 28, 1941. The memo is not in the files.

[69]Courant's department at New York University, Brown University's summer schools in applied mechanics, in addition the more engineering-related institutes like GALCIT.

considerable problems dealing with the new demands.[70] Europeans, in contrast, did have experience with state bureaucracies, and thus Courant joined Weaver's Applied Mathematics Panel during the war.[71] Both parties, Europeans and Americans, had much to learn from each other in this respect, and not everything went smoothly.[72]

Finally, toward the end of the war, the immigrant John von Neumann, the quintessential creative mathematician, "la figure symbolique"[73] for the rise of applied mathematics in the United States, did much to facilitate the acceptance of the German traditions, especially of Courant's and Ritz's methods, in the United States. He also contributed fundamentally to the development of digital computing. In the late 1940s, this continued under the special conditions of the hydrogen bomb project at Los Alamos when calculating devices were much more developed than before the war.[74] At that time, the massive amounts of money pumped into mathematics, even into pure research, by the Department of Defense (Office of Naval Research) and the conditions of the Cold War helped American mathematicians to adapt—at least temporarily—to the "German ideal" of state-financed science.[75]

10.4. The Inner-Mathematical Impact of German-Speaking Immigration on the United States

On the level of "pure mathematics," the influence of "German abstract algebra" is a frequently cited phenomenon of mathematical immigration requiring historical reflection. In retrospect, it was not the "abstractness" of the new methods in algebra and number theory brought by immigrants such as Emmy Noether, Emil Artin, Hans Rademacher, and Carl Siegel that was an innovation. The American tradition in algebra had been very "abstract" prior to immigration.[76] Rather it was, as in applied mathemat-

---

[70]Owens (1989).

[71]Other immigrants took official positions as well. Von Kármán was sent to Europe after the war as an officer in order to survey applied research in Germany. Von Neumann's high rank in the Atomic Energy Commission is well known.

[72]For instance, the new and first journal for applied mathematics, the *Quarterly of Applied Mathematics* (since 1943, Brown University) was clearly modeled after the German *ZAMM*, the more so, since the majority of the faculty at Brown was foreign born. But "individualistic" Europeans such as Richard von Mises had problems to accept the huge editorial board of the *Quarterly* and therefore resigned from that project, as was documented above.

[73]Dahan (1996), p. 172.

[74]Goldstine (1972).

[75]Dahan (1996), p. 181. The fact that the cleavage between pure research and applications was widening again in the USA and many other countries of the West in the years after the war is another story not to be told here.

[76]Artin (1950), p. 65, on Wedderburn. See below under studies (S).

ics, the new conceptual approach, and the variety of themes the immigrants brought with them, the bridge building to topics of classical nineteenth-century European mathematics that had received little notice in the United States and that made immigration effective for and in American mathematics. As late as 1940 Weyl talked about a "dormant interest in number theory" in the United States; he wanted to revive the theory that had long traditions in Europe.[77] Rademacher's and Siegel's immigration proved crucial to the development of number theory in the United States.[78]

In other areas of mathematics, the immigrants H. Weyl and Antoni Zygmund (1900–1992), the latter from Poland, were particularly successful as "bridge builders" to the European traditions. Even today (2008) voices being raised in the United States deem European training in classical analysis and mechanics as superior and encourage the appointment of foreigners with the appropriate education.[79]

10.5. The Impact of the "Noether School" and of German Algebra in General

The origin of the so-called abstract algebra with its revolutionary "structural" approach to mathematics is strongly connected to the work of Emmy Noether in Göttingen. This work had considerable influence internationally during the 1930s, and it had a deep impact on algebraic research in the United States and other countries, such as France.[80] In their public and private commentaries of the 1930s, Americans such as Bell, Lefschetz, Garrett Birkhoff (D), and others almost equated "German algebra" with "Noether's algebra" and saw in it the leading paradigm of

[77]H. Weyl to P. Bernays, March 25, 1940, ETH, Zurich, Weyl Papers 91:23. I thank Liliane Beaulieu (Nancy) for drawing my attention to this letter.

[78]O. Veblen to Warren Weaver, November 19, 1934: "Professor Siegel's line of work is so important and has as yet had so little attention in this country that I feel sure the visit will turn out to be really important." RAC, R.F. 1.1. 200 D, box 143, f. 1769. Rademacher founded an influential number-theoretic school in the United States that, in an article devoted to one of his most important students, A. Whiteman, has recently been called the "Rademacher Tree." See Golomb et al. (1997) and the facsimile from it reproduced here.

[79]W. L. Duren (1989), p. 436. A very critical report on the situation in American mathematics around 1990 says: "A critical shortage of qualified mathematical sciences researchers still looms, held at bay for the moment by a large influx of foreign researchers, an uncertain solution in the longer term" (Renewing U.S. Mathematics 1990, p. 543).

[80]For a confirmation of this assertion consult various articles in the three-volume collection, edited by P. Duren (1988/89), that stress the influence of B. L. van der Waerden's book (1930–31), which is based on Noether's and E. Artin's lectures. See also Bell (1938). For a general discussion of the international success and influence of the new kind of structural thinking in algebra see Corry (1996).

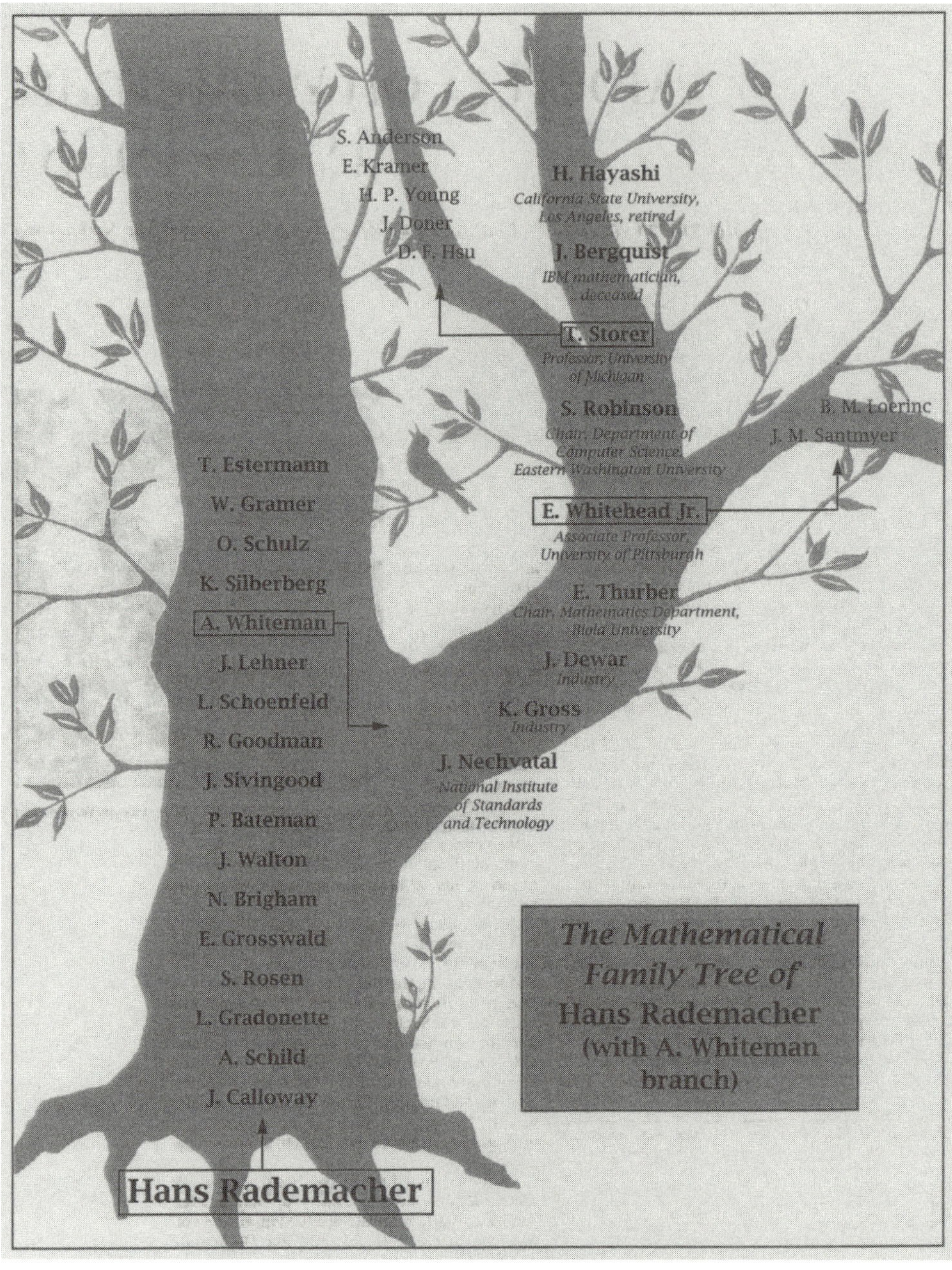

The Mathematical
Family Tree of
Hans Rademacher
(with A. Whiteman
branch)

S. Anderson
E. Kramer
H. P. Young
J. Doner
D. F. Hsu

H. Hayashi
California State University,
Los Angeles, retired

J. Bergquist
IBM mathematician,
deceased

T. Storer
Professor, University
of Michigan

S. Robinson
Chair, Department of
Computer Science,
Eastern Washington University

B. M. Loerinc
J. M. Santmyer

E. Whitehead Jr.
Associate Professor,
University of Pittsburgh

E. Thurber
Chair, Mathematics Department,
Biola University

J. Dewar
Industry

K. Gross
Industry

J. Nechvatal
National Institute
of Standards
and Technology

T. Estermann
W. Gramer
O. Schulz
K. Silberberg
A. Whiteman
J. Lehner
L. Schoenfeld
R. Goodman
J. Sivingood
P. Bateman
J. Walton
N. Brigham
E. Grosswald
S. Rosen
L. Gradonette
A. Schild
J. Calloway

Hans Rademacher

Figure 48 *Rademacher Tree. The family tree of a famous number-theoretic school, founded by Hans Rademacher in American emigration.*

worldwide algebraic research. The questions to ask here are, in what way and to what extent was the influence of Noether's algebra connected to the forced mathematical emigration during those years? In the literature, questions of a similar nature have been raised and partly discussed by Chandler/Magnus (1982) and Hoehnke (1986).[81]

It is not as though the international influence of Noether's algebra started only after her emigration. If one looks at the situation before 1933, one should have expected the international propagation and reception of Noether's algebra to be impeded by the inferior social position Noether occupied in Göttingen—this mainly due to sexist prejudices in the ministry and among nonmathematicians at that time.[82] Despite these impediments there was, already in the late 1920s, a growing recognition of Noether's algebra. This was based on the high degree of internationalization in Göttingen's research practice, on the respect and benevolence of Hilbert and Weyl for Noether, and on the close connections of Richard Courant to the Springer publishing house, responsible for publishing van der Waerden's book *Moderne Algebra* (1930–31). In the late 1920s, P. S. Aleksandrov's and H. Hopf's theory of homology in topology had been markedly influenced by Noether during visits by the two topologists to Göttingen. In a similar way, Noether had an influence on Hasse's and Richard Brauer's work in the theory of algebras. This led to the famous three-person work Brauer, Hasse, and Noether (1932), where the "main theorem" of the arithmetical theory of algebras is proved. This was another triumph of German algebraic research in competition with American attempts at the proof, such as by Abraham Adrian Albert (1905–1972), the student of Leonard E. Dickson (1874–1954).[83]

Now one could imagine at least *four reasons* why the destruction of Göttingen as a mathematical world center in 1933 and the ensuing

---

[81]The German article by Hoehnke, which—unlike my reflections—argues more on a technical mathematical level, deserves to be better known. The author discusses in a chapter "The 'Critical Mass' after Enforced Emigration" (p. 58) briefly "alternatives to the widespread assumption of the Steinitz-Noether tradition as the real origin of modern algebra" and refers in this respect to Birkhoff (1976a), which will be used below as well.

[82]Tollmien (1991). Noether was not a full professor and not elected to the Göttingen Academy of Sciences. The American George David Birkhoff did not even mention Noether in a detailed report on his trip to Göttingen, which he delivered in September 1926 to the International Education Board (IEB). Noether's application to the IEB to give her student Heinrich Grell a fellowship failed, and in her efforts to support Heinz Hopf for the same purpose Noether relied on the prestige of full professors, such as Erhard Schmidt in Berlin. See Siegmund-Schultze (2001).

[83]The German-American race for the proof at that time is described in Fenster (1997), p. 17. In a recent article, Roquette gives Albert equal merit for the proof, and relates the absence of Albert's name from the title of the paper on the main theorem to communication problems in the weeks leading up to publication (Roquette 2004, p. 75).

emigration should have led to difficulties for the future international reception of Noether's algebra.

First, it is a fact that most of Emmy Noether's students remained in Germany, and German mathematics became increasingly isolated in the years to come. Indeed, it makes no sense to talk about an "emigration of the Noether school" from Germany after 1933. Emmy Noether herself went to the United States, but her closest students Max Deuring (1907–1984), Heinrich Grell, B. L. van der Waerden, and Ernst Witt (1911–1991) remained in Germany. Algebraists such as H. Hasse, Wolfgang Krull (1899–1971), and Emil Artin, who were strongly impressed by Noether's thinking or shared similar ideas either did not emigrate or did so (Artin in 1937) at a much later date.

Second, Emmy Noether only worked for eighteen months in the United States before her untimely death at the age of fifty-three in April 1935, and she had no opportunity of creating a new circle of students comparable to that in Göttingen. At the women's college Bryn Mawr near Philadelphia, Noether had limited her teaching to more elementary subjects. Even at occasional seminars in Princeton, before research fellows of the IAS and the university, Noether had to take care not to alienate her listeners:

> I'm beginning to realize that I must be careful; after all they are essentially used to explicit computation and I have already driven a few of them away.[84]

Third, as documented in the book by Chandler and Magnus (1982), one has to consider that emigration brought competing points of view in algebraic research to foreign countries. As a matter of fact, many more German algebraists of the school around Issai Schur in Berlin than those from the Noether school in Göttingen fled the regime. The Schur school was in a certain sense methodically more "concrete" than the Noether school and focused more on the particular problems of group representations.[85] With opportune timing, leading British mathematicians such as Philip Hall (1904–1982) realized around 1933 certain drawbacks in British algebra, particularly in group theory, with the result that eleven German-speaking

---

[84]The translation follows Dick (1981), p. 82. The German original is on p. 204 in the recent edition (Lemmermeyer and Roquette 2006). This quotation should not be taken to mean that the well-educated Americans did not "understand" Noether in a simple sense. One has rather to take into account the revolutionary nature of Noether's conceptual approach, which was not intuitive in a traditional sense and was largely devoid of calculations.

[85]Of course, also the algebraists of Berlin, who organized seminars on Noether's algebra were affected by the modern structural approach. Ledermann (1983), p. 103. Schur's student Richard Brauer—with other emigrants such as Reinhold Baer—thus brought Schur's and Noether's points of view into American and Canadian mathematics. On Brauer's relation to Noether's algebra see the first case study (S) below.

Figure 49 *Walter Ledermann (born 1911). The versatile mathematician, who left Berlin after the state exam with the help of the Weltstudentenwerk (later International Student Service), in 1983 wrote the authoritative article on the school of Issai Schur in Berlin. Until his retirement he was a professor at the University of Sussex (England) and is now (2008) living in London.*

algebraists obtained at least temporary positions in Great Britain, among whom the majority stemmed from the Schur school in Berlin.[86]

Fourth, one could also wonder if the overall trend in the sciences and mathematics to English as the lingua franca raised problems, at least as far as the main textbook for Noether's approach, van der Waerden's *Moderne Algebra* was concerned.

In the following I will argue that—contrary to these expectations—the growing international reception of Emmy Noether's peculiar approach to modern algebra in the years after 1933 was positively correlated with the concrete circumstances of emigration in mathematics and the breakdown in and reconstruction of the international communication network in algebra at that time. In the opinion of this author one can even argue that the expulsions and their concomitant psychological effects contributed to the creation of a "myth" of a homogeneous German algebra—considered to be fundamentally connected to Noether's approach—a myth that exaggerated the importance of Noether's approach in the overall development of algebra.

---

[86]Rider (1984), p. 167. As a result the group-theoretic tradition of Schur's school gained particular influence in the UK. See Ledermann (1983).

Her most famous student, van der Waerden, wrote the following about the uniqueness of her approach in his obituary in the *Mathematische Annalen* of 1935, the publication alone constituting a courageous act in the years of Nazi Germany. Interestingly enough, van der Waerden counted himself—probably with tongue in cheek—among the "ordinary mathematicians" who had still to learn and fully understand Noether's new approach:

> Her thinking deviates indeed in several respects from the majority of other mathematicians. We all like to build on figures and formulas. To her these auxiliary means are worthless, disturbing rather. She was exclusively interested in concepts, not in intuition or calculation.[87]

In fact, Noether's approach differed markedly from the approaches of most contemporary algebraists even in Germany (H. Hasse, I. Schur, E. Artin, R. Brauer). It was precisely the uniqueness and radical nature of her approach that made her name most representative of the new structural approach wthin abstract or modern algebra.

The international communication structure in mathematics changed considerably in 1933, by which time Emmy Noether was at Bryn Mawr, from where she occasionally visited Princeton. In this context a direct, personal impact of her algebraic thinking is palpable, especially her impact on the White Russian–Italian immigrant to the United States, Oscar Zariski (1899–1986). Zariski's new rigorous approach to algebraic geometry was triggered and fundamentally shaped by his encounters and discussions with Noether in Princeton.[88] Also at that time the American Adrian Albert was influenced by Noether, as reported by Roquette:

> Noether's ideas had not yet penetrated everywhere. Albert himself had his training with Dickson, and his papers in those first years of his mathematical activity were definitely "Dickson style." It was only gradually that Albert started to use in his papers the "Modern Algebra" concepts in the sense of Emmy Noether and van der Waerden. In 1937 he published the book "Modern higher algebra" which was a student textbook in the "modern" (at that time) way of mathematical thinking.[89]

Another American mathematician influenced by Noether at that time was apparently Nathan Jacobson (1910–1999).[90] Noether's point of view was

---

[87]Van der Waerden (1935), p. 474 (T).

[88]Parikh (1991), pp. 67–76. I thank David Rowe (Mainz), Norbert Schappacher, and Silke Slembek (both Strasbourg) for informing me about the impact of Noether on Zariski.

[89]Roquette (2004), p. 79. See also the recent German edition of the correspondence between Helmut Hasse and Emmy Noether, which contains Noether's reports on her seminars in Princeton: F. Lemmermeyer and P. Roquette, eds. (2006).

[90]In a personal communication Peter Roquette mentions Jacobson's work on simple algebras of 1945 as probably influenced by discussions with Noether in Princeton.

partly stressed also by another non-German immigrant, the Frenchman André Weil,[91] although Weil had certain reservations about the work of Noether's followers in algebraic geometry, B. L. van der Waerden and M. Deuring.[92] Indeed, it has been argued that between 1932 and 1935 several American and French mathematicians witnessed a kind of quasi-religious "conversion" toward abstract algebra in the Noetherian sense.[93]

Americans such as Garrett Birkhoff (1911–1996, lattice theory) and Saunders MacLane (1909–2005, category theory) based their work deliberately on modern abstract algebra. In 1941 their English textbook *Survey of Modern Algebra*, the so-called Birkhoff-MacLane, was published. It was, according to one of the authors, an "Americanized 'modern algebra.'"[94] At least in its axiomatic spirit and its discussion of Galois theory, it bore similarities to the "original," van der Waerden's *Moderne Algebra* book (1930/31).[95] However, the latter publication continued to exert considerable international influence in the decades to come. This even led to a temporary increase in the use of German as a mathematical language—at least during the 1930s in an algebraic context, the more so since the Dutchman van der Waerden's German was so simple and easy to read.[96]

I will now summarize (by partly adding further arguments) my hypothesis of the emergence during the 1930s of a myth of a homogeneous German algebra, basically represented by Noether's approach. I see the main reasons in the following facts:

[91]Weil had two stipends at the IAS between 1935 and 1937, and he came to the United States for good in 1941, after the German occupation of France. See Weil (1992).

[92]Schappacher stresses Weil's "subtle intersection theory which lay beyond van der Waerden and Deuring" (Schappacher [2006], p. 11). He also indicates that it required Weil's special efforts to draw full profit of the classical theory of geometric correspondences that had partly been lost in the algebraic approach to algebraic geometry particularly by Deuring, as supported by H. Hasse. Compared to the latter two, Schappacher (2007) sees, however, more flexibility in the approach by van der Waerden. See also Slembek (2007).

[93]This conversion was described much later by Dieudonné (1970) and by Garrett Birkhoff (1976a), where they also reported that up to 1930–32 they had known almost nothing about algebra in Göttingen and had not heard about it from their academic teachers either. On the beginnings of the French group of young mathematicians, Bourbaki, which later became the quintessential propagandist of the structural approach in various mathematical disciplines, cf. Corry (1996), pp. 293ff.

[94]See Birkhoff (1976b), p. 69.

[95]Birkhoff stresses also the differences, particularly the "primacy of the real and complex fields," in the American textbook (Birkhoff 1976b, p. 69).

[96]The American Paul R. Halmos (1916–2006) wrote in 1988 on van der Waerden's book: "The German is . . . not as difficult as German can be, perhaps because Van der Waerden is Dutch . . . (Did you ever read something by Hermann Weyl?) For many students the book served a double purpose: you learned German from it at the same time that you were learning algebra." (Halmos [1988], p. 145).

Figure 50 *Emmy Noether (1882–1935). With her "structural" approach to algebra she made one of the most fundamental contributions to mathematics in the twentieth century. She was—with Hermann Weyl—a founder of the German Mathematicians' Relief Fund. Through her seminars and talks in Princeton she influenced the new and rigorous approach to algebraic geometry of the Russian-Italian émigré Oscar Zariski.*

Inner-mathematical stimuli toward further abstraction and the rather late and
   sudden "conversion" of non-Germans to abstract algebra,
the unavoidable loss of concrete, nationally, and linguistically colored scientific
   content resulting from radical changes within the communication structures,
Noether's immigration to the United States, and the defamation of Noether's
   theories as "Jewish" by the Nazis,[97]
Noether's intimate relation to the tradition of the German[98] mathematician
   Richard Dedekind (1831–1916), which made her theories appear "Germanic,"
van der Waerden's book, written in clear and understandable German,
the immigration of most students of the competing school of Issai Schur not to
   America but to the British Commonwealth.[99]

It is plausible to assume that the algebraic work of some other immigrants from Germany and Europe, even of those who had not been Noether's students, was often considered from the point of view of how
much it contained of the relatively new (in America) Noetherian spirit.
It seems as though the influence of certain methodically alternative research standpoints, such as the one of the Schur school in Berlin, was
eclipsed—at least in America—by the new fame of the Noether school.[100]
Schur himself had only little contact with the international research
community after his dismissal in Berlin in 1935, and his most important
student, Richard Brauer, was hindered by the circumstances of his emigration in writing his textbook on algebra for the Springer Grundlehren
series.[101]

[97]This defamation Noether shared with many other emigrants. However, the careers of
Noether's students in Germany, who partly stressed the "Aryan" thinking of their teacher,
do not seem to have suffered under these allegations, which were of a political and quite
voluntary nature. See above chapter 4. Heinrich Grell's later persecution by the Nazis was
not primarily caused by his loyalty to Emmy Noether, although he represented his persecution in that light after the war. This does not mean, however, that he was disloyal to her.

[98]One might add in this context that Richard Dedekind (1831–1916) was non-Jewish. He
was probably the most important predecessor of Noether in respect of algebra and he had a
great influence on her work in detail. See below for his influence on Garrett Birkhoff's work
on lattices.

[99]The most important student of Schur's, Richard Brauer, originally went to Canada
(Toronto).

[100]This impression has certainly to be nuanced as to physics and non-American countries,
such as the UK and Canada, where Richard Brauer would develop a successful group-
theoretic school in Toronto after 1935, on the recommendation of Emmy Noether (see below). The theory of group representations gained importance within quantum mechanics,
which was disseminated in the United States by Weyl, Brauer, and the physicist Eugene
Wigner (1902–1995). See Coleman (1997).

[101]See the case study (S) below. Brauer was, according to his biographer, "the one among
all of Schur's students, . . . who continued Schur's grandiose acting as a teacher and researcher in the most perfect way" (Rohrbach 1981, p. 126 [T]).

The overwhelming influence of abstract algebra and of Noether's school of thought led to a backlash in the 1970s, not entirely independent of the waning Bourbaki euphoria. In this context even some Americans, who had spread Noether's spirit in the 1930s, were critical in their remarks.[102] Forty years after her death Emmy Noether was criticized for allegedly having ignored early British and American sources and results, particularly in the field of the theory of algebras (S). This was connected to the previous distorted (as I believe) image of representing Noether's approach to algebra as *the* "German algebra." As will be briefly discussed in a case study below, Garrett Birkhoff, whose theory of "lattices" in the 1930s had been greatly influenced by the Dedekind-Noether way of thinking,[103] made such accusations against the Noetherian spirit in an exchange of letters with van der Waerden in the beginning of the 1970s.[104] In future decades, other American mathematicians (G.- C. Rota) even criticized the alleged overemphasis on Galois theory in algebra textbooks, which they blamed on the tradition of "German algebra" as well (S).

10.6. Differences in Mentality, the History and Foundations of Mathematics

In addition, several "intangibles" in the sense noted above by Chandler and Magnus (1982), as well as certain traditions and attitudes toward research that differed between Europe and America had their effect during immigration. George David Birkhoff, father of Garrett Birkhoff and in his time the leading American mathematician, highlighted one of these differences when he wrote with some critical intent that "we tend to take our mathematics as serious business rather than as a means of exercising our talent for free invention."[105] The emigrant, Hermann Weyl, saw it as one of his major tasks in the United States to maintain or to reintroduce "re-

---

[102]In Garrett Birkhoff's case this is apparently connected to the fact that under the influence of the war he had turned to more applied mathematical fields, such as hydrodynamics.

[103]Dedekind had proposed the notion of "Dualgruppe," which is basically the "lattice" of the 1930s, and which was retranslated into German as "Verband." See Mehrtens (1979).

[104]Birkhoff was then preparing a historical article on the sources of van der Waerden's book *Moderne Algebra*. As a result of this discussion the article never appeared. Instead a historical paper by van der Waerden (1975), which exculpates Noether, was published. Birkhoff, however, published his standpoint in Birkhoff (1976a/b). See the case study (S) on this discussion below.

[105]G. D. Birkhoff (1938), p. 307. As one of possible fields of "free invention" Birkhoff mentioned "special analysis." In it he saw work by N. Wiener on Tauberian theorems; by Hille, Tamarkin, and Widder on Laplace Transformations; and by L. L. Silverman on summation of divergent series as valuable exceptions to the alleged restrictions of American research.

flection" ("Besinnung") in the ever faster growing world of modern mathematics (D). Connected to these feelings of the more conservative mathematicians such as G. D. Birkhoff and Weyl were leanings toward the foundations and the history of mathematics, where traditionally the European background was stronger.[106] Weyl, supported by IAS director A. Flexner, pleaded for the employment of European mathematicians with knowledge of the history of mathematics, such as Dehn, Hellinger (D), Blumenthal, and Neugebauer (D). But he was mostly unsuccessful. Neugebauer only came to the United States in 1939, and then more on the grounds of his rather "modernist and nonhistorical" activity as editor for *Zentralblatt für Mathematik*, and as founder of the latter's new American counterpart, *Mathematical Reviews*. Also, efforts to organize guest lectures on the history of mathematics for Otto Blumenthal came to nothing because of lack of money and interest[107]—Blumenthal could not be saved and he perished in a Nazi camp.

In any case, emigration with its radical break with the past combined with the ascendancy of English also led to a loss of some historical traditions. Certain people in Europe today criticize the unhistorical method of thinking in American mathematics, because it can be counterproductive for the organic development of the theory itself.[108] However, there still remains a great deal of historical work to be done in order to gain a fair evaluation of American mathematics prior to 1933, and to correct the European perspective on it, one that is often biased and one-sided.[109] Specifically, there has been criticism of the opinion "that mathematical logic in America did not, basically, exist, before their practitioners got logic from studying in Göttingen." This opinion is, according to the refugee Bernays, nothing other than "inappropriate ethnocentrism."[110] The impact of emigration on the historiography and philosophy of mathematics and on research on the foundations is still largely unexplored.

---

[106]In an internal comparison between mathematics in Europe and the United States the International Education Board stated in May 1927 a particular weakness of the Americans in history and philosophy of mathematics, and in applied mathematics (Siegmund-Schultze [2001], pp. 52–53).

[107]Efforts by von Kármán and C. Carathéodory failed. Von Kármán to Blumenthal, April 29, 1936, copy in CPP. See also chapters 5 and 9.

[108]The Dane, O. I. Franksen, stressed the importance of historical retrospection (in the particular case of G. Boole) for the development of "array-based logic," which, in his opinion, is lacking in American logic: "The European scientific tradition with its emphasis on the contributions of the predecessors seems to have no place in this development" (Franksen 1997, p. 188).

[109]An important newer work is Parshall and Rowe (1994). See also Siegmund-Schultze (1998a).

[110]Quoted from Dahms (1987), p. 103 (T). See also Scanlan (1991).

10.D. Documents

10.D.1. The Heterogeneity of the "German-Speaking" Emigration, in Particular Differences between German and Austrian Traditions in Mathematics

The Göttingen logician, Paul Bernays, said of the philosophical work of the Göttingen mathematician and physicist, Paul Hertz, in the early 1920s:

> H. was a leading representative of the philosophy of science at a time when this branch of research was barely appreciated in the domain of the German language.[111]

The Vienna topologist, Karl Menger, writes in his memoirs:

> In Germany, in the 1920's, Abraham Fraenkel was familiar with Polish set theory but was less versed in Polish logic; the logicians in Göttingen were not yet fully familiar with the results obtained in Warsaw; nor had the relations of the Polish logicians with Heinrich Scholz and his group yet developed. The majority of Germans were intensely hostile to the restored Polish nation because of the loss, in the peace of Versailles, of the territories inhabited by Poles, especially the so-called Polish Corridor which joined Warsaw to the sea while separating Berlin from Königsberg, the city of Kant. Even many German intellectuals had an idiosyncratic aversion to Poles, which the latter, mindful of one hundred and fifty years of oppression by Prussia, reciprocated.[112]

By way of contrast, according to Menger, Austria was leading Poland out of its isolation in logic and philosophy:

> Carnap went to Poland in the spring of 1930. Carnap's relations with the Polish logicians were to have a considerable influence on this development; and Tarski's visit in Vienna was certainly one of the first important steps out of isolation for Polish logic and philosophy.[113]

Menger's student, Franz Alt, wrote to me in 1993:

> Menger, although the youngest, was the magnet to whom the foreign guests to Vienna made the pilgrimage. Polish . . . Americans, . . . also Frenchmen, Japanese, but only a few Germans (Nöbeling),[114] maybe because in Germany there was less interest in the fields, which were cultivated mostly in Vienna—logic, foundations, set-theoretic topology, etc.[115]

---

[111]Bernays (1969), p. 172 (T).
[112]Menger (1994), pp. 144–45.
[113]Ibid., p. 156.
[114]Georg August Nöbeling, as mentioned in chapter 3.
[115]F. Alt to R. Siegmund-Schultze, July 12, 1993 (T).

10.D.2. Losses for Germany

Courant wrote to his former student Helmut Ulm in December 1933:

> Germany's best friends such as Hardy, Flexner, Lord Rutherford, the Rockefeller Foundation become alienated while our institutions, which were unequalled in the world, are destroyed—even Cambridge cannot compare to the old Göttingen. Foreign countries take advantage of the situation and employ people, particularly physicists and chemists, who will in the long run give science and its applications there a huge boost.[116]

In 1935, in a letter to the chairman of the mathematics department at Harvard University, W. Graustein (1888–1941), Wilhelm Blaschke of Hamburg recommended his student Erich Kähler (1906–2000) for a position in the United States, but he added:

> Hopefully you do not intend to call Kähler permanently to America. Unfortunately we have suffered great losses to America in the recent past of really good mathematicians, in particular Mr. Siegel from Frankfurt.[117]

The head of the research department of the Reich Aviation Ministry in Berlin, A. Baeumker, wrote in May 1937 from Italy to von Kármán:

> Our research begins to flourish again. But we are still lacking your heretical [ketzerisch] attitude towards the scientific field.[118]

10.D.3. The Profits of Emigration for International Communication

Richard Courant wrote in November 1934 from New York City to the psychologist Heinrich Düker (1898–1986), who would be dismissed from Göttingen in 1936:

> It is sad to see how German intellectual influence is receding here. In this respect the German emigrants, professors etc., who have been forced to work here, are without any doubt a hope for the gradual revival of intellectual relations.[119]

A. Flexner, 1935, to W. Weaver of the Rockefeller Foundation about Carl Ludwig Siegel's talk at the Institute for Advanced Study in Princeton:

[116]R. Courant to H. Ulm, December 22, 1933. CPP (T).
[117]Blaschke to Graustein, February 5, 1935 (T). HUA, Math. Dept. UAV. 561, Correspondence and Papers, 1911–62, UAV 561.8, box 1930–39, f. personnel considered for positions at Harvard 1934. In the event not Kähler but the Finn Lars Ahlfors (1907–1996) was appointed.
[118]Baeumker to von Kármán, May 22, 1937. Kármán Papers, Caltech, Pasadena, 1.40 (T). Von Kármán's unconventional scientific approach, for instance in the statistical theory of turbulence, is discussed in Hanle (1982).
[119]Courant to Düker, November 20, 1934, CPP D (T).

Siegel . . . made a very deep impression upon the mathematicians here. They obviously knew of him while he was still at Frankfurt, but I don't think that they realized how able he was until they had the opportunity for closer personal contact.[120]

10.D.4. Impact of the Institutional Side of German Mathematics (Educational System, Libraries)

Hermann Weyl said in 1940 about Alfred Brauer (the brother of Richard Brauer), who had organized the mathematical library at the University of Berlin in the 1920s and early 1930s and was forced to begin anew as an assistant during emigration:

Alfred Brauer is my assistant, and he is proving very helpful in the building up of a new library . . . in our new quarters called Fuld Hall.[121]

Hermann Weyl wrote to the American A. Dresden in December 1944 responding to Dresden's's reaction to Weyl's obituary (Weyl 1944) of Hilbert:

I was particularly happy about the friendly lines you wrote me about my Hilbert article. You seem to have sensed that it was composed not without an educational side thought, by stressing features and qualities in which the tradition of mathematical training in this country is somewhat deficient.[122]

In January 1945, Weyl added the following in another letter to Dresden:

And to remove my main grievances I am afraid nothing less than an overhauling of the complete educational system—not only of the teaching of mathematics—would be needed.[123]

10.D.5. The Development of New Mathematical Centers in the United States

The Institute for Advanced Study in Princeton, founded in 1932, expanded conspicuously under the influence of immigration. Oswald Veblen's refusal to accept a European-style reform, which would boost applied

---

[120]April 8, 1935. RAC, R.F. 1.1. 200 D, box 143, f. 1769.
[121]Weyl to Bernays, March 25, 1940 (T). Weyl Papers, ETH, Zurich, Hs 91:19a.
[122]Weyl to A. Dresden, December 13,1944. Weyl Papers, ETH, Zurich, Hs 91:185.
[123]Ibid. Weyl to Dresden, January 2, 1945. Hs 91:187. There was a subsequent discussion among Dresden, Courant, and Weyl on mathematical education in the United States. A committee was founded to which belonged, in addition, H. W. Brinkmann, J. R. Kline, Ø. Ore, and H. Rademacher. In the Weyl Papers in Zurich there is a six-page memo written by that committee, whose activities, however, apparently came to nothing.

mathematics and result in more formally organized teaching, was connected to a fear of Courant's "organizing power": "I fear that this power would do harm in our institute though it is just what would be needed in many an American university."[124]

The emigration coincided with a considerable upswing of institutional expansion particularly on the West Coast, which was at least partly stimulated by the influx of new personnel. AMS secretary Richardson wrote in January 1936 to immigrant Hans Lewy of the University of California at Berkeley, who had been temporarily at Brown University before:

> We miss you here but are happy to know that you are fitting into the situation out there in California. There is a possibility of making that one of the great centers of mathematics in the country such as Harvard, Princeton, and Chicago already are. It would be an eye-opener to Easterners to go West for post-doctoral training. I hope that the National Research Fellowships of the new vintage will encourage this migration to different regions of the country.[125]

Peter Lax (born 1926) on Gabor Szegö's impact at Stanford University since 1938:

> Szegö used his powers to turn the provincial mathematics department that Stanford had been under Blichfeldt and Uspensky—both remarkable mathematicians—into one of the leading departments of the country that Stanford is today. He appointed four senior mathematicians from Europe: Pólya, Loewner, [M. Max] Schiffer . . . and Bergman, and half a dozen brilliant young Americans.[126]

In June 1935 Richard Courant wrote to Szegö, at that time holding guest lectures at Stanford:

> I have the feeling that given the relative overcrowding of the East it is perhaps a great fortune to be able to begin the building up work [Aufbautätigkeit] in a somewhat less developed area. Here in the East there have apparently been at least subliminally a lot of difficulties and strains in view of scientific immigration.[127]

In 1986 Wolfgang Wasow remembered the situation in Courant's institute at New York University in the year 1939:

> The graduate department had one small room, a mathematics library consisting mostly of Courant's own reprint collection and one secretary. Courant

---

[124]Veblen to A. Flexner, August 30, 1933, IAS Archives, Faculty Files, box 32, f. Oswald Veblen (1933).

[125]Richardson to Lewy, January 17, 1936. Richardson Papers, Correspondence, BUA, box: German-Jewish Situation, f. H. Lewy.

[126]Quoted from Askey and Nevai (1996), p. 18.

[127]Courant to Szegö, June 8, 1935 (T). Szegö Papers, Stanford University, box 5, f. 15.

recognized me, and he talked in a rather discouraged way about his position at NYU.[128]

10.D.6. Inner-Mathematical Impact on Individual Disciplines

APPLIED MATHEMATICS

A representative of the Department of Genetics of the Carnegie Institution of Washington wrote in May 1935 to the Emergency Committee about Felix Bernstein:

> It seems to be a fair statement that Dr. Bernstein is the leading biological statistician in the United States; and that this country needs the best available biological statisticians to train a body of statisticians. Somebody has recently written that biometrics in this country is undergoing a masked decline—this is regrettable.[129]

However, the integration of Bernstein's field into the American system met with difficulties, which were partly due to Bernstein's personality, and in particular, his age.[130] Bernstein explained his problems as having to do with the interdisciplinarity of his field within applied mathematics. In a letter to Albert Einstein he wrote in April 1935:

> These problems lie in the atomization of the Am. department system. . . . But I have, nevertheless, the feeling that I should not give up and simply return to pure mathematics, as much as this would make things easier.[131]

Hermann Weyl, 1940, in a letter to the American consul in Lisbon in favor of his former assistant in Zurich, Alexander Weinstein:

> In the years gone by he has done some remarkable and outstanding work in mathematics, especially in applied mathematics. I believe that there will be an increasing demand in this country for this type of mathematics, in which Europe has specialized, because the European nations have for a long time been forced to squeeze the last five per cent of efficiency from their natural resources.[132]

The Russian-born theoretical mechanic, Stephen Timoshenko, who had heard lectures by Ludwig Prandtl in Göttingen in 1905 and had immigrated to the United States in the 1920s, confirmed in a letter to Courant in 1937, that mechanics could not develop without parallel progress in applied mathematics. He therefore supported the work of the immigrant Kurt O. Friedrichs:

---

[128]Wasow (1986), p. 213.

[129]C. Davenport (?) to EC, May 30, 1935. EC, box 2, f. F. Bernstein, 1933–35.

[130]On Bernstein's problems see above, particularly chapter 9.

[131]F. Bernstein to A. Einstein, April 5, 1935 (T). Einstein Papers, Jerusalem, 49-269-1-2.

[132]December 3, 1940, EC, NY Public Library, box 103, f. Weinstein.

I agree entirely with you that there is a great need in this country in developing applied mathematics of the kind as presented some time ago by Professor C. Runge at Goettingen University. The development of mechanics, in which I am especially interested, is impossible without a simultaneous development of applied mathematics. . . . I know Dr. Friedrichs, and know some of his publications in the field of bending of plates, which are excellent. I am sure that such a man as Friedrichs can contribute immensely to the development of applied mathematics in this country.[133]

Michael Golomb, who had been a student of Adolf Hammerstein (1888–1941) in Berlin in theoretical analysis (integral equations), was— similar to other immigrants—forced into applications immediately upon immigration to the United States in 1939. He wrote to me on July 19, 1993:

In order to obtain and hold the first position I found in this country, I was compelled to work eight hours a day in a field quite remote from that of my interest. I became a "research associate" to a professor of electrical engineering, who was writing a textbook on circuit analysis, but did not have the knowledge of the relevant mathematics.[134] I was engaged in this work for three years. Two other Jewish mathematicians, émigrés from Germany, Stefan Warschawsky [sic] and Fritz Herzog, preceded me in this position and were compelled to do work of little interest to them. Also my first appointment to an assistant professorship in a mathematics department (Purdue 1942) was in no small measure due to the fact that, as a mathematician educated at a European university, I was expected to be able to give lectures on Theoretical Mechanics and other applied fields, in fact I became a "captive applied mathematician." I know of others to whom this happened, too.

The Viennese mathematician Franz Alt, student of Karl Menger, the topologist interested in economics, wrote to me about his special connections with American research in economics prior to his emigration in 1938:

To me the close relations between the Vienna school of [national] economics [Nationalökonomie] and America were of particular importance. Josef Schumpeter, originally professor in Vienna and finance minister in the first Austrian government after World War I, was appointed at Harvard [in 1932; R. S.]. Shortly upon my arrival in New York in 1938 he wrote to me about an article, which I had published in the Austrian Journal for [National] Economics [*Zeitschrift für Nationalökonomie*]. . . . And of course, Oskar Morgenstern,[135]

[133]Timoshenko to Courant, April 29, 1937, EC, box 9, f. Friedrichs, K.
[134]Golomb is talking about the engineer Malti who has been discussed in chapter 8.
[135]As is well known, Morgenstern (1902–1976) published with John von Neumann the classical monograph *Theory of Games and Economic Behavior* (1944).

Figure 51 *Franz Alt (born 1910). The student of Karl Menger in topology and geometry left Vienna only a few months after the Anschluss in 1938 and worked in the United States in econometry and computing. He lives today (2008) in New York City.*

professor in Vienna and head of the Austrian Institute for Business Cycle Research [Institut für Konjunkturforschung], was appointed at Princeton University.[136]

Gustav Bergmann, who shifted from mathematics to philosophy during emigration, was registered in the files of the Emergency committee as a specialist for "Applications of Mathematics to Psychology, Philosophy, and Sociology."[137]

Felix Bernstein, in 1949 in letters to Albert Einstein, deplored the fact that Weyl and von Neumann at the IAS did not support a position for him at the Institute. Judged by the following letter his desperation seems to have impaired his insight:

> Von Neumann . . . is indeed a terrible egoist. It is only embarrassing to see how he is now, following the fashion, turning to numerical computing.[138]

Feodor Theilheimer, who in 1936 had earned his PhD on invariant theory as the last doctoral student of Issai Schur in Berlin, came to the United States in 1937, after his brother—who had arrived in the country earlier—had written him an affidavit. Theilheimer had to survive several years as a teacher in Jewish studies before his return to mathematics through the summer school of the University of Chicago in 1941. But this meant at the same time a move to applied mathematics, a field of rapid expansion during the war. After the war (1948–77), Theilheimer worked for the U.S. Department of Defense, contributing with his mathematical work to the theory of spline functions. The conditions of his emigration prevented him from returning to work as an academic teacher.[139]

PROBABILITY THEORY AND STATISTICS

Willy Feller wrote in a 1934 letter to Courant about the influence of Stockholm, his temporary place of refuge, on his view of stochastics:

> It was for me a great revelation that beside mathematics proper there exists a very complicated insurance mathematics, where for instance Cramér throws around Dirichlet series with great effect [dass alles kracht].[140]

Feller's publications from the late 1920s and early 1930s belonged to differential equations and measure theory, his first explicitly probabilistic paper appeared in 1936, still in German, in the *Mathematische Zeitschrift*.[141]

---

[136]F. Alt to me, July 12, 1993 (T).
[137]EC, box 148, f. Rockefeller Foundation, 1940–45.
[138]Bernstein to Einstein, undated [1949]. Einstein Papers, Jerusalem, 57 650-4 (T).
[139]Letter by Theilheimer's daughter Rachel to me, December 17, 1977. See also the history of the family Theilheimer in Gunzenhausen in Bavaria at http://www.gunnet.de/stephani/step_p27.
[140]Feller to Courant, November 11, 1934, CPP (T).
[141]See Feller's bibliography in Birnbaum, ed. (1970).

Figure 52 *Feodor Theilheimer (1909–2000). Theilheimer was Issai Schur's last doctoral student. In American emigration since 1937 he worked mostly in military research (Photo 1992).*

The early Polish immigrant and mathematical statistician, Jerzy Neyman (1894–1981), was an adherent of von Mises's concept of probability,[142] which was about to lose out against the new paradigm of the Russian A. N. Kolmogorov (1903–1987). Neyman saw a chance to cultivate von Mises's standpoint after his immigration to the United States in 1939. He also supported the immigration of Hilda Geiringer, whom he found to be an important asset in the dissemination of von Mises's theory,[143] and wrote accordingly a letter to the Rockefeller Foundation on her behalf:

> The von Mises Theory constitutes a great and real advance both in the theory of probability proper and especially in the philosophy of the subject. It is known but little, there are many regrettable misunderstandings about it and it is certainly in the interest of American science to have here one more competent person who could successfully teach it.[144]

Neyman's hope for the dissemination of von Mises's theory did not come true, in spite of the later renaissance of von Mises's works in the 1960s in the context of Kolmogorov's theory of complexity.

Experience in stochastics was not an automatic guarantee for jobs for the emigrants—it may even have been a disadvantage, as long as the theory was not yet fully recognized among mathematicians in the early 1930s. Courant sympathized with Wolfgang Sternberg's problems in a letter from 1934 in which he stated that he was one of the few representatives of analysis still keeping in touch with the applications (probability and statistics).[145]

Apparently there were also tendencies among American statisticians not to take immigrants from Europe too seriously, at least the non-English ones. This becomes evident in the following rather arrogant letter from Harold Hotelling (1895–1973) to Weyl from 1947, in which he refused to consider the immigration of Robert Frucht from South America to the United States:

> The great difficulty about placing European mathematicians in statistics is that they are all almost ignorant of the modern theory of statistics. There is just one outstandingly successful mathematical statistician of European birth in this country today. He is Abraham Wald, and before emerging as a good statistician, he had to have several years of work in mathematical statistics as my assistant in spite of having previously published contributions in differential geometry and other parts of mathematics.[146]

---

[142]See Siegmund-Schultze (2004b).

[143]See Siegmund-Schultze (1993b), p. 366.

[144]Neyman Papers, Berkeley, cont. 47, f. 33, Neyman to W. Weaver, April 22, 1940.

[145]Courant to the Notgemeinschaft deutscher Wissenschaftler im Ausland, September 23, 1934, CPP. On Sternberg's career problems in the United States see chapter 8.

[146]Hotelling to Weyl, February 10, 1947, OVP, box 30, f. Frucht, Robert, 1938–47.

STATISTICAL MECHANICS, ERGODIC THEORY, CELESTIAL MECHANICS

In 1940 William C. Graustein of Harvard University recommended Artur Rosenthal, quoting the Greek-German Constantin Carathéodory:

> [Rosenthal] was one of the first to understand the ergodic problem solved by Birkhoff (his work is about 15 years before that of Birkhoff).[147]

In 1939 Weyl wrote more generally about the lack of suitably qualified German immigrants in celestial mechanics,[148] in a letter to the astronomer H. Shapley (Harvard):

> I roamed through our files for a man who combines astronomy and mathematics in the same way as Freundlich does and found none. This combination, very common in the first half of the nineteenth century, has become much less frequent in my generation in Germany, although we used to have a fair amount of analytic mechanics and many of us studied Poincaré's Mécanique Céleste.[149]

In the early 1930s, Eberhard Hopf, during his Rockefeller Fellowship with Shapley and Birkhoff, had already stressed the shortcomings of most German mathematicians in this field.[150]

GERMAN ABSTRACT ALGEBRA AND NUMBER THEORY IN THE UNITED STATES

Toward the end of 1934 (after the immigration of Emmy Noether), Solomon Lefschetz of Princeton revealed his somewhat biased opinion of her work, at the same time revealing a rather disturbing lack of insight into the work carried out by the Schur school in Berlin:

> As the leader of the modern algebra school, she developed in recent Germany the only school worthy of note in the sense, not only of isolated work but of a very distinguished scientific work. In fact, it is no exaggeration to say that without exception all the better young German mathematicians are her pupils.[151]

The American Eric T. Bell (1883–1960) underlined in 1938 the international dominance of German algebra and of the Noether school:

> This latest phase of algebra is distinctly European in origin, and practically all German. . . . Its roots are in Dedekind's work. . . . Steinitz' paper of 1910 and

---

[147]Shapley Papers, box 6C, file R, Graustein to Shapley, undated [1940].

[148]Note also C. L. Siegel's work on the three-body problem and his lectures on celestial mechanics in the United States (Siegel 1956), which his student, Jürgen Moser (1928–1999), published in German, based on a lecture script.

[149]Weyl to Shapley, May 30, 1939, Shapley Papers, box 6F, f. Weyl, Veblen, Einstein, IAS.

[150]See Siegmund-Schultze (2001), p. 114. See also Hopf's commentary to the Rockefeller Foundation as quoted in chapter 3 and the case study on Hopf's return to Germany in 1936 in chapter 7.

[151]Kimberling (1981), p. 35.

Emmy Noether's abstract school, trained by her either personally or through her writings from about 1922 to her death in 1935 at the age of 53.[152]

B. L. van der Waerden in Leipzig wrote to Courant in 1936:

By the way, my works on algebraic geometry find much applause primarily in the United States.[153]

Veblen from the IAS wrote in November 1934 to Weaver of the Rockefeller Foundation on behalf of Siegel:

Professor Siegel's line of work is so important and has as yet had so little attention in this country that I feel sure the visit will turn out to be really important.[154]

According to Saunders MacLane in 1988, Garrett Birkhoff's introduction in 1933 of "general algebras" was "a natural development of the German idea of modern algebra, and is the starting point of the whole field of 'universal' algebra and its relation to model theory."[155] Moreover, Garrett Birkhoff considered it, in hindsight in 1989, pure good fortune that the German algebraists had left him with some work to do in his theory of lattices in the 1930s:

In retrospect, I think I was very lucky that Emmy Noether, Artin, and other leading German algebraists had not taken up Dedekind's 'Dualgruppe' concept before 1932.[156]

HISTORY, PHILOSOPHICAL REFLECTION, FOUNDATIONS OF MATHEMATICS

In September 1933 the director of the Institute for Advanced Study, Abraham Flexner, wrote to the leading American mathematician of the Institute about the noted historian of ancient mathematics Otto Neugebauer, currently facing political problems in Göttingen:

My disposition would be to invite Neugebauer, because he would bring to this country something absolutely new, namely the historical and humanistic side of mathematics. The success of the Institute of the History of Medicine at the Johns Hopkins with its liberalizing influence over the faculty as well as the students encourages me to try this novelty. Mathematics is something more than an affair of today and yesterday. It is a part of the cultural history of the race.[157]

---

[152]Bell (1938), p. 32.

[153]Van der Waerden to Courant, March 28, 1936, CPP (T). Van der Waerden was probably alluding to the work of Zariski, which would later on supersede his own approach, using Wolfgang Krull's general theory of valuations. See Slembek (2002) and Schappacher (2007).

[154]O. Veblen to W. Weaver, November 19, 1934, RAC, R.F. 1.1. 200 D, box 143, f. 1769.

[155]MacLane (1988), p. 331.

[156]Birkhoff (1989), p. 41.

[157]A. Flexner to O. Veblen, September 8, 1933, IAS Archives, Faculty Files, box 32, f. Oswald Veblen (1933).

In February 1934, Weyl of the IAS wrote to F. H. C. Northrop of Yale University on behalf of the logician Paul Bernays at the time visiting from Zurich where he had no position:

> Dr. Flexner allowed me to read your memorandum concerning the plan for a philosophical academy in this country. . . . In one respect I thoroughly agree with you: that it is highly important to broaden the philosophical outlook of the young scientists in this country, to open their eyes which they keep focused too exclusively on the technicalities of a special subject, to the more philosophical implications of science, its methods and its attitude.[158]

In the same vein, Weyl wrote to his friend Erich Hecke (1887–1947) in Hamburg in April 1936. He saw it as his main goal to educate the young American mathematician in "reflection" ("Besinnung"). In particular, Weyl deplored the exaggerated abstractness shown in recent topological research:

> Personally I will hardly have the energy left for huge mathematical accomplishments. . . . When I look around I realize that the Americans have fully learned the techniques of science. Enough and more than enough mathematics is being produced. If the Americans are lacking something—and here I can possibly be of some help—it is "reflection" ["Besinnung"]. For myself I am aiming at a balance in the essential and healthy tension between "creation" and "reflection." . . . I am opposed to topology's increasing abstractness which bears the danger of degenerating into an axiomatic game. Unfortunately Siegel has not come over. Through his personality and teaching he would have acted strongly for a substantial mathematics.[159]

In 1939 Weyl viewed Ernst Hellinger's knowledge of the history of mathematics as a positive point in favor of employing him, although admitting that Hellinger had been less active recently in mathematical research:

> An important point seems to me his familiarity with the history of mathematics, which is solidly founded on the actual study of the sources;. . . . a little more education in these respects is quite definitely wanted for our young American mathematicians.[160]

In August 1940, George Pólya in Zurich wrote to his former boss Weyl, reporting on problems with the American consul in Zurich, as he had only a one-year offer from Stanford University. Alluding to certain tendencies in the United States to discourage the immigration of older mathematicians such as himself (he was fifty-two at that time), Pólya said:

---

[158]IAS Archives, School of Mathematics, f. P. Bernays (1935–65).

[159]H. Weyl to E. Hecke, April 21, 1936, Weyl Papers, ETH Zurich, Hs 91:258 (T). Thanks to Liliane Beaulieu (Nancy) for bringing this letter to my attention.

[160]EC, box 13, f. E. Hellinger (1933–41), Weyl to EC, November 11, 1939.

The argument "we have enough bright young men," which certainly has some truth to it, causes me to come back to a point in an earlier letter. . . . I have thought more thoroughly about some points of teaching (everything which has to do with "heuristics") than most other people. . . . Given that the USA are so prone to "education," would it not be possible to find a position for me there which is not prestigious [glänzend] but useful and could not be occupied by a "bright young man"?[161]

TRANSFER AND MEDIATION OF THE EUROPEAN SCIENTIFIC TRADITION

Hermann Weyl, who did so much for stimulating "reflection" ("Besinnung") in modern mathematics, was also known for familiarizing Americans with the European scientific tradition. In retrospect in 1977, Garrett Birkhoff saw a kinship of mind between his father George David Birkhoff and Weyl and added about Weyl:

When he came to our country, he brought with him a substantial piece of European scientific tradition, and from him flowed wisdom and creative contributions to many aspects of mathematics until his retirement.[162]

This bridge-building function of immigrants was not restricted to German-speaking immigrants. In 1940, Oswald Veblen pleaded for the immigration of the Italian geometer Beniamino Segre (1903–1977), who was then in England, with the following words:

I should say that, as compared with some of his contemporaries, he tends to be very geometrical in the classical sense of the word. This tendency has fallen somewhat into decay in the United States, and I think that it would be very desirable to revive it.[163]

A non-German-speaking immigrant to the United States, the Polish analyst Antoni Zygmund, also contributed greatly to upholding the European scientific spirit. Two of his American students wrote the following about him in 1989:

It is important to realize the following unique features of this school. When Zygmund came to Chicago, the "trend" in mathematics was very much influenced by the Bourbaki school and other forces that championed a rather abstract and

---

[161]Pólya to H. Weyl, August 5, 1940 (T). OVP, cont. 32, f. Pólya, George, 1940–41. Pólya finally went first to Brown University, which offered a two-year contract, before getting a permanent position at Stanford in 1942.

[162]Birkhoff (1977), p. 73.

[163]O. Veblen to W. Weaver, October 8, 1940., OVP, cont. 32, f. Segre, Beniamino, 1938–41. As the same folder reveals, Segre finally stayed in England, due to resistance by S. Lefschetz in Princeton, who was more prone to the modern approach to algebraic geometry in the sense of Zariski.

algebraic approach for all of mathematics. Zygmund's approach toward his mathematics was very concrete. He felt that it was most important to extend the more classical results in Fourier analysis to other settings. . . . He realized that fundamental questions of calculus and analysis were still not well understood. In a sense, he was "bucking the modern trends."[164]

10.S. Case Studies

10.S.1. The Failure of Richard Brauer's Book on Algebra in 1935, or the Paradoxical Victory of "Talmudic Mathematics" Due to Nazi Rule

There are three participants in this case study: Richard Courant in New York, Richard Brauer in Princeton and Toronto (where he had moved in the fall of 1935, after a recommendation by Emmy Noether), and Bartel van der Waerden in Leipzig.[165] Emmy Noether, who died on April 14 that same year at Bryn Mawr College, and Issai Schur, dismissed in Berlin, act in the intellectual background.

In the summer of 1935 the emigrant Courant, who had close contacts with Springer in Germany, was obviously still determined to publish Brauer's *Algebra* (an older project). He wrote to him on July 17:

> From Germany come cries for help both from Springer and from my new co-editors van der Waerden und F. K. Schmidt because of your Algebra. For reassurance I would like to send Mr. Schmidt the present plan of your book how you imagine it.

Brauer chose to interpret this letter apparently as a reminder of his tardiness and immediately submitted a detailed plan of contents with fifty-six chapters. Courant, however, was aware of principled resistance toward the project in Germany. He had already written to van der Waerden on July 16, 1935:

> Brauer's book has been discussed last year repeatedly also with Emmy Noether and will certainly not be a superfluous enterprise.

Apparently Courant reacted to objections raised by the young, but already famous author of *Moderne Algebra*, van der Waerden, a student of Emmy Noether's, as van der Waerden replied on August 10, 1935:

---

[164]Coifman and Strichartz (1989), p. 347.

[165]The archival source for the following study is the private correspondence of Richard Courant (CPP), in particular with van der Waerden. Brauer's letters are quoted from copies and originals in that correspondence. The quotations are translated from German.

Figure 53 *Richard Brauer (1901–1977). Brauer, who was expelled from Königsberg in Eastern Prussia, was, through his work on the theory of algebras and groups, the most important student of Issai Schur's in Berlin. The publication of a textbook on algebra, for which he had a contract with the Springer publishing house in Berlin, did not materialize due to the circumstances of emigration and the discrimination of Jewish authors in Germany. After a stopover of several years in Toronto, Brauer received an appointment as a professor at Harvard University since 1952.*

I am sorry you found my letter aggressive. . . . With regard to Brauer (R.) I was of the opinion that he himself was not really keen on writing the book. If this is not the case and even Emmy Noether agrees with the book I withdraw my reservations for the time being.

In another letter to van der Waerden on August 20, 1935, Courant explained how Brauer's book came into being:

An ancient plan by Schur for the edition of Frobenius' lectures on algebra had been converted for long into a plan for the edition of Schur's lectures. Schur then had nominated Richard Brauer as a collaborator and gradually shifted the responsibility onto the latter. After very careful deliberations also with Emmy Noether the contract was completed thereby clearly indicating it should be a "concrete algebra" and in a way a complement of your book. . . . Over here Brauer has worked much on the book—in very intimate contact with Emmy Noether to whom he was closer than any other human being.

However, in the following weeks, Courant seems to have gotten cold feet and lost his courage. He wrote to Brauer on October 28, 1935:

One of the weak points of Springer is his close contact to Jewish emigrant-authors. Springer is decided to continue business in an objective way, but he has to be cautious not to leave himself open to attack. Now there is danger of a concentric attack within Germany with respect to your Algebra. Many people are putting into question the objective need for the book on the German market given the existence of many algebra books (Perron, Hasse, Haupt, van der Waerden, Bieberbach-Bauer, Dickson-Bodewig, Steinitz, Weber-Fricke, [Maxime] Bôcher). Apparently van der Waerden and Schmidt have asked Springer to postpone the publication of the book to more quiet times, and it seems Springer has become a little bit anxious. I have fought a long battle in this matter in which I have come to the conclusion that compromises are not possible in this case and that—given the changed situation—you have to be free to publish the book, in the form as you want it, with an English or American firm. This would be by the way in the long run more profitable for you . . . A repayment of the advance is out of the question.

Brauer was no doubt hurt that even Courant had let him down. He wrote to van der Waerden on November 24, 1935, stating that the whole idea of a more general algebra book (as opposed to a more special one on groups of linear substitutions) was not his or Schur's, but had come from Springer and in particular from Courant:

Courant wanted me to write an elementary book on algebra rather. . . . It was supposed to address a different readership than your book, younger students, which had not yet gained understanding for axiomatic, abstract thinking. Addressed to people whose interest was more in analytical direction and who shared the prejudices against the modern "abstract" (or "Talmudic")[166] algebra which have never died out. To use Courant's words the book was supposed to build a bridge to the "abstract" algebra, such that the regions on both banks be given due attention.

In the same letter to van der Waerden, Brauer also tried—though rather belatedly—to dispel reservations in the latter by quoting him indirectly in the following passage:

> ... Emmy Noether never co-operated actively contrary to what you seem to assume, but she was warmly interested in the whole project. This might imply that it was not something which "was opposed to the entire tendency of her work."

The correspondence between Courant, Brauer, and van der Waerden shows that there was more at stake than political caution on Courant's side and professional jealousy on van der Waerden's part. They possessed different points of view regarding "structural mathematics," where Courant in particular, known for his interest in applications, erred more on the side of conservatism and caution, while Brauer, the follower of the Schur school, found himself somewhere in between.

Thus, paradoxically, "talmudic" mathematics à la Emmy Noether as expressed in van der Waerden's *Moderne Algebra* and certainly not liked by the Nazi fighters for "Deutsche Mathematik," was partly "strengthened" by the Nazi seizure of power. This did not mean, however, that van der Waerden's book now became the standard for mathematical beginners in Germany. Rather to the contrary, van der Waerden felt forced to leave out some "dubious set-theoretic methods" ("bedenkliche mengentheoretische Schlussweisen")[167] in the second edition of 1937, and a fully developed structural algebraic method was reintroduced in the German curricula only after the war. But van der Waerden remained in Leipzig in his influential chair, and the proponent of a somewhat "more concrete" concept of algebra, Richard Brauer, was driven out of the country and his book project was destroyed. In fact, Brauer would never publish a book on algebra. He experienced another disappointment at the same time, when his contribution to the second edition of the German *Enzyklopädie der mathematischen Wissenschaften* was refused by the mathematicians remaining in Germany.[168]

This example shows that the Nazi rule not only led to the oppression of "abstract" mathematics but—dependent on the specific historical constellation—also to the destruction of more "concrete" directions in

---

[166]This is of course a strongly ironic use of words by the Jewish mathematician Brauer.
[167]B. L. van der Waerden "Moderne Algebra," 2nd ed., vol. 1, Berlin 1937, p. vi.
[168]See Tobies (1994).

mathematical research, as emigration included representatives of all different fields of mathematics.[169] This not only underscores the demagoguery of Nazi propaganda, supported by mathematicians like Bieberbach, and claiming a correlation between "abstract" mathematics and "Jewish" mathematics, it also demonstrates the considerable role played by National Socialism in the destruction of the *overall tradition* of German mathematics. Against this backdrop one might even understand a paradoxical quotation such as the following by Carl Ludwig Siegel, who returned to Germany after the war. In 1968, over three decades after those destructive happenings and the consequent emigration, he came with the following concerns about prevailing trends in algebra, seemingly too abstract for him, showing a disturbing move away from the concrete classical problems of the subject:

> The current threatening situation in mathematics reminds of the times of National Socialism, when they were marching until everything fell in broken pieces [bis alles in Scherben fiel].[170]

Siegel's quote—so it seems to me—is a good example of how, in the memory of scientists, parallel historical processes, which in the time may even have partially contradicted one another, can amalgamate. In the case of Siegel's quotation it was the parallel rise of structural mathematics and the decline of the German tradition in mathematics under the Nazis, not least due to the emigrations.

My previous arguments may have indicated that the partial destruction of the German mathematical culture by the Nazis and the rise of "Modern Algebra" were not one-sidedly in contradiction to each other in a mutually exclusive or even always opposite sense. Even within mathematical disciplines—and in abstract algebra in particular—there were potentials for one-sidedness in the development of mathematical objects or for the suppression of certain alternative standpoints of research (such as the one adopted by the Schur school and Brauer). Therefore it seems to me that Siegel's grotesque quotation also has some grain of historical truth in it, at least in the sense of a personally experienced reality. Siegel's remark from a period when exaggerated belief in the omnipotence of abstract "structural mathematics" in the sense of Bourbaki was already on the wane,

---

[169]One might also think of the two leading "applied mathematicians," in a more narrow sense, who had to go: Richard Courant and Richard von Mises.

[170]Siegel (1968), p. 6 (T). The quotation, which alludes to an infamous Nazi marching song, is cited and commented upon in more detail in Siegmund-Schultze (1992), p. 727. It is basically contained already in a letter by Siegel to L. J. Mordell, dated Göttingen March 3, 1964. Siegel hoped to find in Mordell an ally against recent modernist books by Serge Lang and others. Siegel added there: "I am afraid that mathematics will perish before the end of this century if the present trend for senseless abstraction—I call it: Theory of the empty set—cannot be blocked up." Mordell Papers, 28.17.

leads to a controversy that can be interpreted as a partly ill informed counterreaction to the dominance of Noether's algebra in the 1930s and 1940s. This controversy will now be described briefly.

10.S.2. Late American Criticism of "German Algebra,"
a Controversy between Garrett Birkhoff
and B. L. van der Waerden in the 1970s
and Commentary by G.-C. Rota in 1989

A debate of historic interest, but arising rather late (1973) between two main actors who had been promoting Noether's algebra in the 1930s, B. L. van der Waerden and Garrett Birkhoff, indicates, once again, that the absorption of mathematical ideas during emigration did not go uninfluenced by questions of national prestige and national jealousy. Moreover, global changes in mathematics during the 1970s, a reconsideration of more "constructive" (in the sense as this word is used within work on the foundations of mathematics), of discrete and applied mathematics, contributed to that discussion, one to which, occasionally, other mathematicians such as C. L. Siegel (see above), A. Weil,[171] and G.-C. Rota contributed as well. The present case study does not give a thorough analysis, as this would have to be based on the actual mathematical developments. The case study has therefore to be regarded as a mere stimulus for further discussion.

Garrett Birkhoff said the following in 1976 on Emmy Noether's algebra, which had influenced him so much in the 1930s:

> Emmy Noether and her students owed more to British algebraists than they recognized.[172]

For, in one particular instance, having refrained from quoting the British-American algebraist Joseph H. M. Wedderburn (1882–1948), Birkhoff attacked Noether (and implicitly also van der Waerden) in 1973, long after her death, with the following words:

> This seems like an example of German 'nostrification': reformulating other people's best ideas with increased sharpness and generality, and from then on citing the local reformulation.[173]

Van der Waerden, in his reply to Birkhoff, flatly denied that this criticism was justified:

---

[171]For instance with his criticism of the one-sidedness of van der Waerden's algebraic geometry as mentioned above. See Schappacher (2006).

[172]Birkhoff (1976a), p. 48.

[173]Birkhoff to van der Waerden, November 1, 1973. Van der Waerden Papers, ETH, Zurich, Hs 652:1056, Manuscript: "The Sources for van der Waerden's Moderne Algebra," p. 2, footnote.

As far as Wedderburn is concerned, this impression is definitely not true. I heard two courses of lectures of her, both on Group Theory and Algebras. The contents of the second course was published in Math. Zeitschrift 30 (1929), p. 641; it included the Theory of Representations. In the first course (Winter 1924/25) the culmination point was the structure theorem of Wedderburn. . . . Wedderburn was for her "la crème de la crème." As for Dickson I cannot be so positive. His "Algebras and their Arithmetics" was translated into German with valuable additions by Speiser, but it was hardly mentioned in her lectures. . . . Wedderburn was the man who created the structure theory, and Molien and Peirce were his main precursors, not Dickson.[174]

As a result van der Waerden took issue with Birkhoff's planned article, "The Sources for van der Waerden's Moderne Algebra," which Birkhoff wished to publish in *Historia Mathematica*. Instead, in 1975, van der Waerden published his own version of the events leading to his book in that journal, briefly alluding in the publication to his discussion with Garrett Birkhoff.

Birkhoff refrained from publishing his article, but wrote the following one year later on Noether's talk on "Hypercomplex Systems" given at the International Congress of Mathematicians in Zurich in 1932:

> Reading it today, one sees clearly, how completely German algebraists had taken over: no reference to Wedderburn, passing references to Dickson and Chevalley, and the rest is Germanic.[175]

In addition to the changed global situation in mathematics at that time, in the 1970s, one might be forgiven for suspecting traces of remorse in Garrett Birkhoff, given the fact that American mathematics in his era often took the glory for foreign results, that is, practiced "nostrification" (= "make it ours") in a similar vein as Göttingen mathematics had once done in the 1920s.[176] As a matter of fact, Birkhoff admitted on another occasion that some Americans presented a distorted picture of the history

---

[174]Van der Waerden to G. Birkhoff, November 7, 1973, ETH, Hs 652:1057. Winfried Scharlau (Münster), in a Göttingen talk on the history of the theory of algebras on July 15, 1997, remarked that Dickson's book of 1923 on algebras was "putting off" ("abschreckend") the readers, because it announced proofs but did not always deliver them. However, according to Scharlau, Noether quoted Dickson in her lectures. Also David Rowe (Mainz), who has intimate knowledge of Noether's lecture scripts, underscores that she generally quoted conscientiously (personal communications).

[175]Birkhoff (1976a), 55. Indirectly alluding to the letter by van der Waerden, mentioned before, and to an article by Noether in the *Mathematische Zeitschrift* in 1929, Birkhoff then said in a footnote: "Three years earlier, she was more generous" (Noether 1929).

[176]On reproaches of nostrification within Germany in the 1920s, particularly from mathematicians outside Göttingen, see Kowalewski (1950), p. 193.

Figure 54 *Bartel Leendert van der Waerden (1903–1996). Through his book* Moderne Algebra *(1930/31) the Dutch mathematician (who stayed in Leipzig during the time of Hitler), became the most influential disseminator of the structural thinking of his teacher Emmy Noether.*

of computing by ignoring contributions by Germans such as Konrad Zuse (1910–1995).[177]

Up until quite recently resentment about the dominance of the Noether/van der Waerden algebra has been reiterated by other Americans as well. Gian-Carlo Rota complained, in 1989, about how the one-sidedness of van der Waerden's algebra book, which also included among its sources Emil Artin's lectures, continued to exert an influence on mathematical training in the United States:

> The table of every algebra textbook is still, with small variations, that which Emil Artin drafted and which van der Waerden was the first to develop. (How long will it take before the imbalance of such a table of contents—for example,

[177]Birkhoff (1980), p. 29.

the overemphasis on Galois theory at the expense of tensor algebra—will be recognized and corrected?)[178]

Rota, appreciated as he is for his often-polemical and sharp commentary said in a tone of disapproval the following about Artin's influential lectures in Princeton after the war:

> He inherited his mathematical ideas from the other great German number theorists since Gauss and Dirichlet. To all of them a piece of mathematics was the more highly thought of, the closer it came to Germanic number theory. This prejudice gave him a particularly slanted view of algebra. He intensely disliked Anglo-American algebra, the kind one associates with the names of Boole, C. S. Peirce, Dickson, the late British invariant theorists . . . and Garrett Birkhoff's universal algebra (the word "lattice" was expressly forbidden, as several other words).[179]

In 1950, Emil Artin, an immigrant to the United States in 1937, but who would return to Germany in 1958, had the following to say about Wedderburn's work from the early twentieth century. Stressing the importance of the American sources in German algebraic work, Artin seemed to be eager to fend off possible American resentment:

> The most striking fact is the difference in attitude between American and European authors. From the very beginning the abstract point of view is dominant in American publications whereas for European mathematicians a system of hypercomplex numbers was by nature an extension of either the real or complex field. While the Europeans obtained very advanced results in the classification of their special cases with methods that were not well adapted to generalization, the Americans achieved an abstract formulation of the problem, developed a very suitable terminology, and discovered the germs of the modern methods.[180]

---

[178]Rota (1989), p. 233.

[179]Ibid., pp. 231–32. Rota is probably alluding here to C. S. Peirce's father, Benjamin Peirce, the author of "Linear Associative Algebra" of 1870.

[180]Artin (1950), p. 65. In 1927, Artin had used his chain conditions in order to prove fundamental theorems of Wedderburn's.

ELEVEN

Epilogue

THE POSTWAR RELATIONSHIP OF GERMAN
AND AMERICAN MATHEMATICIANS

> In us, the ones directly and indirectly affected, there must be
> enough love to let pale the evil pictures of the past.
> —Max Dehn c. 1950[1]

THE postwar situation[2] brought with it a multitude of partly contradictory developments in mathematics in Germany. Almost all of them were
the result of the war, and most of them were connected to consequences
of emigration, in particular to the relationship between mathematicians
who had remained in Germany and emigrants. Some of these developments led to a further weakening of German mathematics due to further
emigration of established mathematicians and promising students[3] and restrictions on research imposed as a result of decisions by the Allied Control Council in Germany. In addition, the division of Germany into zones
of occupation by the victorious powers impeded communication within
Germany. Other developments, however, eased the almost total international isolation of German mathematics: the help by emigrants such as
Courant (see below), the return of some (if only a few) emigrants,[4] and
the rapidly increasing number of German students in the United States
that led to the importation and reimportation of certain mathematical
subdisciplines to Germany. In recent decades a number of (West-) German
mathematicians who maintained particularly close connections with the

---

[1]Max Dehn in a private letter to an unknown addressee in Germany around 1950 (T).
Quoted from E. Lissner, Dehn Papers, Austin (see below).

[2]The most important source for postwar relations is, once again, the Richard Courant Papers at the New York University Archives. Courant was at the center of efforts to renew international relations in mathematics between America and Germany.

[3]Among the latter was, for instance, Jürgen Moser. Among the former, Wilhelm Magnus,
Eberhard Hopf, and Hans Zassenhaus (1912–1991) were particularly prominent. Several
postwar emigrants are named in Bers (1988), p. 241.

[4]Among them were R. Baer, H. Hamburger, F. Levi, and C. L. Siegel. E. Artin returned
rather late in 1958. In 1952, at the age of sixty-seven, H. Weyl returned to Zurich, where he
had worked prior to 1930.

United States have produced important results, among them Friedrich Hirzebruch (born 1927, algebraic topology) and Gert Faltings (born 1954, proof of Mordell's conjecture in 1983).

The last remark, regarding West Germany, also applied mutatis mutandis to the relationship between East German and Russian mathematicians. Although there were only a few immigrants from Germany to the Soviet Union, communication between East German and Russian mathematicians developed rapidly. As a result of the increasing isolation of East Germany (which from 1949 became the German Democratic Republic [GDR]), partly self-chosen, most immigrants to the West directed their attention toward the West German Federal Republic, unless otherwise interested in the GDR for political reasons.[5] The vastly superior economic strength of West Germany, aided by the American Marshall Plan, ensured that all emigrants could count on material compensation (Wiedergutmachung) from West Germany, even if the place of their dismissal had been in the eastern part of Germany.[6] This, in turn, reinforced the West German claim for the exclusive right to represent "Germany" ("Alleinvertretungsanspruch" and "Hallstein-Doktrin") in the international arena. For this reason, this chapter is largely confined to West Germany, even though mathematicians from both parts of Germany were members of the Deutsche Mathematiker-Vereinigung (DMV) until the foundation of the East German Mathematische Gesellschaft der DDR in 1962.[7] The sporadic contacts between emigrants and East German mathematicians or administrators, which during the Cold War were often politically prejudiced or tainted, require further examination.[8] The postwar relationship between American and Austrian mathematicians, which has to be considered against the background of the Anschluss of 1938, is not discussed in this chapter, due to lack of space and insufficient sources.

After 1945, largely due to the catastrophic working and living conditions in Germany, another wave of immigration to the United States and to the Soviet Union began. Another factor contributing to emigration was a

---

[5]This applies to two Jewish émigrés who were open to Marxist ideology, namely the mathematics teacher Wilhelm Hauser, and the former mathematics student in Göttingen, Ludwig Boll. Cf. Wirth (1982), Oswald (1983), Arnold (1986), and an interview with Boll by the author in November 1983.

[6]This applies for instance to refugees from the University of Berlin, which was located in East Berlin and named after the Humboldt brothers in 1949. Among those from the University of Berlin who received compensation was Richard von Mises, although he did not live to see it.

[7]Siegmund-Schultze (1996).

[8]For material on the relationship of Richard von Mises and of Richard Courant to the East German Academy of Sciences see Siegmund-Schultze (1993c). Richard Courant's Papers at NYU are also an important source with respect to the relations between emigrants and East German mathematicians.

strong demand for applied mathematicians in the United States.[9] In addition, initial stipulations excluding former Nazis from immigration were gradually relaxed during the Cold War. This softening of attitudes was also apparent in the so-called Operation Paperclip through which the United States secured the help of German specialists in rocketry and aerodynamics. However, victims of National Socialism frequently received preferential treatment when applying for immigration to the United States; for some it was a first and for others a second chance at a career.[10] The start of the Brain Drain from Europe to the United States not only affected German mathematics. Richard Courant's Institute in New York City became a center of attraction for immigrants from several European countries.[11]

Several of the emigrants[12] and also some American mathematicians[13] visited Germany. In Germany the Mathematical Institute in Oberwolfach (Black Forest), which had been founded in the Nazi period, became a place for the resumption of international contacts in mathematics.[14] The Nazi victim, the mathematician Erich Kamke (1890–1961), played an important role in the revitalization of the International Mathematical Union (IMU) around 1950, ensuring the inclusion of Germany into the work of

[9]Courant helped to facilitate the employment of Germans in the U.S. Navy (CIP, file: German scientists 1946–). Weyl warned Courant against the immigration of F. Ringleb (1900–1966), who was in his opinion incompetent and a former Nazi (ETH, Hs 91:164, August 30, 1946. I am grateful to L. Beaulieu, Nancy, for this information). The applied mathematician Ringleb succeeded in getting to the United States. See Tobies (2006), p. 274. See also Th. v. Kármán's postwar correspondence with the former head of the research department in Göring's aviation ministry, A. Baeumker, who also went to the United States (Kármán Papers, Caltech, 1.40). In May 1945, v. Kármán had written a report, titled "German Scientists Recommended for Evacuation to U.S." (Kármán Papers, Caltech, 91.1).

[10]The latter applied, for instance, to the aerodynamicist K. Hohenemser, who failed in his efforts to be rehabilitated and reinstated into his previous position in Germany (D). See also Rammer (2002).

[11]In an obituary of Lipman Bers one reads: "After the move to the New York City area, the Berses lived in the enclave of European-born mathematicians in New Rochelle, New York, that had formed around Richard Courant. Members of that community included, at various times, Fritz John, Kurt Friedrichs, Cathleen Morawetz, Harold Grad, Jürgen Moser, Wilhelm Magnus, and Joan Birman." (Abikoff [1995], p. 13). One could add the names of Hungarian born Peter Lax, and of Wolfgang Wasow.

[12]There were visits to Germany by von Kármán (1945), by the nonmathematician Michael Weyl (1945), by B. H. Neumann and O. Taussky (1945), and by Richard Courant (1947 and 1950). G. Szegö's report on his stay in Germany from July 1948 until January 1949 is contained in the Szegö Papers, Stanford University, box 5, f. 22.

[13]S. MacLane traveled to Paris and Germany, in particular Göttingen, in 1947–48. See his letters in the C. R. Adams Papers, Brown University Archives, Providence (D).

[14]The Austrian emigrant Olga Taussky and her British husband John Todd visited Oberwolfach in 1945 as members of the Intelligence Corps of the British Army (letter B. H. Neumann, who was himself in Germany as a member of that Corps, to me, August 22, 1993). See also Süss (1967), p. 32.

the IMU from the very beginning.[15] An important project for the resumption of international contacts was the "FIAT Reviews." These Reviews provided not only proof of the Allies' interest in mathematics done in Germany during the war, but were also crucial in providing a job-creation scheme for German mathematicians. In these valuable and little-known printed reports, secret Nazi war documents were used to analyze progress in several disciplines.[16] There were also initiatives taken by emigrants to improve the libraries in Germany (and other European countries) that had been devastated during the war.[17] German mathematicians such as Franz Rellich and Wolfgang Franz (1905–1996) kept emigrants and Americans informed of the state of mathematical institutes and personnel in Germany. In September 1945, Courant received a four-page list on the whereabouts of German mathematicians written by Franz Rellich, who had taken over as director of the mathematics institute at Göttingen.[18]

The focus on Germany of many emigrants led others among them (Harald Bohr) to issue warnings not to forget the difficult situation of mathematics in Eastern Europe.[19]

Immediately after the war, emigrants such as Richard Courant worked for a gradual rapprochement between American and German mathematicians. He was without doubt the key figure in this process, as shown in the following examples.[20]

In October 1945 Courant wrote a letter to the chief of the American Economic Intelligence Division in which he expressed his desire not to impose on Germany scientific reparations that would impede possible recovery.[21] Courant was concerned not just about mathematics in this context

[15]See Lehto (1998) and, for more details, the files of the AMS Archives (AAMS) in Providence, RI and the M. H. Stone Papers in the same archives, located at Brown University.

[16]*FIAT Review of German Science, 1939–1945*, 84 volumes, published by the Office of Military Government, Field Information Agencies Technical (British, French, U.S.), Wiesbaden: Dieterich, 1948–53. The volumes on Applied Mathematics were edited by Alwin Walther, the volumes on pure mathematics by Wilhelm Süss. Courant in his letter of 1945 to the Economic Intelligence Division had proposed such work very early after the war. See Appendix 5.1.

[17]"Report of Committee on Aid to Devastated Libraries" (1948), AAMS (Providence), 75.2, box 13, f. 64.

[18]Göttingen, September 26, 1945. AAMS (Providence), 75.5, box 30, f. 45. Communicated by Courant to J. R. Kline the secretary of the AMS. Franz wrote a six-page German "Report on the Development since 1933 and the Current State of the Mathematical Institute of the University Frankfurt" (W. T. Reid Papers, AAM, Austin, box 13, f. 7).

[19]See CIP, file: Trip to Europe 1947, twenty-six-page report by Courant: "Harald stresses that psychologically necessary not to forget Poland and other victims over German problems" (p. 23).

[20]The examples are all documented in the Courant Papers at the New York University Archives.

[21]Published as Appendix 5.1 below.

but also about other scientific disciplines. In the summer of 1947 Courant traveled to Germany for the first time since the war. During the trip he kept a twenty-six-page diary and wrote several reports. Among the reports a four-page "Memorandum on Scientific Rehabilitation in Germany,"[22] was followed by a seven-page "Some General Impressions of the Present Situation of Science in Germany" (August 31, 1948), and the three-page memorandum "Rehabilitation in Germany and the Marshall Plan" (December 15, 1948).

Needless to say there was plenty of misunderstanding between emigrants and the mathematicians who had chosen to remain in Germany. One problem concerned the publication rights for German books seized by the Americans, who then republished the books during and after the war, generally without paying royalties to the German authors.[23]

The harsh judgment passed by some emigrants on the political behavior of German mathematicians who had remained in Germany was often induced by their own sufferings, personal losses, and disappointments and therefore was sometimes exaggerated (D). A man such as Richard Courant was, on the one hand, well informed about the developments in Germany during the Hitler years and therefore was able to show more discernment in the matter decades before some professional historians of that period.[24] On the other hand, Courant could not easily forget and disregard the humiliations and sufferings he had personally experienced in Germany after 1933.[25] Nevertheless, Courant, along with many of the emigrants was prepared—in the interest of their science in general—to do his utmost in helping colleagues in the "denazification" procedures.

Apart from Courant, testimonies by emigrants were often based on judgment from a distance and showed little familiarity with circumstances within Nazi Germany. Instead, personal relationships and scientific interests were the main basis for their testimonies. The so-called Persilscheine (whitewash certificates) of the emigrants thus contributed, at least in the Western zones of occupied Germany, to the general failure of denazification, a process that was soon to be distorted and curtailed anyway by the developing Cold War. There was a general lack of critical self-reflection on the part of

[22]October 29, 1947, written with N. Artin. Courant Papers, New York University.

[23]For an admission of this continuation see a letter by the Office of Alien Property to J. R. Kline, April 11, 1949 (AAMS, Stone Papers, box 37, f. 5). See also remarks above in chapter 10.

[24]On March 14, 1947, Courant wrote to Franz Rellich about G. Doetsch's "horrifying behavior" during the Nazi period and added: "It is not the party membership that really matters." (CIP, file Rellich, 1945–47). At the same time Courant defended M. Deuring against unjustified Polish attacks, according to which Deuring had destroyed Polish mathematical literature while in Poznan during the war (CIP, file: K. Reidemeister, 1947–60).

[25]See Courant's letter on Hasse to W. Hanle in (D) below.

German mathematicians about their past under Hitler's regime,[26] which proved a liability in the relations between German and American mathematicians at least until the 1960s.[27] Some of these subliminal conflicts are still palpable today.[28] The often bureaucratic and callous treatment of former refugees in their compensation claims against the German authorities led to further friction between the emigrants and the new Germany. The pseudo-legality of the Nazi laws was often used as a basis for decisions, and thus the true terrorist nature of Nazi rule was misjudged.[29] In some cases, including even mathematicians of Jewish descent, German postwar authorities alleged "voluntariness" of emigration, causing a temporary hold-up in the granting of compensation, although most claims were apparently settled after a while.[30] In fact, during the 1960s the group of people eligible for compensation was widened to include, for example, those whose application for civil service as a teacher had been turned down for political (including racist) reasons after 1933.[31] However, no general and official invitation was ever extended to the emigrants to return to Germany or Austria,[32] nor was it always made easy for them to regain their German or Austrian citizenship.[33] In at least in one case there are reports about in-

[26]One of the few self-critical mathematicians was, of all people, Oskar Perron, who had repeatedly tried to oppose the Nazis. As late as 1951 he felt "personally ashamed" and advised against electing foreigners to the Bavarian Academy of Sciences, because "we should not try to elevate ourselves before them" (Georgiadou 2007, p. 5).

[27]This became very clear in connection with Helmut Hasse's invitation in 1963 to be the main speaker at a Mathematical Association of America memorial symposium to honor the recently deceased Emil Artin. One emigrant, L. Bers, protested vehemently against the invitation, if in vain, while Courant wrote a critical opinion on Hasse. Courant's disappointment over a lack of critical discussion in Germany on the Nazi past is also palpable in his rather negative evaluation in 1960 of the German plans for a Max Planck Institute for mathematics. See Siegmund-Schultze (1993c), p. 35, and the letter by Courant to W. Hanle, quoted below in (D).

[28]For example, there have been discussions about the Brain Drain in mathematics, and on the development of mathematical reviewing, notably the competition between *Zentralblatt* and *Mathematical Reviews* with clear reference to the past. See Siegmund-Schultze (1994).

[29]Dehn was humiliated in his claims by the authorities in Frankfurt (D), and so was Rosenthal in Heidelberg (Mußgnug 1988, 274–75). In order to obtain pensions some expelled Jewish mathematicians had to rely on expert opinions by German mathematicians, including some who were incriminated and involved in the Nazi regime, such as Bieberbach (D).

[30]This applied to Richard von Mises (D) and to Otto Neugebauer (CIP, file: Neugebauer, 1938–53). Schwerdtfeger received no compensation (D).

[31]This applied to the emigrant Wolfgang Wasow (D).

[32]There were, however, offers to return that were extended on a more local level, such as by the philosophical faculty of the University of Köln, which wrote to Hans Hamburger to this effect in May 1946. In this particular case Hamburger first proposed to return in autumn 1947, but then decided to accept a professorship in Ankara (Turkey). He returned to Köln in 1953. See SPSL, box 279, f. 6 (Hamburger, H.), folios 442, 447.

[33]Strauss (1991), p. 12. According to Stadler (1988), p. 121, Menger's wish to return to Vienna "failed miserably," while several former Nazis continued to blossom in their careers

trigues against inviting refugees to return.[34] Of course, such invitations would hardly have been accepted on a broad scale, given the high percentage of Jewish emigrants. As a rule, within the overall cultural-scientific emigration the wish to return was greater, the more political the motive for emigration had been.[35] However, explicit political motives had not been typical for the emigration of mathematicians. Also, the pressure for naturalization, which had been exerted upon the immigrants in the United States and the legal measures that threatened immigrants with the loss of their American citizenship if they returned to Europe, did nothing to strengthen the desire to return.[36] In several cases honorary doctorates (von Mises, Hempel, etc.) were extended to emigrants by German and Austrian universities, and accepted. However, invitations to join societies and academies in Germany often received negative responses from former refugees, owing to old wounds and disappointments.

The beginning of the remilitarization of Germany and the Cold War added to the problems of mathematicians both from the East and the West of politically and scientifically "coping with the past" ("Vergangenheitsbewältigung"). A letter by Max Dehn sent to the refounded Deutsche Mathematiker-Vereinigung in 1948, in which he explains his refusal to rejoin the DMV, is a particularly impressive document (Appendix 5.2). Discussions on the political history of mathematics under the Third Reich have continued until quite recently, as the exchange of letters to the editor in the *Notices of the American Mathematical Society* on the quality of obituaries in the *Jahresbericht der Deutschen Mathematiker-Vereinigung* shows.[37]

---

at their original institutions after the war. In the case of A. Rosenthal, compensation was originally given only on the condition of his return to Heidelberg. See Mußgnug (1988), p. 275. On P. Thullen's problems to regain his German citizenship see (D) below.

[34]This concerns the case of Ernst Jacobsthal and the Technical University in Berlin (D).

[35]Möller (1992), p. 608. According to Möller, about one-third of the cultural emigration returned after the war (ibid., p. 610).

[36]These measures (McCarran-Walter Act of 1952), which affected Weyl's return to Switzerland, prompted Courant to write to U.S. Senator S. Kilgore (CIP, file: H. Weyl, 1947–57) on May 3, 1955. They also threatened F. Bernstein's opportunity to return to Europe under reasonable conditions. See above 9.S.2.

[37]See in particular the letter by Kahane, Krickeberg, and Lorch (1994), which justly criticizes several less than sensitive biographies in the *JDMV* but slightly exaggerates the point by not acknowledging the merits of more recent contributions, for instance the biography, Schappacher/Scholz, eds. (1992), of the Nazi-mathematician O. Teichmüller.

Figure 55 *Carl Gustav Hempel (1905–1997). Hempel was an emigrant from Berlin and became a leading figure in the mathematically inspired analytical theory of science in the United States. The honorary doctorate he received from the University of Konstanz in 1991 after the German reunification of 1989–90 came at a time when the wounds of the past had begun to heal.*

11.D. Documents

11.D.1. *The New Wave of Emigration after the War*

On March 28, 1950, the American Federal Security Agency sent a round-robin letter to college and university administrators with the following information:

> If you need an outstanding mathematician, physicist, chemist, biologist or scientist in other fields, either for teaching or research, the attached statements of qualifications of selected German and Austrian scientists who are available may be of interest to you.[38]

Involved in this initiative were also the Institute of International Education (New York), which had housed the Emergency Committee during the war and which now offered services to foreign students, and the International Science Foundation (San Francisco), which created and supported reception centers for foreign engineers and scientists.[39] Emil Julius Gumbel wrote to G. Radbruch (Heidelberg), on June 29, 1947:

> A scientific friend of mine, Dr. Hermann von Schelling, has recently tried to habilitate [i.e., tried to obtain a teaching permit; R. S.] for statistics in Heidelberg. Schelling continued to write to me, even after my denaturalization [Ausbürgerung], in order to keep scientific contact with me. This was not without danger to him. His scientific qualification is without doubt. However, the poor man is stricken with polio and tied to a wheel chair which is bound to impede his teaching. As he wrote, to me his chances in Heidelberg are small.[40]

11.D.2. *Remigration and Obstacles to It*

Ernst Jacobsthal, who had been expelled from Berlin and become a refugee to Norway and Sweden in 1939 and 1943, wrote to the British SPSL in July 1947, when he was sixty-five years old:

> Obstacles have been put in the way of my return to Berlin. My former colleague Professor A. Timpe has reached a decision by the faculty to create a full professorship for me. However, the rector and Professor Mohr have schemed against this decision. According to my information the anti-Semitic sentiment there is still as virulent as it used to be. I am wondering why the [British]

---

[38]MITA, AC4, box 100, f. 1.

[39]The related correspondence (1954) of the International Science Foundation with the American Mathematical Society is in AAMS 75.5, box 39, f. 99.

[40]Gumbel Papers, Chicago, box 2, f. 1 (T). See also von Schelling's correspondence with Richard von Mises in 1947, which shows that he finally went to the United States (Mises Papers, HUA, HUG 4574.5, box 4, f. 1947).

Military Administration does not intervene! Personally I am no longer that much interested since the Norwegian government has created a post for me at the Technical University in Trondheim.[41]

Bernhard H. Neumann wrote in a letter to me on August 22, 1993:

I never was approached officially about returning to Germany, though many friends asked me if I would. I have often been back to Germany, a country I love to visit because I still have family and friends there, and because I can speak the language, but where I could not envisage to live.

In a review of the book by Wirth (1982) on the fate of the Freiburg teacher of mathematics, Wilhelm Hauser, who went back from England to the German Democratic Republic (GDR) and who was sixty-two years old in 1945, one reads:

When Wilhelm Hauser went to the GDR after World War II, it was partly as a result of the communist convictions of his son Harald. Harald Hauser had sympathized with the KPD before 1933 and was connected to communist groups of the résistance, when he was in French exile. But other motives also played a role: for instance the desire to teach again in Germany. His application, written from the Soviet occupation zone, to be re-admitted as a teacher in Baden [in the French zone, R. S.] was turned down due to the age limit. In the GDR he was allowed to teach until he was 74 years old.[42]

Hans Samelson, who had an impressive career in the United States as a topologist, wrote in a letter to me in 1994:

Finally I want to say that the USA was very good to me and that I am very grateful and feeling at home. I never had the wish to return to Germany, although I have many good memories of my young years there, especially the time of the youth movement.[43]

11.D.3. Resumption of Scientific Communication

Saunders MacLane wrote in 1948 to Clarence Raymond Adams (1898–1965) in Providence about his visits to Göttingen and Frankfurt:

The mathematicians at Göttingen (Kaluza, Herglotz, Magnus, Rellich, Arnold Schmidt) are practically starving for lack of scientific contacts with the outside

[41]SPSL, box 280, f. 5 (Jacobsthal, E.), folio 428v/429 (T). Ernst Mohr was a non-Jewish victim of the Nazis, sentenced to death for anti-Nazi propaganda in 1944, but saved from execution by involvement in war research. He was at that time, in 1947, still waiting for the official acknowledgment of his status as a victim (Litten 1996). Maybe his case is indicative of an occasional clash of interests of Jewish and non-Jewish victims.

[42]Schmolze (1983), p. 54 (T).

[43]H. Samelson to R. Siegmund-Schultze, May 2, 1994 (T).

world. . . . [Wolfgang] Franz and I traveled from Frankfurt to Heidelberg in an overfilled German train. . . . In spite of the crowding, we enjoyed ourselves discussing topology and class field theory. It still seems a minor miracle that one can find, in the midst of European confusion and German disorganization, someone who talks the same language on the most recondite aspects of algebra and topology. <u>This</u> much international communication is still possible.[44]

The emigrant Michael Golomb (Purdue University, Indiana) wrote retrospectively to me on July 19, 1993:

> As more and better mathematics was being produced in Germany the relations with the American mathematical community developed rapidly. We had several visiting professors and also graduate students from Germany in our department and the same is true for many other American universities, possibly with the exception of some eastern schools, where there was considerable mistrust against all middle-aged Germans. As an illustration of the respect American mathematicians have paid to mathematical achievements in Germany I mention the following fact. One of the requirements for an advanced degree in science is demonstration by the candidate of reading knowledge of two foreign languages, chosen for their usefulness in his field of specialty. For mathematics students the choice has almost always been French and German.[45] Finally as you surely are aware, American mathematicians have been frequent and enthusiastic participants in the mathematical seminars and conferences in Oberwolfach.

11.D.4. Compensation for the Emigrants

Hans Schwerdtfeger's widow wrote to me in 1993:

> Compensation was applied for, but refused, because for a non-Jew there was no compulsion to emigrate.[46]

When Richard von Mises applied for compensation, the Federal Ministry of the Interior alleged that his departure from Berlin in November 1933 had been voluntary. In his response on February 10, 1953, shortly before his death, von Mises made it clear that the Nazi mathematician and ministerial officer, Theodor Vahlen, had deceived him in 1933:

> Dr. Vahlen emphatically assured me that my departure abroad to accept the offered chair, would not change anything with regard to my pension rights accrued during my previous years of service. Later, on December 1, I received a

[44]S. MacLane to Adams, March 29, 1948. C. R. Adams Papers, Brown University, Providence.

[45]This is probably based on tradition and does not reflect the growing importance of Russian in mathematics in the second half of the twentieth century [R. S.].

[46]Hanna Schwerdtfeger to the author, undated, received July 21, 1993 (T).

demand to relinquish my claims, whereupon I replied that I was not prepared to comply with such a demand.[47]

Compensation for von Mises was approved only after his death. In the process of making their decision the ministry even asked von Mises's former colleague and prominent Nazi-mathematician, Ludwig Bieberbach, for advice. Hilda Geiringer-von Mises wrote a grateful letter to Bieberbach in November 1957, thanking him for his support:

> I believe what you said about von Mises' departure immediately after the seizure of power is exactly true: It was his pride and his intelligence + instinct which caused him to leave a position which had become untenable.[48]

Peter Thullen had not obtained his habilitation when he immigrated to South America in 1935. For this reason he did not receive any material compensation after the war. He was not offered a professorship at a German university either and was happy to receive offers from Swiss universities late in his life. As Thullen had obtained Ecuadorian citizenship during emigration he had to go through a complicated procedure to regain his German citizenship. On this issue, his son Georg said in a letter to me:

> My father insisted that in his case it was not a question of naturalization but of recovering his German citizenship, which he never renounced. All this, however, was to no avail. In the end he had to subject himself to the naturalization procedures. There was no material benefit in doing so; on the contrary, he had to pay a rather hefty fee.[49]

Wolfgang Wasow wrote in his memoirs in 1986 about his late and successful application for compensation after the law had been changed at the beginning of the 1960s:

> After the Second World War, the new German government, conscious of the horror and the lasting revulsion the revelations about the German attempt at exterminating everybody of Jewish descent had caused in the world, did pay considerable sums to surviving victims of racial and political persecution. I doubted that I could succeed with such an application to the German authorities. The sum total of my life experiences had been improved, not damaged by the Nazi period. While I had no objection to getting money from Germany, I felt I should give precedence to real victims of the Nazis. . . . Finally, I did send in documentation for my claims, which were rejected—not to my surprise.

[47]Mises Papers, HUA, HUG 4574, box 8, f. Antrag auf Wiedergutmachung (T). See also above chapter 4.D.6.

[48]Hilda Geiringer to L. Bieberbach, November 13, 1957 (T). Estate L. Bieberbach, Oberaudorf, Germany.

[49]Quoted in Siegmund-Schultze (2000), p. 62. Dated December 5, 1999. Translation from German by Georg Thullen. See also Thullen's memories below in Appendix 6.

A few years later, in 1961, the Germans changed the law governing these matters. . . . The new law extended to persons who had fulfilled all professional conditions for a government position and had applied but were rejected for such [racial or political] reasons. I fell precisely into that category: I had passed the State examination for a career in the German public secondary school system and had applied for a position as "Studienreferendar." . . . My grades and qualifications were good, and I was excluded because of my Jewish ancestors. The point that in 1933 the teaching profession was so overcrowded in Germany that I might not have been hired even under normal conditions, fortunately was not raised.[50]

11.D.5. Political "Coping with the Past" ("Vergangenheitsbewältigung")

Heinz Hopf (Zurich) wrote in 1945 to Hans Freudenthal about his visit to Germany:

Hasse dismissed, as well as Blaschke and Udo Wegner. Süss reinstated recently (what he seems to deserve, according to everything I heard).[51]

Theodor von Kármán described his mixed feelings about the situation in Germany in a letter to Courant on February 4, 1946:

I am somewhat reluctant to interfere with things in Germany. To be quite frank, I am rather disgusted with affairs in that country.[52]

The applied mathematician Kurt Hohenemser, who had been dismissed in Göttingen in 1933 as a "half-Jew," suffered further humiliation when asking—quite categorically—for compensation after the war. He wrote to Courant in May 1946:

I received a letter from Prof. Prandtl saying that Herr Schuler is very angry with me and I should apologize before taking up my appointment in Göttingen. Furthermore—according to Prandtl—it was inappropriate of me to try to get access to Prof. Schuler's institute by exerting pressure through the administration. It was really too much for me that I was supposed to ask this man most courteously for a position in his institute. After all, he occupies the professorship of a dismissed Jewish colleague (Bernstein), he gained his position through

---

[50]Wasow (1986), pp. 368–69.

[51]ETH, Zurich, manuscript division, papers Hans Freudenthal. H. Hopf to Freudenthal, December 11, 1945 (T). I am grateful to L. Beaulieu for this information. Wilhelm Süss (1895–1958) was the president of the DMV during much of the Third Reich and collaborated with the regime without being as fanatic a Nazi as Bieberbach was. See Remmert (2004). Segal (2003) tends to agree with Hopf's view of Süss.

[52]Kármán Papers, Caltech, 6.16.

Figure 56 *Kurt Hohenemser (1906–2001). The applied mathematician and aerodynamicist was a victim of anti-Semitism in Göttingen in 1933. His claims for rehabilitation immediately after the war failed and he went to the United States, where he continued to work in helicopter research. His application for compensation from Germany was successful in 1958.*

the political pressure of his Nazi students, and treated me as an undesirable beggar when I went to see him after the war. And I was supposed to apologize to him because I dared to mention his less than impeccable past. I vehemently refused to do so and got further serious admonitions from Prof. Prandtl who wrote that my integration into the faculty was being jeopardized by my behavior. But I won't flinch until I am fully rehabilitated. I'd rather give up Göttingen and play at being a school teacher [den Schulmeister machen] somewhere else.[53]

Courant, who saw Hasse during his visit to Göttingen on 21 July 1947, wrote in his diary: "Hasse. Mixed feelings."[54] Hermann Weyl commented in early 1947 about his feelings concerning the possibility of resuming his relationship with the Göttingen Academy of Sciences:

I would feel not too happy to be associated with men like Hasse in the same learned body. On the other hand I really do not wish that they now take repressive actions against these men, of the same type as were used against us.[55]

Ruth Moufang (1905–1977), who had been discriminated against as a woman-mathematician during the Nazi period, wrote in 1948 in a letter from Frankfurt to Max Dehn about the delays in Dehn's and Ernst Hellinger's applications for compensation (full retirement benefits). She did not think that the old anti-Semitism was responsible for the delays but admitted:

Not all of the fixed ideas of the Nazi time have disappeared. And denazification was a failure, it resulted neither in moral nor in actual cleansing (for instance 90 percent of the briefly dismissed have been reinstated). The Germans will never manage to behave reasonably . . . it will get worse as the influence of occupation gets less.[56]

Emil Julius Gumbel wrote in 1950 to Norbert Wiener about the repercussions of the Cold War on his career as an emigrant in the United States:

I was and am a convinced antifascist and antimilitarist. And I had to leave Germany + France for this reason. As an unrepentant sinner I am not welcome at a

[53]OVP, Library of Congress, cont. 4, f. Courant (1939–48). Hohenemser to Courant, May 21, 1946, T). K. Hohenemser wrote to me on December 23, 1997 that he did not have proof for the statements in his letter, particularly with respect to the Nazi past of Max Schuler (1882–1972). Partly for this reason his rehabilitation failed. He immigrated to the United States in 1947 and continued working on the aerodynamics of helicopters as he had done previously in the German aviation industry from 1935. From 1958 he received a pension at the level of a professor as compensation from Germany. See also Rammer (2002).

[54]Courant's twenty-six-page diary on his trip to Europe, June–August 1947, p. 22 (CIP, NYU).

[55]Weyl Papers, ETH, Zurich, Hs 91:167, Weyl to Courant January 27, 1947. I thank L. Beaulieu for this information.

[56]R. Moufang to M. Dehn, September 16, 1948 (T). Dehn Papers, AAM, Austin.

time when the Nazis and Militarists are put back into power in Germany. Consequently + rightly I am without a job.[57]

In 1950, Ernst Jacobsthal, the immigrant to Trondheim (Norway), wrote the following to the Springer publishing house about Wilhelm Blaschke's lack of sensitivity about the past:

> In 1950 Springer published a book by Herr Wilhelm Blaschke, which has the title Introduction into Differential Geometry [Einführung in die Differentialgeometrie]. . . . Herr Blaschke still cannot give up his Nazi methods. . . . Quoting mathematicians of Jewish descent as "Jewish" while mentioning other mathematicians of non-Jewish descent only by their names and without nationality is clearly tendentious and inexcusable in Germany in 1950. Even the addendum "outstanding" ["hervorragende"] does not conceal the agenda, which Blaschke pursues with such quotations.[58]

Hans Rademacher wrote to Gumbel in July 1952:

> But already now the old Nazi-leaders, feeling to be indispensable in the scheme of rearmament devised by the Allies, are reappearing everywhere on the scene, in particular in education and civil administration.[59]

The immigrant to Australia, Felix Behrend, said the following about his conversation with Thomas Mann on June 26, 1954 in Zurich:

> "And now we have found asylum here again," said Thomas Mann, who spoke out against McCarthyism sharply, almost passionately. "It began in the year 1949," he continued, amused by the memory, "when I played a trick on the Americans, going to Weimar, to East Germany, to the celebration of Goethe's 200th birthday." Thomas Mann expressed an almost child-like joy over this successful trick. "The persecution of the physicist Oppenheimer," I said, "reminds me of Einstein's similar fate in Berlin. This was the time of anti-Semitic riots at the University of Berlin."[60]

Richard Courant, having been asked for his opinion about a possible doctorate honoris causa for Helmut Hasse, wrote in May 1963 to the director of the Physical Institute of the University Gießen, W. Hanle:

---

[57]Gumbel to Wiener, December 22, 1950; MIT Archives, Wiener Papers, box 3, f. 132.

[58]Jacobsthal to F. Springer, September 17, 1950 (T). Copy in ETH, Zurich, van der Waerden Papers, Hs 652:11988c. Springer reacted immediately and terminated Blaschke's collaboration in the Grundlehren series, the "yellow collection." For Blaschke's political stance during the Third Reich see chapter 4.

[59]H. Rademacher to E. J. Gumbel, July 8, 1952. Gumbel Papers, Chicago, box 4, f. 3.

[60]ETH, Zurich, Thomas Mann Archives, F. Behrend, ms. "Die Fahrt zu den Vätern" ["The Journey to the Fathers"], p. 18 (T). Behrend is apparently alluding to an episode in Berlin in 1929.

92 — Jacobsthal
51 (092)

Ernst Jacobsthal

AV

SIGMUND SELBERG

(Minnetale i Fellesmøtet 5te april 1965)

Den 6. februar i år gikk vårt medlem Ernst Jacobsthal bort, i en alder av 82 år. Da dødsbudskapet kom, var det mange av oss som følte det som en av våre nærmeste hadde forlatt oss. Jacobsthal hadde en merkelig evne til å knytte dem han kom i berøring med, til seg med vennskapsbånd. Hans venneskare ble også som en stor familie, hvor de enkelte medlemmer ikke behøvde å komme i personlig kontakt, men likevel kjente til hinannen gjennom den sentrale skikkelse — Ernst Jacobsthal. Hans interesser spente over et så stort register, fra vanskelige matematiske problemer til små dagligdagse gleder og bekymringer, at han kunne nå inn i alles sinn. Ja så menneskelig enkel og naturlig var han, at man måtte bli glad i ham.

Professor ERNST JACOBSTHAL ble født 16. oktober 1882 i Berlin av foreldre Dr. Martin Jacobsthal og Ida, f. Rosenstern. Hans far og også hans farfar var lege. Det er vel grunnen til at professor Jacobsthal alltid var så interessert i den medisinske forskning. Likevel valgte han ikke denne vei selv.

I 1906 tok Jacobsthal embetseksamen og doktorgrad ved Berlin Universitet. Hans lærere i hovedfaget matematikk, var de kjente matematikere Frobenius og Herman Amandus Schwarz. Doktoravhandlingen hadde tittelen: «Anwendungen einer Formel aus der Theorie der quadratischen Reste». Det er et arbeide som nå er blitt klassisk, og som finnes omtalt i de fleste større lærebøker i tallteori.

Figure 57 *Ernst Jacobsthal (1882–1965). The number theorist and student of G. Frobenius and H. A. Schwarz immigrated in 1939 to Norway, from where he had to escape for a period in 1943, as described in a letter to H. Weyl quoted in chapter 6. After the war he complained about the lack of political sensitivity on the part of his colleagues in Berlin and in Hamburg.*

Mister Hasse is doubtless a mathematician who is enthusiastic about his science. I have known him since World War I, and I have been friendly with him, although even then I found his wild, and quite open pan-German convictions [alldeutsche Gesinnung] quite sinister. Nevertheless I was deeply disappointed when, in 1933, he forgot about all his loyalty to his Jewish teachers and mentors and followed the Nazis with full enthusiasm. . . . To my knowledge, he has not behaved nastily to anybody. . . . It is well known among mathematicians that he was dismissed from office after the Collapse [Zusammenbruch], that he then became a professor at the Humboldt University in Berlin and was finally appointed by the Nazi Blaschke in Hamburg. Gradually, of course, the past is being forgotten by the younger generation; this apparently explains that Hasse was invited to give a talk in Colorado this summer (which provoked passionate protests).[61]

In 1970 the postwar emigrant Wilhelm Magnus (New York University) sent critical remarks about Max Pinl's report "Kollegen in einer dunklen Zeit" ("Colleagues in a Dark Time," Pinl 1969) to the editor of the *Jahresbericht der Deutschen Mathematiker-Vereinigung*:

At the end of the Introduction (p. 168) one reads:

"In addition we want to avoid the dubious [anrüchige] terminology, which can be found in reports on the infamous period." I do not know what the word "dubious" means here, but I assume it relates to the fact that in the entire report the words "Jew" and "Nazi" have been avoided. But they do belong in it precisely where it should have been said that the overwhelming majority of the colleagues mentioned in the report were persecuted because they were Jewish (in the sense of the Nazi legislature) or because they had close connections to Jews, and that, in addition, some colleagues were persecuted because of their political opinions, which they had articulated publicly before 1933. One could possibly give the reasons including the concrete names at the end of the report. This would also render superfluous the following, most unfortunate sentence (p. 168): "To mention reasons for the persecutions is not necessary, because there were none."[62]

---

[61]R. Courant to W. Hanle, May 25, 1963 (T), CIP, NYU, file: L. Bers. The protests against Hasse's invitation to honor Artin have been mentioned above. Courant's judgment on Hasse and Blaschke is probably somewhat exaggerated and unbalanced. See Segal (1980) and Hasse's obituary of his Jewish teacher Kurt Hensel (Hasse 1936). Courant's letter appears to have killed the attempt to make Hasse "Ehrendoktor" in Gießen, although the title was awarded to Hasse in Kiel on another occasion.

[62]W. Magnus to J. Tits, January 27, copy, 1970 (T). Dehn Papers, AAM, Austin, box 4. The later English report, Pinl and Furtmüller (1973), gives the "reasons" in detail and thus seems to take into account the criticism by Magnus that had been made in agreement with an unnamed "American mathematician."

11.S. Case Study

11.S.1. A Case of Failed Compensation: Max Dehn

An anonymous article in *Aufbau* (New York) on Friday, August 29, 1952, pointed to a campaign in favor of—in the meantime deceased—Max Dehn[63] in the daily newspaper *Frankfurter Rundschau* and in the University of Frankfurt. Dehn had been forcibly retired in 1935 prior to the Nuremberg Laws, using the obscure "rubber paragraph" 6 of the Law for the Restoration of the Professional Civil Service[64] of April 7, 1933. "Dehn should have become Professor Emeritus, according to the wishes of the university, which would have secured him the undiminished payment [Nachzahlung] of his entire lost salary." Erich Lissner wrote in the *Frankfurter Rundschau* of July 5, 1952:

> For ten years he—who could no longer expect a full professorship due to his advanced age—has worked at several smaller universities over there, finally at the Black Mountain College in North Carolina, which upheld true humanistic traditions. . . . His rights did not consist in a payment of his retirement benefits [in Frankfurt] but in his emeritization [Emeritierung], which means the undiminished payment of his salary—the legitimate [wohlerworbene] right of every Ordinarius of a German university.[65]

However, the ministry did not support this campaign. Also, Lissner's article came only after Dehn had already died from a lung embolism on June 27, 1952. The historian of science, Willy Hartner (1905–1981),[66] wrote in Dehn's obituary in the *Frankfurter Allgemeine Zeitung* on July 8, 1952 rather angrily:

> Many years before [his dismissal], Dehn, who had the highest standards for his own work and for the work of others, had sharply criticized an inferior publication by a colleague.[67] When the latter was appointed in spring 1935 to the

[63]For the following see the Max Dehn Papers at the AAM in Austin, box 3, no. 56. The events are also documented in the University Archives in Frankfurt am Main.

[64]See chapter 4.

[65](T). Lissner also says that Rector and Senate of the University referred to the case of the orientalist Gotthold Weil (1882–1960), then in Jerusalem, where this procedure and interpretation had been applied. Lissner's article also contains the quotation from Dehn that serves as the epigraph of this chapter.

[66]See Cohen (1983) and Siegel (1965) for Hartner's courageous stance in the Third Reich and on the purely financial/existential motives for his return to Germany from the United States.

[67]Hartner alluded to Dehn's scathing review of Theodor Vahlen's book *Abstrakte Geometrie* (1905), criticism that seems to have stimulated Vahlen's "flight" from pure into applied mathematics. See Siegmund-Schultze (1984), pp. 20–21. It was also Vahlen who deprived Richard von Mises of his rights to a pension when he immigrated to Turkey. Vahlen was,

Figure 58 *Vienna Exhibit. This picture was taken in front of the poster dedicated to the refugee from Vienna, Franz Alt, at the Vienna exhibition titled "Cool Good Bye from Europe," in September 2001. From right to left: Franz Alt, Robert M. Wald (the well-known theoretical physicist and cosmologist, son of the emigrant, Abraham Wald), the author of the present book, and the organizer of the exhibit, Karl Sigmund.*

ministry of education, Dehn knew that his days at the University of Frankfurt were over. What he predicted soon became reality: with effect of April 1, 1935, even before the Nuremberg Laws, he was retired under the flimsy pretext of austerity measures. . . . As late as April [1952] the authorities in Wiesbaden, responsible for compensation, adopted the hypocritical argument of the Rust ministry that Dehn was dismissed because of the cancellation of his chair and not because he was a Jew and that therefore his right to emeritization [Emeritierung] was not proven. The authorities stuck to this in spite of all petitions stressing the urgency the matter; even pointing out the possibility of his death did not alter the situation. He did not live to see his right granted. Nevertheless he was never shaken in his belief in a new Germany. He had even considered returning to Frankfurt University the next winter without attaching any conditions to it. The carelessly neglected opportunity for compensation [Wiedergutmachung] imposes a severe moral liability for our young Republic.

---

since 1933, in the Nazi ministry, somewhat contrary to Hartner's remarks. See also Dehn's position to the DMV in 1948. See Appendix 5.2.

Lists of Emigrated (after 1933), Murdered, and Otherwise Persecuted German-Speaking Mathematicians (as of 2008)

PRELIMINARY REMARKS ON THE SOURCES AND SPECIAL
SYMBOLISM USED FOR THE LISTS IN APPENDIX 1

THE FOLLOWING lists are based on published and unpublished sources. The sources are quoted as before in the text from the bibliography or from archives or left papers with the abbreviations introduced (such as OVP for Oswald Veblen Papers, Library of Congress). Mentioning one of the latter implies that a file on the respective person is kept there. EC-N means there is a file at EC on the refugee, however, as a nongrantee. Among the published sources addressing the fate of the persecuted mathematicians collectively, Pinl (1969/72) and Pinl/Dick (1974/76) are still the most detailed, despite of some errors. Further biographical information can be gathered from *IBD*, from Pinl/Furtmüller (1973), and Tietz (1980). Almost all mathematicians in the lists 1.1 and 1.2 are named in at least one of these printed sources, which are therefore as a rule not quoted again, if additional sources are available. The few mathematicians not named in them, are labeled NSS (=not in standard sources). In these cases, other, mostly unpublished sources are quoted. The *Lists of Displaced Scholar*s of 1936/37 (*LDS*), are rather unreliable and incomplete and contain only names for first orientation. Information about unpublished sources in American archives is given in Spalek (1978) and *AC*. In addition to the archives, an occasionally used unpublished source are the *IBD* microfilms containing the original questionnaires for the *IBD*, which are available in several libraries, for instance in the "Zentrum für Antisemitismusforschung Berlin." The films comprise some victims who were not included in the printed version of the *IBD*. A particular degree of the prominence of the respective mathematician is indicated when his/her career was honored by the publications *DSB*, *NDB,* and *CW*. In addition, some dates have been taken without specific reference from the *Biographisch-literarisches Handwörterbuch zur Geschichte der exakten Wissenschaften* (*Poggendorff*, Leipzig 1863–2003). The latter, which is now also available on CD-ROM, was supplemented in 2004 by a printed

"Ergänzungsband Mathematik" (Weinheim: Wiley, ed. H. Kühn) in two parts, to which it will be referred below as Pogg-Erg. Temporary addresses and short biographies of living mathematicians can be found in several editions of *American Men and Women of Science* (AMWS). Reference to the *Notices of the AMS* means usually only a short notice of death. Appendix 1.3 comprises non-émigrés, many of them teachers of mathematics, which are as a rule not mentioned in the standard sources. A useful newer source pertaining to this appendix is Tobies (2006) = Tob. Further conjectures on the fates of German mathematicians can be gathered from Dresden (1942) and Toepell, ed. (1991) = Toep.

For the special purpose of the Appendices 1.1 through 1.3 the following abbreviations are used: "Letter" means a private letter to the author of this book exists. Special symbols used are NJ = not Jewish, PR = partner was victim of racist persecution, NU = not originally from the university system, RM = remigration after 1945, Interview = interview with author, Photo = the quoted source contains a photograph of the mathematician. "Places of Emigration" in the first list do not include countries that—as often France—served only as countries of transit without offering job opportunities.

For more applied fields of mathematics the following information is given on the mathematicians' subject, which in some cases was only one of several others or changed during emigration: Econ = close to mathematical economics, Eng = close to engineering mathematics, Hist = historian of mathematics, Ped = mathematics teacher, Phil = mathematician close to philosophy, Stat = mathematical statistics or insurance mathematics.

List of German-Speaking Mathematicians
Who Emigrated during the Nazi Period
(First Generation)

No.	Name	Dates	Place of expulsion	First place of emigrat.	Final place until 1945	Sources and Remarks
1	Alt, Franz	1910–	Vienna	1938 US	1938 US	NU, AC, Pinl/Dick (1974), letters (1993/97, photo), partic. ICM Berlin 1998, lives in New York City
2	Artin, Emil	1898–1962	Hamburg	1937 US	1937 US	NJ, PR, RM, DSB, Spalek, R. Brauer (1967, photo), Reich/Kreuzer (2007)
3	Artzy, Rafael, born Deutschländer	1912–2006	Königsberg	1933 Palestine	1933 Palestine	IBD, letter (1997, photo), died in Haifa (Israel)
4	Baer, Reinhold	1902–1979	Halle	1933 GB	1935 US	RM, SPSL, EC-N, Gruenberg (1981, photo)
5	Baerwald, Hans G.	1904–1987	Berlin	1933 GB	1939 US	NU, EC-N, IBD-Microfilm, Spalek, AMWS 11 (1965), Eng
6	Basch, Alfred	1882–1958	Wien	1939 US	1939 US	RM, SPSL, HSP, OVP, EC-N, Wunderlich (1959), Eng
7	Behrend, Felix	1911–1962	Berlin	1934 GB	1940 Australia	SPSL, Cherry/Neumann (1964), ETH Thomas Mann Archives
8	Bergmann, Gustav	1906–1987	Vienna	1938 US	1938 US	NU, HSP, OVP, EC, Thiel (1984); Stadler (2001, photo)
9	Bergmann, Stefan	1895–1977	Berlin	1934 SU	1939 US	DSB (Suppl), Schiffer/Samelson (1979, photo), EC-N; OVP, SPSL
10	Bernays, Paul	1888–1977	Göttingen	1934 Switzerl.	1934 Switzerl.	DSB (Suppl), SPSL, CPP, HSP, OVP, Müller (1978, photo), IASA
11	Bernstein, Felix	1878–1956	Göttingen	1933 US	1933 US	DSB, EC, EP, OVP, SPSL, Bernstein Papers, further papers at Göttingen (not used) and

No.	Name	Dates	Place			Sources
12	Bers, Lipman	1914–1993	Prag	1940 US	1940 US	Basel (not seen), Frewer (1981), Stat, Appendix 4.4
13	Bochner, Salomon	1899–1982	Munich	1933 GB/US	1933 US	OVP, Albers et al. (1990), Abikoff (1995), Bers (1988)
14	Boll, Ludwig	1911–1985	Frankfurt	1934 Netherl.	1934 Netherl.	DSB (Suppl), EC-N, IASA, Bochner Papers, Rice University (not seen by this author); Appendix 2
15	Brauer, Alfred	1894–1985	Berlin	1939 US	1939 US	RM, NSS, Boll (1977), Arnold (1986), Interview 1983
16	Brauer, Richard	1901–1977	Königsberg	1933 US	1935 Canada	HSP, EC, OVP, IASA, Rohrbach (1988)
17	Breuer, Samson	1891–1974	Karlsruhe	1933 Palestine	1933 Palestine	DSB (Suppl), SPSL, CPP, EC, IASA, Szegö Papers, CW (1980, photo), Green (1978), Rohrbach (1981)
18	Breusch, Robert	1907–1995	Freiburg	1936 Chile	1939 US	Pinl (1971/72), SPSL, IBD-Microfilm, Deutsche Biographische Enzykl. 2 (1995), 126; Stat
19	Busemann, Herbert	1905–1994	Göttingen	1935 Denmark	1936 US	NJ, PR, NU, IBD, *Notices AMS* 42 (1995), 779; Tob 70/71
20	Caemmerer, Hanna von (married to B. Neumann)	1914–1971	Berlin/ Göttingen	1938 GB	1938 GB	NJ, AC, CPP, Jessen (1993, photo), Beyerchen (1982), *Notices AMS* 41 (1994), 472
21	Carnap, Rudolf	1891–1970	Prague	1935 US	1935 US	NJ, PR, IBD-Microfilm, CW (1988), Newman/Wall (1974, photo), letter B. Neumann (1993), Pogg-Erg

(*continued*)

No.	Name	Dates	Place of expulsion	First place of emigrat.	Final place until 1945	Sources and Remarks
22	Cohn, Arthur	1894–1940	Berlin	1940 Palestine	1940 Palestine	NU, NSS, doctor student I. Schur 1921, Ped., death after shipwreck in Haifa 1940, accord. to inf. S. Hank (Berlin)
23	Cohn-Vossen, Stefan	1902–1936	Cologne	1935 SU	1935 SU	Pinl (1971/72), *Uspechi Mat.Nauk* 2 (1947), no. 3, 107–41
24	Courant, Richard	1888–1972	Göttingen	1933 GB	1934 US	EC, SPSL, OVP, CIP, CPP, CP, Spalek, Reid (1976, photos), Weaver (1972), Appendices 4.2 and 5.1
25	Dehn, Max	1878–1952	Frankfurt	1938 Norway	1941 US	DSB, NDB, Dehn Papers, SPSL, CPP, OVP, EC, AC, Spalek, Siegel (1965), Sher (1994), Appendix 5.2
26	Dubislav, Walter	1895–1937	Berlin	1936 Czechosl	1936 Czechosl	NJ, Thiel (1984), Danneberg et al. (1994), NDB, Pogg-Erg, Phil
27	Fanta, Ernst	1878–1939	Vienna	1939 Brazil	1939 Brazil	Einhorn (1985), Eng
28	Fanta, Werner	1905–?	Vienna	1939 Brazil	1939 Brazil	NSS; son of E. Fanta, Tob 101; letters to A. Dick 1971 in Archives Austrian Academy of Science, Dick Papers, kind information by Ch. Binder (Vienna)
29	Feller, Willy	1906–1970	Kiel	1933 Denmark	1939 US	DSB (Suppl), Birnbaum (1970, photo); SPSL, CPP, EC
30	Fenchel, Werner	1905–1988	Göttingen	1933 Denmark	1942 Sweden	CPP, OVP, SPSL, Jessen (1993, photo), *Notices AMS* 35 (1988), 644; married to refugee mathematician Käte Sperling

No.	Name	Dates	City			References
31	Fraenkel, Adolf A.	1891–1965	Kiel	1933 Palestine	1933 Palestine	DSB, CPP, SPSL, EC, Fraenkel (1967)
32	Freudenberg, Karl	1892–1966	Berlin	1939 Netherl.	1939 Netherl.	RM, IBD, Stat
33	Freudenthal, Kurt (Fulton)	1910–1995	Munich	1938 Colombia	1938 Colombia	IBD, letter (1994)
34	Fried, Hans	1893–1945	Vienna	1940 US	1940 US	NU, NSS, IBD-Microfilm, EC, OVP, HSP, Duggan/Drury (1948), Dresden (1942), 426; Ped
35	Friedrichs, Kurt	1901–1982	Brunswick	1937 US	1937 US	NJ, PR, CPP, EC, Spalek, Reid (1983)
36	Froehlich, Cecilia	1900–1992	Berlin	1937 Belgium	1941 US	IBD, EC-N, Tob 115; Eng
37	Frucht, Robert	1906–1997	Berlin/Trieste	1930 Italy	1939 Chile	NU, HSP, OVP, Frucht (1982), expelled from Italy after racial laws 1938, Notices AMS 44 (1997), p. 1113
38	Geiringer, Hilda married Pollaczek, 2nd marr. v. Mises	1893–1973	Berlin	1934 Belgium	1939 US	HSP, SPSL, EC, OVP, Frank (1988), Richards (1987), Binder (1992), Siegmund-Schultze (1993b), Appendix 4.3
39	Gödel, Kurt	1906–1978	Wien	1940 US	1940 US	NJ, DSB (Suppl), AC, EC, OVP, CW, IASA, Einhorn (1985), Dawson (1997)
40	Golomb, Michael	1909–2008	Berlin	1933 Yugoslavia	1939 US	IBD, SPSL, HSP, OVP, letter (1993), particip. ICM Berlin 1998
41	Gumbel, Emil J.	1891–1966	Heidelberg	1932 France	1940 US	EC, SPSL, OVP, CPP, EP, Spalek, Gumbel Papers, Gumbel (1991), Jansen (1991), expelled by Nazis before 1933
42	Hamburger, Hans	1889–1956	Cologne	1939 GB	1939 GB	RM, NDB, SPSL, HSP, EC-N, OVP, Golczewski (1988), Grimshaw (1958)

(*continued*)

No.	Name	Dates	Place of expulsion	First place of emigrat.	Final place until 1945	Sources and Remarks
43	Hartley, Herman O. (born Hirschfeld)	1912–1980	Berlin	1934 GB	1934 GB	Pinl (1969), David (1982), until 1937 Hermann Otto Hirschfeld; Pogg-Erg, Stat
44	Hauser, Wilhelm	1883–1983	Freiburg	1938 France	1939 GB	NU, NSS, Archives Konrad Adenauer Foundation St. Augustin (Bonn), Wirth (1982), Oswald (1983), Ped
45	Heilbronn, Hans	1908–1975	Göttingen	1933 GB	1933 GB	DSB (Suppl), SPSL, EC, OVP, CW (1988 = Kani/Smith), Cassels/Fröhlich (1977, photo), Rider (1984)
46	Heller, Isidor	1906–?	Vienna	?	1943 Switzerl.	NSS, Stone Papers, AMWS 12 (1972)
47	Hellinger, Ernst	1883–1950	Frankfurt	1939 US	1939 US	DSB, Spalek, SPSL, HSP, EC, OVP, Hellinger Papers Evanston, Siegel (1965), Rovnyak (1990)
48	Helly, Eduard	1884–1943	Vienna	1938 US	1938 US	DSB (Suppl), NDB, HSP, OVP, EC-N, Einhorn (1985), Sigmund (2004)
49	Helly, Elisabeth (born Bloch)	1892–1992	Vienna	1938 US	1938 US	NSS, OVP, Sigmund (2004), Ped., wife to E. Helly
50	Helmer-Hirschberg, Olaf	1910–?	Berlin	1934 GB	1937 US	IBD, SPSL, OVP, Thiel (1984), 233; Rescher (1997), Phil, Econ
51	Hempel, Carl G.	1905–1997	Berlin	1934 Belgium	1937 US	NJ, IBD, Thiel (1984), Rescher (1969, photo), Rescher (1997), Phil
52	Hermann, Grete	1901–1984	Göttingen	1934 Denmark	1938 GB	NJ, RM, NSS, Hermann (1985, photo); inform. by S. Slembek Strasbourg 1998, Phil

No.	Name	Dates	Place	Emigration 1	Emigration 2	References
53	Hertz, Paul	1881–1940	Göttingen	1934 Switzerl.	1938 US	NDB, HSP, EC-N, Thiel (1984), 235; Phil
54	Herzberger, Max	1899–1982	Jena	1934 Netherl.	1935 US	IBD, SPSL, *The Annual Obituary* 1982 (New York), 174–76 (photo), Eng
55	Herzog, Fritz	1902–2001	Berlin	1933 US	1933 US	IBD-Microfilm, Dresden (1942), AMWS 19 (1995/96), Michigan State University News Bulletin, January 10, 2002
56	Hirsch, Kurt	1906–1986	Berlin	1934 GB	1934 GB	SPSL, Gruenberg (1988, photo), Hirsch (1986)
57	Hopf, Ludwig	1884–1939	Aachen	1939 Ireland	1939 Ireland	NDB, HSP, Sommerfeld/Seewald (1952/53), Eckert (1993); Pinl (1969, photo); Eng
58	Ille, Hildegard (marr. E. Rothe)	1899–1942	Breslau	1937 GB	1938 US	NJ, NSS, Pinl (1969, p. 209); Tobies (ed., 1997), p. 41, Ped
59	Jacobsthal, Ernst	1882–1965	Berlin	1939 Norway	1943 Sweden	NDB, SPSL, HSP, OVP, Lorenz (1992), Knobloch (1998), Selberg (1965)
60	Jacobsthal, Walther (Bruns)	1876–?	Berlin	?	1939 US	NSS, OVP, Dresden (1942), 423; Toep 177; Ped, later University of Syracuse
61	Jacoby, Walter	ca. 1905– ca. 1968	Berlin	? after 1938	1939 US	NSS, Dresden (1942), 424; Ped, letter Lutz Noack (Stroud, GB), February 25, 1999
62	John, Fritz	1910–1994	Göttingen	1934 GB	1935 US	CPP, SPSL, EC, OVP, CW (1985), Hildebrandt (1988), *Notices AMS* 42 (1995), 256/57
63	Karger, Ilse (marr. R. Brauer)	1901–1980	Königsberg	1933 US	1935 Canada	NSS, Rohrbach (1981), p. 126; Green (1978), p. 318; Ped

(continued)

No.	Name	Dates	Place of expulsion	First place of emigrat.	Final place until 1945	Sources and Remarks
64	Kaufmann, Boris	1904–?	Heidelberg	1933 GB	1933 GB	NSS, SPSL, Mordell Papers, IBD-Microfilm, Toep 190; Tob 179; Fletcher (1986)
65	Kober, Hermann	1888–1973	Breslau	1939 GB	1939 GB	NU, NSS, Fuchs (1975, incl. photo)
66	Kober (Silberberg), Käte	1908–?	Breslau	1939 GB	1939 GB	NU, NSS, Ped., Fuchs (1975), Tob 315; married to H. Kober
67	Korn, Arthur	1870–1945	Berlin	1939 US	1939 US	NDB, EC-N, Korn (1945), Litten (1993), Weiher (1983, photo), Eng
68	Kürti, Gustav	1903–1978	Wien	1938 GB	1939 US	NU, AC, Kürti Collect. Case Western Reserve University; Pogg-Erg; Eng
69	Kuhn, Paul	1901–?	Prague	1939 Norway	1943 Sweden	Pinl/Dick (1974), OVP, 1970 alive
70	Lasker, Emanuel	1868–1941	Berlin	1935–? SU	1937 US	NSS, NDB, Hannak (1952), chess world champion until 1921
71	Ledermann, Walter	1911–	Berlin	1934 GB	1934 GB	IBD, SPSL, Ledermann (1983, photo), Gaines/Laffey (1985, photo), letters (1997/98), particip. ICM Berlin 1998, lives in London
72	Leibowitz (Winter), Grete	1907–?	Heidelberg	1934 Palestine	1934 Palestine	NSS, doctorate 1933 Heidelberg (Tob 207), Hist
73	Levi, Friedrich	1888–1966	Leipzig	1936 India	1936 India	RM, NDB, SPSL, CPP, EC, OVP, Fuchs/Göbel (1993, incl. photo), Kegel/Remmert (2003)
74	Levin, Victor	1909–1986	Berlin	1933 GB	1938 SU	NSS, LDS 1936, SPSL, *Uspechi Mat. Nauk* 25, no. 1, 205–10

No.	Name	Dates	Place			References
75	Lewy, Hans	1904–1988	Göttingen	1933 US	1933 US	CPP, EC, Albers et al. (1990, photo), *Notices AMS* 35 (1988), 1152
76	Lichtenstein, Leon	1878–1933	Leipzig	1933 Poland	1933 Poland	*Wissenschaftliche Zeitschrift* University Leipzig 29 (1980), series math./science, no. 1 (special issue for Lichtenstein)
77	Löwner, Karl	1893–1968	Prag	1939 US	1939 US	DSB, CW (1988), Pinl/Dick (1974), SPSL, HSP, EC, OVP, Appendix 3.2
78	Lotkin, Michael (Max, Mark)	1911–?	Berlin	1937 US	1937 US	NSS, HSP, EC-N, Dresden (1942), 420; AMWS 14 (1979)
79	Lüneburg, Rudolf	1903–1949	Göttingen	1934 Netherl.	1935 US	NJ, NSS, LDS 1936, SPSL, CPP, Beyerchen (1977), Lüneburg (1964, photo)
80	Lukacs, Eugen	1906–1987	Wien	1939 US	1939 US	NU, AC, OVP, SPSL, EC-N, Lukacs (1982, photo); *Notices AMS* 35 (1988), 473
81	Mahler, Kurt	1903–1988	Göttingen	1933 GB	1937 GB	SPSL, OVP, Poorten (1991, photo)
82	Mann, Heinrich	1905–2000	Vienna	1938 US	1938 US	NU, OVP, Pinl/Dick (1974), Zassenhaus (1977, photo and biogr. sketch, pp. xx–xxv), letter Henry Mann (1993), www.osu.edu
83	Marx, Arnold	1905–?	Königsberg	1934 South Africa	1934 South Africa	NSS, Termination DMV membership 1938 (Toep 248), doctorate Königsberg 1932, to Capetown 1934 (Tob 227)
84	Mayer, Anton	1903–1942	Vienna	1938 GB	1938 GB	OVP, Pinl/Dick (1974), Einhorn (1985)
85	Mayer, Walther	1887–1948	Vienna Berlin	1933 US	1933 US	Einhorn (1985), Pinl/Dick (1974), Pais (1982), 492–94

(*continued*)

No.	Name	Dates	Place of expulsion	First place of emigrat.	Final place until 1945	Sources and Remarks
86	Menger, Karl	1902–1985	Vienna	1937 US	1937 US	NDB, OVP, Spalek, Menger (1994), Kass (1996); Einhorn (1985)
87	Mises, Richard von	1883–1953	Berlin	1933 Turkey	1939 US	DSB, NDB, CW (1963/64), SPSL, Mises Papers Harvard, Spalek, Siegmund-Schultze (2004b), Stat, Eng, Appendices 3.3 and 4.3
88	Nemenyi, Paul	1895–1952	Berlin	1934 Denmark	1938 US	NSS, LDS 1936, SPSL, EC-N, IBD-Microfilm, Dresden (1942), 422.; Nicholas/Benson (2002), Eng
89	Neugebauer, Otto	1899–1990	Göttingen	1934 Denmark	1939 US	NJ, EC, SPSL, OVP, Spalek, Swerdlow (1993)
90	Neuhaus, Albert (Newhouse)	1914–?	Hamburg	1937 US	1937 US	IBD, AMWS 19 (1995/96)
91	Neumann, Bernhard	1909–2002	Berlin	1933 GB	1933 GB	OVP, SPSL, CW (1988), letters (1993/98)
92	Neumann, Johann von	1903–1957	Berlin	1933 US	1933 US	DSB, CW, AC, OVP, Biermann (1988), Appendix 2
93	Noether, Emmy	1882–1935	Göttingen	1933 US	1933 US	DSB, CW (1982), EC, OVP, Dick (1981), Kimberling (1981, photos), Lemmermeyer/Roquette (eds. 2006), Appendix 4.1
94	Noether, Fritz	1884–1941	Breslau	1934 SU	1934 SU	EC-N, HSP, OVP, Schappacher/Kneser (1990), Schlote (1991), Pinl/Furtmüller (1973, photo), murdered by Stalinism

#	Name		City			
95	Oppenheimer, Friedrich	1904–?	Frankfurt	1933 ??	1933 ??	Ped, doctorate 1929 Marburg, dismissed Frankfurt due to racial laws 1933 (Tob 249), possibly emigration
96	Ornstein, Wilhelm	1905–?	Berlin	1933 Poland	1943 Egypt/US	NU, IBD, AMWS 19 (1995/96), technical advisor to US in Egypt 1943–46, Eng
97	Peltesohn, Rose	1913–1998	Berlin	1938 Palestine	1938 Palestine	Pinl/Furtmüller (1973), information by W. Ledermann
98	Pollaczek, Felix	1892–1981	Berlin	1933 Austria	1939 France	NU, NSS, SPSL, HSP, OVP, EC-N, Cohen (1981, photo), Schreiber/ LeGall (1993, photo), Pogg-Erg
99	Pólya, Georg(e)	1887–1985	Zurich	1940 US	1940 US	Pólya Papers, ETH, Zurich and Stanford, Alexanderson (2000)
100	Prag, Adolf	1906–2004	Frankfurt	1933 GB	1933 GB	NU, NSS, Ped, Hist, Scriba (2004, incl. photo)
101	Prager, Wilhelm	1903–1980	Göttingen/ Karlsruhe	1934 Turkey	1941 US	HSP, OVP, Prager (1972), Neumark (1980)
102	Pringsheim, Alfred	1850–1941	Munich	1939 Switzerl	1939 Switzerl	DSB, Perron (1953), Toepell (1996)
103	Rademacher, Hans	1892–1969	Breslau	1934 US	1934 US	NJ, DSB, CW (1974), AC, SPSL, CIP, EC, OVP, Grosswald (1974), Berndt (1992, photo)
104	Rado, Richard	1906–1989	Berlin	1933 GB	1933 GB	SPSL, Rogers (1991, photo), Lenz et al. (1991)
105	Reichenbach, Hans	1891–1953	Berlin	1933 Turkey	1938 US	DSB, AC, Spalek, Reichenbach Papers, Pittsburgh, Kamlah (1993), Danneberg et al. (eds., 1994); Phil

(continued)

No.	Name	Dates	Place of expulsion	First place of emigrat.	Final place until 1945	Sources and Remarks
106	Reissner, Eric(h)	1913–1996	Berlin	1936 US	1936 US	Reissner (1977), E. Reissner-Papers San Diego, letter (1994); *Notices AMS* 44 (1997), 352; Eng
107	Reissner, Hans	1874–1967	Berlin	1938 US	1938 US	H. Reissner-Papers San Diego, Kármán Papers, Ebner (1959, photo); Reissner (1977); Eng
108	Reschovsky, Helene	1907–1994	Vienna	1938 US	1938 US	NU, NSS, Dresden (1942), 422; IBD-Microfilm, AMWS 14(1979); A. Dick Papers, Austrian Academy Sciences, Sigmund (2001), 76
109	Riess, Anita	?	Hamburg?	1939 US?	1939 US?	NSS, Dresden (1942), 425; Ped
110	Rogosinski, Werner	1894–1964	Königsberg	1937 GB	1937 GB	SPSL, OVP, Szegő-Papers, Hayman (1965, photo), Rider (1984)
111	Romberg, Werner	1909–2003	München	1933 SU	1938 Norway (1944 Sweden)	NJ, NSS, Lorenz (1992), 360, letter (1998), Hemmer (2005)
112	Rosenthal, Artur (in emigrat: Arthur)	1887–1959	Heidelberg	1939 Netherl.	1940 US	EC, SPSL, OVP, HSP, Haupt (1960), Mußgnug (1988), Appendix 3.4
113	Rothberger, Fritz	1902–2000	Vienna	1937 Poland	1940 Canada	NSS, SPSL, Sigmund (2001), 76; deported from GB to Canada; inf. by Tom Archibald (Vancouver)
114	Rothe, Erich (Eric)	1895–1988	Breslau	1937 GB	1938 US	OVP, SPSL, EC-N, Richardson Papers; *Notices AMS* 35 (1988), 544

No.	Name	Dates	City			References
115	Sadowsky, Michael	1902–1967	Berlin	1934 Belgium	1938 US	NJ, PR, NSS, LDS 1936, EC-N, SPSL, OVP, CP, Knobloch (1998), Dresden (1942), 422; Eng
116	Samelson, Hans	1916–2005	Breslau	1936 Switzerl.	1941 US	IBD, CPP, CIP, letters (1994/98), Tamari (2007)
117	Scherk, Peter	1910–1985	Göttingen	1936 Czechosl	1943 Canada	OVP, SPSL, EC-N, *Notices AMS* 32 (1985), 818; Pogg-Erg
118	Schiffer, Max	1911–1997	Berlin	1933 Palestine	1933 Palestine	IBD, Dinghas (1945), Schiffer (1986/98)
119	Schilling, Otto	1911–1973	Marburg	1934 GB	1935 US	NJ, IBD, AC, Schilling Papers, Purdue Univ., Pogg-Erg
120	Schur, Issai	1875–1941	Berlin	1939 Palestine	1939 Palestine	DSB, EC-N, CW (1973), Pólya Papers, Stanford; A. Brauer (1973), Ledermann (1983, photo), Schiffer (1986/98)
121	Schwerdtfeger, Hans	1902–1990	Göttingen/Bonn	1936 Czechosl.	1939 Australia	NJ, EC-N, SPSL, OVP, HSP, Schwerdtfeger, P. et al. (1991, photo); letters by Hanna Schwerdtfeger (1993/97)
122	Seckel, Alfred	?	Freiburg?	1939 US	1939 US	NSS, Dresden (1942), 425; Ped
123	Siegel, Carl L.	1896–1981	Frankfurt/Göttingen	1940 US	1940 US	NJ, RM, DSB (Suppl); CW (1966), EC, CPP, OVP, Spalek, Weyl Papers; Siegel (1965)
124	Simon, Heinz	?	Frankfurt	1940 US	1940 US	NSS, Dresden (1942), 427; Ped
125	Sperling, Käte (marr. W. Fenchel)	1905–1983	Berlin	1933 Denmark	1942 Sweden	NU, NSS, Jessen (1993, photo), Høyrup (1987)

(continued)

No.	Name	Dates	Place of expulsion	First place of emigrat.	Final place until 1945	Sources and Remarks
126	Steinhaus, Heinz	1908–?	Göttingen	1933 US	1933 US	NSS, LDS 1936, EC-N, Tob, Becker (1987), F. Bernstein to Einstein, April 25, 1933 (Einstein Papers), probably at The Equitable Life Assurance (NYC), Stat
127	Sternberg, Wolfgang	1887–1953	Breslau	1935 Palestine	1939 US	CPP, SPSL, HSP, EC-N, OVP, Schaefer (1954)
128	Szász, Otto	1884–1952	Frankfurt	1934 US	1934 US	EC, SPSL, CW (1955, bust), Siegel (1965), Szegö (1954)
129	Szegö, Gabor (Gabriel)	1895–1985	Königsberg	1934 US	1934 US	Szegö Papers; CPP, EC, OVP, CW(1982), Askey/Nevai (1996, photos)
130	Tamari, Dov (Teitler, Bernhard)	1911–2006	Frankfurt	1933 Palestine	1933 Palestine	NSS, Tamari (2007)
131	Taussky, Olga	1906–1995	Göttingen/ Vienna	1934 US	1937 GB	AC, OVP, Taussky-Todd (1985, 1988), Notices AMS 43 (1996), Zassenhaus (1977, photo), Luchins (1987), letter (1993)
132	Theilheimer, Feodor	1909–2000	Berlin	1937 US	1937 US	NSS, OVP, Dresden (1942), p. 421; letters Rachel Th. (1997/98), Tob, Eng
133	Thullen, Peter	1907–1996	Münster/(Rome)	1934 Italy	1935 Ecuador	NJ, LDS 1936, CIP, Notices AMS 43 (1996), Thullen (2000), Siegmund-Schultze (2000), Appendix 6
134	Tintner, Gerhard	1907–1983	Vienna	1936 US	1936 US	NU, IBD, Fox et al. (1969, photo), Sigmund (2001), 76; Pogg-Erg; Econ

135	Toeplitz, Otto	1881–1940	Bonn	1939 Palestine	1939 Palestine	Toeplitz Papers, DSB, CPP, HSP, Behnke/Köthe (1963)
136	Vajda, Stefan	1901–1995	Vienna	1939 GB	1939 GB	Bather (1996), letter J. Bather (1997) with unpubl. interview
137	Wald, Abraham	1902–1950	Vienna	1938 US	1938 US	NU, DSB, CW (1957, photo), AC, Stat
138	Warschawski, Stefan	1904–1989	Göttingen	1933 Netherl.	1934 US	EC-N, SPSL, Becker (1987), Notices AMS 36 (1989), 781
139	Wasow, Wolfgang	1909–1993	Göttingen	1933 France	1939 US	IBD, OVP, Wasow (1986), O'Malley (1995), Becker (1987)
140	Weinberg, Josef	1909–1943	Freiburg	1936 Belgium	1936 Belgium	NSS, doctorate 1935 Freiburg, Hist (Tob 356)
141	Weinstein, Alexander	1897–1979	Breslau	1933 France	1941 Canada	HSP, SPSL, EC-N, OVP, CW (Diaz 1978)
142	Weyl, Hermann	1885–1955	Göttingen	1933 US	1933 US	NJ, PR, DSB, CPP, EC-N, OVP, IASA, Weyl Papers, CW (1960), Spalek, Müller (1986), Appendix 4.4
143	Winternitz, Artur	1893–1961	Prague	1939 GB	1939 GB	SPSL, Pinl/Dick (1974), OVP
144	Zatzkis, Henry	1915–prior 2007	Heidelberg	1939 GB	1940 US	IBD, letter (1994), was living 1998, "deceased" 2006 according returned letter
145	Zorn, Max	1906–1993	Hamburg	1933 US	1933 US	NJ, Pinl (1971/72), Notices AMS 40 (1993), 640

List of German-Speaking Mathematicians Who Were Murdered or Driven to Suicide by the Nazis

No.	Name	Dates	Location in 1933/38/39	Sources, remarks
1	Berwald, Ludwig	1883–1942	1939 Prague	NDB, OVP, Pinl (1965), murdered
2	Blumenthal, Otto	1876–1944	1933 Aachen	NDB, CPP, OVP, HSP, Butzer/Volkmann (2006), 1939 tempor. emigr. to Netherlands, murdered
3	Eckhart, Ludwig	1890–1938	1938 Vienna	NJ, Einhorn (1985), Wunderlich (1948), suicide
4	Epstein, Paul	1871–1939	1933 Frankfurt	Siegel (1965), Schappacher/ Kneser (1990), suicide
5	Fröhlich, Walter	1902–1942	1939 Prague	OVP, HSP, Pinl/Dick (1974), murdered
6	Grelling, Kurt	1886–1942	1933 Berlin	NSS; EC-N, Peckhaus (1994), 1939 temporary emigration to Belgium, murdered
7	Haenzel, Gerhard	1898–1944	1933 Karlsruhe	NJ, NDB, driven into suicide by military court
8	Hartogs, Fritz	1874–1943	1933 Munich	Pinl (1971/72), suicide
9	Hausdorff, Felix	1868–1942	1933 Bonn	DSB, NDB, HSP, OVP, Neuenschwander (1996), CW suicide
10	Hurwitz, Charlotte	1889–?	1933 Berlin	Ped., 1933 dismissed as teacher, 1941 still working for Jewish congregation, murdered (Tob 165)

No.	Name	Dates	Location in 1933/38/39	Sources, remarks
11	Kahn, Margarete	1880–1942	1933 Berlin	Tobies (1997), 50; Ped., murdered
12	Lonnerstädter, Paul	1900–?	1933 Würzburg ??	NSS, doctor 1924 Würzburg, unclear whether 1933 still in mathematics, deported and murdered (Tob 216)
13	Neumann, Nelly (temp. N. Courant)	1886–1942	1933 Essen	NSS, Rohsa (1983), Tobies (1997), Stein (1986); Ped., murdered
14	Pick, Georg	1859–1942	1939 Prague	Pinl/Dick (1974), murdered
15	Remak, Robert	1888–1942	1933 Berlin	OVP, HSP, Dinghas (1945), Merzbach (1992), Vogt (1998), 1939 tempor. emigr. to Netherlands, murdered
16	Strassmann, Reinhold	1893–1944	1933 Berlin	NSS, Strassmann (2006), murdered
17	Tauber, Alfred	1866–1942	1938 Vienna	DSB, Binder (1984), Einhorn (1985), Sigmund (2004), murdered

List of German-Speaking Mathematicians Persecuted[1] in Other Manners (Includes Teachers of Mathematics and Is Probably Incomplete)

As in the list 1.1 of emigrants only those who were performing jobs as mathematicians when the Nazis seized power have been included in the following list. Many of them were mathematics teachers (=Ped), which had rather bad chances to emigrate. Their fate after dismissal and the date of their death are usually unknown. Teachers are as a rule not mentioned in the "standard sources," as defined above. The following list is partly an informed guess on the basis of Tobies (2006) and Toepell (1991). The latter source indicates membership in the DMV (German Mathematicians Association), while the former gives information on mathematicians with a doctoral dissertation. In some cases the termination of DMV membership might indicate persecution of mathematicians, particularly after 1937, if they had partners of Jewish origin (PR). As a rule for Jewish mathematicians, persecution has to be assumed. In cases of unknown persecution two question marks are added in the fourth column. We include here victims who possibly emigrated without available documentation so far.

---

[1]On the problematic notion of persecution see chapter 2. The list does not include again the murdered mathematicians of list 1.2.

No.	Name	Dates	Location at beginning of persecution	Sources and remarks
1	Baer, Siegbert	1910–?	1936 Berlin	Ped., doct. 1913 Göttingen, teacher Jewish school until 1936 (Tob, p. 44)
2	Baltz, Hermann	1891–?	1934 Kassel	Ped., doct. 1921 Göttingen, Stat., no longer teacher Kassel 1934 (Tob 46)
3	Barneck, Alfred	1885–1964	1933 Berlin	Pinl/Furtmüller (1973)
4	Baule, Bernhard	1891–1976	1938 Graz	NJ, Pinl/Dick (1976), SPSL, HSP, OVP
5	Beer, Gustav	1906–1945	1938 Vienna	Pinl/Furtmüller (1973)
6	Berlinger, Jakob	1902–?	1933 Frankfurt	Ped., doct. 1928 Frankfurt, 1933 dismissed as Jew in F.a.M. (Tob 54)
7	Duschek, Adalbert	1895–1957	1938 Vienna	NJ, PR, OVP, Einhorn (1985), Rybarz (1957)
8	Fischer, Ernst	1875–1954	1933 Cologne	NDB, Pinl (1971/72)
9	Fleck, Albert	1861–1943	1933 Berlin	NU, Stürzbecher (1997), physician and mathematician
10	Frenzen, Egon	1904–?	1934 Kiel	Ped., doct. Kiel 1932, left teaching in 1934 (Tob 111)
11	Freudenthal, Hans	1905–1990	1940 Amsterdam	Pinl (1969), Bos (1992), Dalen (2005), letters (1983/84)
12	Freund, Eugen	1885–?	1936 Breslau	Ped., doct. 1909 Breslau, dismissed 1936 because of racist laws (Tob 111)
13	Fuchs, Richard	1873–1945	1933 Berlin	NDB, Harnack (1964), Eng., letters B. Jaeckel (1998, incl. photo)
14	Funk, Paul	1886–1969	1939 Prague	HSP, SPSL, OVP, Hornich (1956)
15	Glücksohn, Jakob	1909–?	?? Würzburg	Born in Vilkaviskis (Lithuania), doct. 1931 Würzburg, further fate unknown (Tob 125)

(continued)

No.	Name	Dates	Location at beginning of persecution	Sources and remarks
16	Goldmann, Frieda	1881–?	1933 Breslau	Ped., doct. 1909 Breslau, dismissed because of racist laws (Tob 127)
17	Görke, Lilly (born Buchhorn)	1904–1992	1933 Berlin	NU; interview with author 1986 (photo); Pietzsch (1992, photo); Ped
18	Grell, Heinrich	1903–1974	1935 Halle	NJ, Pinl (1971/72), Schappacher/Kneser (1990), 41
19	Grötzsch, Herbert	1902–1993	1933 Gießen	NJ, Kühnau (1997, photo), Schappacher/Kneser (1990)
20	Halberstadt, Siegbert	1891–?	1933 Berlin	Ped., doct. 1923 Königsberg, 1934 at Jewish girl school (Tob 140)
21	Heesch, Heinrich	1906–1995	1933 Göttingen	NJ, Bigalke (1988)
22	Helms, Alfred	1909–?	??1937 Hamburg	Ped., doct. Erlangen 1935, dismissed Hamburg 1937 (Tob 150)
23	Hensel, Kurt	1861–1941	1933 Marburg	DSB, NDB, Hasse (1936)
24	Hohenemser, Kurt	1906–2001	1933 Göttingen	CPP, Rammer (2002), letter (1997, photo), Eng
25	Hopfner, Friedrich	1881–1949	1938 Vienna	NJ, NDB, Mader (1950)
26	Irrgang, Robert	1908–?	?? 1938 Breslau	Ped., doct. Breslau 1933, dismissed in 1938 Breslau (Tob 167)
27	Jacob, Moses	1900–?	1940 Warsaw	doct. Vienna 1925, insurance math. Warsaw 1940 (Toep 177), Stat
28	Jaks, Erich	1891–?	1933 Berlin	Ped., doct. Königsberg 1914, dismissed due to racist laws (Tob 169)
29	Jolles, Stanislaus	(1857–1942)	1933 Berlin	OVP
30	Kamke, Erich	1890 1961	1933 Tübingen	NJ, PR, NDB, Ehrlich et al. (1968)
31	Kohn, Paul	1895–1954	1939 Prague	NU, Pinl/Dick (1974)

No.	Name	Dates	Location at beginning of persecution	Sources and remarks
32	Kraft, Artur	1891–?	??Vacha	Ped., teacher Vacha (Rhön), left DMV 1933 (Toep 213)
33	Lampe, Ernst	1886–?	?? Elsterwerda	Ped., teacher Elsterwerda, left DMV 1938 (Toep 224)
34	Landau, Edmund	1877–1938	1933 Göttingen	DSB, NDB, Heilbronn/Hardy (1938)
35	Lange, Werner	1893–?	?? Dresden	Ped., teacher Dresden, left DMV in 1933 (Toep 225)
36	Liebmann, Heinrich	1874–1939	1933 Heidelberg	NDB, *Mitteilungen des Bundes der ehemaligen Carolo-Alexandriner in Jena*, Nr. 43 und 44; Appendix 3.4
37	Löbenstein, Klara	1883–?	1936 Landsberg (Brandenburg)	Ped., doctorate Göttingen 1910, dismissed due to racist laws (Tob 213)
38	Löwenheim, Leopold	1878–1957	1933 Berlin	NU, DSB (Suppl), Thiel (1975), Ped., Phil
39	Löwenherz, Arthur	1890–?	1934 Saarbrücken	Ped., Toep 237, doct. 1911 Königsberg, dismissed 1934 due to racist laws (Tob 215)
40	Löwig, Heinrich (Loewig, Henry)	1904–?	1939 Prague	SPSL, OVP, immigration to Australia 1947
41	Loewy, Alfred	1873–1935	1933 Freiburg	DSB, NDB, Fraenkel (1938), Remmert (1995)
42	Magyar, Franz	1894–1958	1938 Vienna	NDB, Wunderlich (1959), Eng
43	Mahlo, Paul	1883–1971	1933 Mansfeld	NJ, NU, Gottwald/Kreiser (1984), Ped, Phil
44	Malsch, Fritz	1890–?	?? Frankfurt	Ped., last in Frankfurt, left DMV in 1938 (Toep 245)
45	Mittag, Walter	1891–?	1933 Berlin	Ped., doct. 1918 Halle, dismissed 1933 in Berlin due to racial laws (Tob 233)
46	Mohr, Ernst	1910–1989	1939 Prague	NJ, NDB, Litten (1996), sentenced to death 1944

(*continued*)

No.	Name	Dates	Location at beginning of persecution	Sources and remarks
47	Moufang, Ruth	1905–1977	1933 Frankfurt	NJ, DSB (Suppl), NDB, Srinivasan (1984, photo), discriminated against as a woman
48	Müller, Oswald	1877–1940	1933 Bonn	NJ, PR, Pinl/Furtmüller (1973)
49	Müntz, Chaim (H.)	1884–1956	Leningrad/ Stockholm	Pinl (1969), SPSL, OVP, Ortiz/Pinkus (2005)
50	Naas, Joseph	1906–1993	Berlin	NJ, NU, Gähler/Gähler (1993, photo), Tob, persecuted as Communist
51	Pinl, Max	1897–1978	1939 Prague	NJ, SPSL, HSP, OVP, Pinl/Dick (1974), Kracht (1981), letters Claudia Pinl, Appendix 3.5
52	Plaut, Hubert C.	1889– 1978 (?)	1936 Berlin	According Tobies (2007) possibly emigration in 1936; Eng
53	Pöschl, Theodor	1882–1955	1938 Karlsruhe	NJ, PR, Pinl (1971/72), Eng
54	Porusch, Israel	1907–?	1933 Marburg	doctorate 1933 Marburg (Tob 262)
55	Reidemeister, Kurt	1893–1971	1933 Königsberg	NJ, DSB, OVP, Szegö Papers, Artzy (1972), transferred for disciplinary reasons
56	Rembs, Eduard	1890–1964	1933 Berlin	NJ, NU, Pinl (1969), Ped
57	Rosenblüth, Emanuel	1901–?	1934 Marburg	doct. 1934 Marburg, Jewish origin (Tob 277)
58	Schlesinger, Ludwig	1864–1933	1933 Gießen	Dunnington (1935)
59	Stern, Antonie	1892–?	?? Göttingen	Ped., left DMV 1938 (Toep 370), doct. 1925 Göttingen (Tob 325)
60	Stessmann, Berthold	1906–?	1934 Frankfurt	Ped., doct. 1934 Frankfurt, Jewish origin, further fate unknown (Tob 326)

No.	Name	Dates	Location at beginning of persecution	Sources and remarks
61	Tewes, Wilhelm	1904–?	?? Kiel	Ped., doct. 1930 Kiel, left school 1934/36 (Tob)
62	Thaer, Clemens	1883–1974	Greifswald	NJ, Ped., Schreiber (1996, photo), Tamari (2007), Hist
63	Thomsen, Gerhard	1899–1934	?? Rostock 1934	NJ, Pinl (1971/72), univ. Prof., Engel (1985), suicide
64	Torhorst, Marie	1888–1989	1933 Berlin	Ped., doct. Bonn 1918, as teacher dismissed, because SPD (Tob 338)
65	Wagner, Karl W.	1883–1953	1936 Berlin	NJ, univ. prof., 1936 dismissed, Weiher (1983), Rürup (1979)
66	Weil, Ilse	1901–1987	1934 Kassel	Ped., doct. 1926 Frankfurt, dismissed probably due to racist laws 1934 (Tob)
67	Weyl, Gertrud	1895–?	1934 Schweidnitz	Ped., doct. Breslau 1921, dismissed in 1934 due to racist laws (Tob 362)
68	Willers, Friedrich	1883–1959	1942 Freiberg	NJ, Sauer/Heinrich (1960), tempor. dismissal, Eng
69	Winkler, Wilhelm	1884–1984	1938 Vienna	NJ, Johnson/Kotz (1997), 322–24; Stat
70	Wolf, Karl	1886–1950	1938 Vienna	NJ, Pinl/Dick (1974), Basch (1950)
71	Zermelo, Ernst	1871–1953	1933 Freiburg	NJ, DSB, Peckhaus (1990b, photo), Ebbinghaus (2007)
72	Zweiling, Klaus	1900–1968	1933 Berlin	NJ, doct. 1923 Göttingen, 3 years jail because Communist activities (Tob 377)

Excerpt from a Letter by George David Birkhoff
from Paris (1928) to His Colleague-Mathematicians
at Harvard Concerning the Possibility of or
Desirability to Hire Foreigners[1]

IT SEEMED to me that our department would get very little advantage from any young man of second-rate qualifications and that any really promising material would be likely to receive strong bids to substantial positions in Europe. Furthermore, I knew that there were always a number of moderately good young men, in general publishing as rapidly as possible, whose mathematical promise was not sufficient to secure early recognition here in a substantial way. These men are generally in a very difficult position insofar as money is concerned, and would naturally welcome a remunerative position in America and the possibility even of staying there for some years, or permanently. I may remark that in four cases out of five these men are Jewish. It was my feeling that it would be better policy to get some promising young American, of about equal standing and achievement, to come to us for half a year at a similar salary. . . .

. . . Bochner . . . His conversation gives the impression of genuine devotion to his science. He says he is interested in the whole field of analysis. Personally I have heard nothing much of Bochner in Vienna, Budapest, Szeged, Göttingen, or Berlin, which seems to me to indicate that he is not the man of outstanding promise. It is easy to understand that C[arathéodory]'s recommendations are slightly tinged by his personal relations to B., and also perhaps, by a feeling (not rare here) that any second-rate European youngster is good enough for us.

In Berlin I met young Van [sic] Neumann, originally from Hungary and also a very attractive young fellow although Jewish. He was tremendously

---

[1]HUA, Department of Mathematics, UAV 561.8, Box 1920s, f. Miscellaneous, G. D. Birkhoff to Julian [Coolidge], Oliver [Kellogg], and Bill [Graustein], Paris, July 7, 1928, 3 pages typewritten. None of the mathematicians discussed was finally hired at Harvard. Von Neumann went to Princeton University for one term per year since 1930. This document was published first in Siegmund-Schultze (1998), pp. 307–8.

recommended by everyone in Hungary among the mathematicians there and also by the men at Berlin. . . . His range of information is astounding and he is enthusiastically devoted to his science. I found him very likeable personally. I think that he is going on with Mathematics but just at present he is working with quantum theory from the mathematical point of view. I regard B.'s preference for pure mathematics as a point in his favor as over against V. N. . . .

Finally there is Van der Waerden in Göttingen. . . . So far as I have been able to find out, he is more promising scientifically than either of the other two men. He is a Dutchman and evidently enthusiastic in devotion to mathematics. I think he ought to be seriously considered.

Report Compiled by Harald Bohr "Together with Different German Friends" in May 1933 Concerning the Present Conditions in German Universities, in Particular with Regard to Mathematics and Theoretical Physics[1]

THE "REVOLUTION" has had a more profound effect on the universities than one would have thought at first sight. A determined attempt is being made to alter the fundamental character of the German universities; the definite character of the changes is not as yet fully determined, but changes are certain. The official slogan is as follows: "The time for objective science is past; the universities now simply have the problem of laying foundations for and building up a national sentiment." Science and mathematics only have a right to exist insofar as they can serve the national aims; beyond that they are, to a certain extent, a dangerous and "disintegrating" factor because, by the teaching of independent objective thinking, they support individualism and liberalism.

The direction of the universities and organization of the higher schools has deliberately been passed into the hands of the humanists [Geisteswissenschaftler] with the clear exclusion of the physical sciences [Naturwissenschaften]. The whole tendency is expressed in numerous official speeches; if there is no vigorous reaction, it will certainly bring about a strong check on mathematical and scientific study in Germany. At present the term "objective, logical science" is almost a designation of contempt.

In the ministries of public instruction [Unterrichtsministerien] all continuity and every tradition is violently broken up by the dismissal of the former officers. The new people in authority are for the most part enthu-

---

[1]The report is attached in German to a letter by H. Bohr to R. G. D. Richardson, dated May 30, 1933 (4 pp. typewritten). Richardson Papers, BUA, box: Correspondence 1933 (German-Jewish situation), f. H. Bohr. On June 13, 1933, the report was sent in English translation by AMS secretary Richardson to the "Members of the Executive Committee of the Graduate Council" of the AMS, among them G. D. Birkhoff, O. Veblen, and N. Wiener. It is also in German copy in the RAC, R.G. 2 (General Correspondence), 1933, box 91, f. 727. The translation given here is partly based on the one by Richardson, which contains some errors. Some words will be explained by quoting the German original in brackets. The original German version is published in Siegmund-Schultze (1998), pp. 308–11.

siastic (but dilettante) national socialists with very vague aims. Their "confidential agents" [Vertrauensmänner] in all universities are not the administrative officers [Kuratoren] but anonymous spying members of their party among the students and chiefly among the young assistants and instructors, especially dissatisfied individuals, held back in their careers, who are filled with hate against the old "system" and who in secret have already been spying on the political opinions of their colleagues for a long time. They are frequently assistants in the Department of Agriculture, of German literature or of other not purely scientific branches, and in many cases they are filled with resentment against the exact sciences that have seen so much success in recent years. These confidential agents inform the ministry about the individual teachers on the basis of personal dislike or on the basis of gossip. It also happens sometimes that people denounce their colleagues as politically unreliable in order to take personal revenge or cleverly to cause difficulties for more able competitors.

The granting of leaves, which have been given by the minister, are directed in part against those of Jewish origin. A single Jewish grandparent is considered sufficient grounds for dismissal. If, however, the respective person was on the faculty already before 1914 or fought in the trenches, or finally, if his father or son was killed in the war, the Jewish origin is not in itself a ground for dismissal. The campaign is also directed against those who are politically "unreliable," that is, people who, according to the confidential agents or the student leaders of the National Socialist party in the university, are not prepared with their whole strength to serve the state. However, so far the decisions of the "civil service law" have not always been applied consistently. Many who are designated as victims by the law have not been given leave or punished, especially if they happened to maintain good personal relations with one of the confidential agents or if one of the student leaders of the National Socialist party wanted to take a course or an examination with the particular instructor. It thus turns out that in many places the philologists and philosophers have been less disturbed than the followers of the exact sciences, which had no connections to those National Socialist circles.

From a legal standpoint the leaves granted are considered to be merely a preparation for the final separation from the professorship. Leaves in some cases may be retracted. In this respect the ministries of instruction have in many cases had considerable pressure exerted on them, but success is very doubtful. Certainly in most cases the leaves will result in eventual dismissal.

There are optimistic observers who hope for a relaxation of the policy embarked upon. At least it seems that the tempo of the "cleansing measures" at the universities is slowing down. However, counting on a fundamental alteration in the new policy would be a dangerous error.

<u>Special List of Leaves of Mathematicians and Theoretical Physicists,</u> <u>with Commentary</u> (i.e., probable reasons for the leave, which in some cases appears totally irrational)

1. <u>University Berlin</u>
   Prof. I. Schur. Reason: Jewish origin (but faculty before 1914).
   Dr. London, theoretical physicist. Reason: Jew.

2. <u>University Kiel</u>
   Prof. A. Fraenkel. Reason: Jewish origin. Fraenkel has already assumed a new position in Jerusalem.
   Dr. Feller, Privatdozent. Reason: Jewish grandmother.

3. <u>University Königsberg</u>
   Prof. Reidemeister. Reason: liberal convictions and sister who is employed in the socialist Vienna town administration.

4. <u>University Frankfurt</u>
   Prof. Szasz. Reason: Hungarian citizen. Jew.

5. <u>Technical University Aix-la-Chapelle (Aachen)</u>
   Prof. Ludwig Hopf. Reason: Jew.
   Prof. Blumenthal. Reason: Jewish origin (but faculty before 1914). He has been in jail for several weeks (Reason: he is member of the Peace Society [Friedensgesellschaft] and has, in addition, made favorable public remarks on Russian scientific institutions after traveling to Russia).

6. <u>University Halle</u>
   Dr. Reinhold Baer, Privatdozent. Reason: Jew.

7. <u>University Köln</u>
   Dr. Cohn-Vossen, Privatdozent. Reason: Jew.

8. <u>University Göttingen</u>. Here mathematics and physics have been persecuted the most, probably due to general jealousy on the part of the National Socialist circles of instructors [Dozentenkreise], which are a priori hostile toward mathematics. On leave are:
   Prof. Born, theoretical physicist. Reason: Jew (but faculty before 1914), liberal convictions, open remarks on government, personal and scientific relations with Einstein.
   Prof. Bernstein, statistician. Reason: Jew (but faculty before 1914), liberal man.
   Prof. Courant. Reason: Jew (but faculty before 1914, in addition in the trenches), as director of the Mathematical Institute particularly exposed. Flunked a female National Socialist student several years ago, this student sued the examination commission for years and used in this connection a National Socialist attorney who has now become Minister of Justice; the student is insane according to a doctor.

Prof. Emmi [*sic*] Noether. Reason: Jew, woman, liberal convictions.

In addition, all the younger theoretical physicists <u>Nordheim</u> and <u>Heitler</u> as well as the mathematicians Prof. <u>Landau</u>, <u>Bernays</u>, <u>Lewy</u>, and <u>Neuge-</u>

bauer have been "temporarily" excluded from instruction by a boycott (the last because of a personal denouncement, he is no Jew).

The greatest German experimental physicist, Prof. Franck in Göttingen, has resigned from his chair out of inner protest against the general situation.

Mathematics and physics in Göttingen are for the time being almost totally blown up [gesprengt].

In Southern Germany one has apparently been more cautious with leaves and does not seem to have disturbed our fields so far.

There are, however, still several young, partly excellent mathematical and physical talents who are threatened, above all: Dr. John, Göttingen, in whom Mr. Morse in Cambridge should be interested, Dr. Heilbronn, assistant to Landau, Dr. Fenchel, Göttingen, Dr. Rado and Dr. Bernhard Neumann, Berlin, students of Schur, and Dr. R. Brauer, outstanding algebraist from Königsberg.

Dr. Bochner (Jew, Pole) from Munich has already left the country for good.

This list is, particularly as far as the younger people are concerned, certainly not complete.

Translation of a Letter from Professor Karl Löwner of the University of Prague to Professor Louis L. Silverman (Dartmouth College) Dated August 2, 1933[1]

Dear Professor Silverman:

I have just returned from a three weeks trip into Germany to get information and I am giving you the report I promised. In the first place you are interested in the cases of Schur and Szegö. I happened to meet the latter yesterday in the Wilson Depot in Prague on his way to Budapest and spoke with him. He still holds his position and hopes that he can hold it. He wishes me very particularly to keep in mind his assistant, Richard Brauer, who according to the new law has lost his position and like all privatdozenten who did not distinguish themselves in the war, has been given leave and was not allowed to lecture during the last semester. I believe that you will hear from Szegö from Budapest and he will give you a fuller account of what has happened in the University of Königsberg.

Professor Schur I found to be quite depressed. He has as I have written to you earlier, been given leave unlawfully since he held his position before 1914. The reason seems to be that there were too few Jewish professors in Berlin who were affected by the law so that some had to be thrown to the hungry wolves and he was chosen. He has marked Jewish features and first name, and he seems to have been sacrificed. In spite of his difficult situation, Schur has been working hard and has completed several of his investigations which will appear in the Berichten of the Berlin academy. He has assured me that he feels energetic and would gladly receive a call from a foreign country.

The possibilities of reinstatement in his position seem to be very slight. Erhard Schmidt has busied himself over this matter, and he has attempted

---

[1]Emergency Committee in Aid of Replaced Foreign Scholars, Box 4, folder: Richard Brauer (1933–41), New York Public Library, Rare Books and Manuscripts Division, English translation of German original as in the files. The Silverman addressed is probably L. L. Silverman (1884–1967) of Dartmouth College, New Hampshire. Löwner was a former Privatdozent at Berlin. Some statements by Löwner, in particular concerning Lewy and F. (?) Pollaczek, proved to be too optimistic. This document was published first in Siegmund-Schultze (1998), pp. 312–13.

to interview a number of persons in authority. Whether he has carried out his plans and with what success, I have not yet ascertained. The plan that the department of mathematics at Berlin collectively should stand behind Schur was defeated by Bieberbach who seems not to have acted fairly. It was evident that Bieberbach was afraid for himself since he had earlier denounced the Nazis.

The pension which Schur will receive in case he is ousted will suffice for a modest existence. However it would be an excellent thing for him to receive a call from another country or at least an invitation to a visiting professorship. Can't you do something for him in America? It may be well for you to get in touch with Von Neumann who has recently assumed duty as a professor in Princeton.

While relatively many physicists have got positions in foreign countries, so far little has been done for the mathematicians. Mr. Lewy of Göttingen has through the good offices of Hadamard received a position (what kind, I do not know) in Spain. Mr. Fenchel of Göttingen has received a stipend in Copenhagen. Dr. Pollaczek has been called to Frankfort [sic].

In case you have an opportunity to place a young mathematician in America I should like to call your attention to Mr. Cohn-Vossen who has lost a position in geometry in Köln. When I get further news I will communicate with you again.

Richard von Mises's "Position toward the Events of Our Time" in November 1933[1]

6. [XI.33] On train all day, read several things, mainly the Braunbuch[2] . . .
Otherwise tried to formulate my position on the events of our time.

I count myself among the class of so-called intellectuals who are so little appreciated today. All my interests lie almost exclusively in the intellectual domain. By my talents I was early directed toward the exact sciences. I thought—when looking at my modest writings—I was a reasonably
useful member of society. Although I never actively engaged in politics, I
believe I have served the state also directly, for instance in the organization
of aviation and of teaching. It seems to me that a government interested
in the well-being of its nation should be able to make use of me. However,
the only really valuable and dignified work, which nobody should be allowed to withdraw from without reason, seems to me the "emendatio intellectus humani,"[3] the preservation and growth of the intellectual property of mankind.

The rulers of the Third Reich instruct me that emphasis on intellectual
interests and on the exact sciences is but an outgrowth of my belonging to
that race that they themselves experience as alien. To them it appears that
physical exercise, sports and training, above all the ability to defend oneself physically and to attack, are at least equally important as intellectual
education. The political benefit to the nation should be the highest measure of worth and dignity, right and wrong of any action, leaving open
whether the nation itself understands that benefit accurately. Humanity in
the sense of an educational ideal is as decidedly rejected as the principled
consideration of humanistic attitudes, prescribed by the Jewish-Christian
religions of all denominations.

I know quite well that my Father's house has many mansions, and I do
not claim to have the key to the only right one. But I am not young enough

---

[1]Excerpt from Richard von Mises's personal diary, written in the German shorthand
Gabelsberger, dated November 6, 1933, in the train from Vienna to Istanbul, the first stop
of his emigration, before he proceeded to the United States in 1939. Translated into English
by the editor with kind assistance by Magda Tisza (Boston). HUA, HUG 4574.2.

[2]Means obviously the anonymous "Braunbuch über Reichstagsbrand und Hitler-Terror"
(Basel 1933).

[3]Apparently following Spinoza's "Tractatus de intellectus emendatione."

to relearn from scratch and not old enough to adapt for my convenience contrary to my conviction. Therefore I cannot do otherwise than resolutely and unequivocally uphold the old principles of civilization, by which empires much more important than the ephemeral "Third" have acquired and maintained greatness: the primacy of the intellect over violence, of freedom over force, of humanity over politics.

Report by Artur Rosenthal (Heidelberg) from June 1935 on the Boycott on His and Heinrich Liebmann's Mathematical Courses[1]

2.6.35

To his magnificence the Rector of the University Heidelberg, with the request for expedited dispatch to the <u>Baden Ministry of Instruction</u> and to the <u>Reich Education ministry</u>

Since May 17, the majority of my audience has been urged by the National Socialist Student Organization [N.S. Studentenbund] to stay away from my mathematical lectures and seminars and to do the same with Professor Liebmann's courses. Shortly afterward, the same action has been started against further five colleagues (three law professors, one economist, one theologian). Since the beginning of last week the actions by the Studentenbund against us mathematicians have been further aggravated by the fact that its scientific section [naturwissenschaftliche Fachschaft] has organized three parallel courses (called "study groups" [Arbeitsgemeinschaften]) during the hours of our courses, which are being led by two assistants from the state observatory [Landessternwarte] and one junior assistant [Hilfsassistent] of the Mathematical Institute. A colleague from the Medical Faculty has offered his auditorium for that purpose.

I wish to stress that my students entertained—as in the past years—the best of personal relationships to me this year. They attended my lectures eagerly until the sudden outbreak of that action. They also followed the practical training and my seminars diligently and with interest. One student told us that a section of the audience greatly regrets being forced to give way to that action; but it is perfectly understandable that those students are not able to resist. The university administration here has not opposed the actions of the N.S. Studentenbund, which according to the Rector was not even possible because the Studentenbund was not subordinate to him. It should be mentioned as well that no personal reproach is being raised against us; all our difficulties result from our being "Non-Aryans."

---

[1]University Archives Heidelberg, personal file Artur Rosenthal, PA 5515. The extant document is a copy. The German original is published in Siegmund-Schultze (1998), pp. 314–15. Translation by the editor.

In spite of the fact that the majority of the audience was prevented from coming, we have until now continued to teach before a small remainder of our students, in my case for about 1/5 of my listeners. I felt obliged, in the interests of the authority of the university, to fight the difficult battle against that action of the Studentenbund, although it was perhaps doomed to failure from the outset. I knew, however, that the fight would be possible only if the Reich Education Ministry supported us. Therefore Herr Liebmann and I asked by telegraph on the evening of May 29th in agreement with the Rector and the ministerial councilor Fehrle, to be received for a consultation by the ministerial director Vahlen or by the responsible officer of the Reich Education Ministry. On the afternoon of May 31st we were informed by telephone that it was not necessary to make the trip to Berlin now. As a result, and because it seemed very unlikely to us that the split between us and the leadership of the students could be resolved, which had occurred from no choice of ours, we decided—first Herr Liebmann on the morning of June 1st, then I as well—to give up the fight and to apply for our retirement [Emeritierung]. This decision was finally confirmed by our Dean, Prof. Vogt, who phoned us on the evening of June 1st telling us that he had talked to ministerial director Vahlen, according to whom our trip to Berlin would be to no avail and he recommended our application for retirement. This happens therefore herewith:

<u>I ask the Reich Education Ministry to grant me retirement [Emeritierung].</u>
Dr. Artur Rosenthal
Full [Ordentlicher] Professor of mathematics at the University Heidelberg

[Appended to Rosenthal's application for retirement was the following statement by the rector, dated June 6, 1935:]
After ministerial director Vahlen, in a personal conversation with the dean of the faculty for science and mathematics, Prof. Dr. Vogt, has approved of the position of the university and faculty leadership in the matter under discussion, it seems superfluous to comment in detail on the application by Professor Rosenthal. I only wish to point out for the sake of clarity that the current number of students of Prof. Rosenthal's, who says that he is lecturing for 1/5 of his former students, is between two and three, and they are Jews and foreigners.

I propose to grant the application for retirement.

Max Pinl—Later Author of the Pioneering Reports (1969–72) on Mathematical Refugees—in a Letter to Hermann Weyl on the Situation in Czechoslovakia Immediately after the Munich Dictate of September 29, 1938[1]

Amsterdam Z, Molenbeekstraat 6/I; 4.10.38 [October 4, 1938]

(Nederland)

Dear Professor [Weyl or, less likely, Veblen],

I take the liberty of sending you copies of my two most recent publications. For the time being I am in Amsterdam in Holland and cannot return to Prague, because due to the political tensions between the German Reich and the Republic of Czechoslovakia the entire railway service is interrupted. I had business here in Holland, because my German translation of Prof. V. Hlavatý's Textbook "Differential Geometry of Curves and Surfaces and Tensor Calculus"[2] will be published here. During that visit I was taken totally by surprise by the recent sad political events.

When I resigned voluntarily from Berlin three years ago,[3] since I could no longer support the bad conduct of Professor Bieberbach [schlechte Benehmen . . . nicht länger unterstützen wollte], I believed I had finally found a quiet and pleasant if financially rather unsatisfactory position at the German University in Prague. During my last term I had, in my lec-

---

[1]Translation by the editor from the German typewritten original. The letter is in the Veblen Papers (OVP, cont. 32, f. Pinl, Max, 1938–39), which contain more letters of Pinl's to Hermann Weyl. Although Weyl does not seem to have been acquainted with Pinl personally (according to a commentary in the same file), Pinl's admiration for Weyl's work is well known, which together with allusions to details in Germany and to the journal *Compositio Mathematica* makes it probable that the following letter was addressed to Weyl, although Veblen's mathematical interests were not far from Pinl's either.

[2]Pinl's translation from Czech into German appeared as "Differentialgeometrie der Kurven und Flächen und Tensorrechnung" with Noordhoff in Groningen in 1939.

[3]This could qualify Pinl also as a refugee from Nazi Germany, but his final country of residence before 1945 was to be—due to the occupation of Czechoslovakia—Nazi Germany, although he had to spend part of his time there in a prison.

ture "Analytical Geometry"—in spite of the already existing political tensions—nevertheless no fewer than 39 listeners and much joy and success with them. While I am writing this, Northern Bohemia[4] is being occupied by the German Army and when I return in one or two weeks' time to Prague, they will have taken away our students. For one can hardly expect from the heavily damaged and mutilated Czechoslovakian Republic that it will provide to a university of their own for the remaining 600,000 Germans. I believe the university will be closed in one semester's time at the latest. Then I will be out of work, once again. Do you happen to know of some help for me, dear Professor? You can hardly imagine how depressing it is for scientists and artists in today's Europe. If it were only for the material hardships! But this constant inner-political police pressure, the constant spying out of racial descent [Abstammungsverhältnisse], of the personal relations one entertains, this constant mistrust vis-à-vis scientific work, that does not lend itself to purposes of armament and preparation of a war!

I was happy to escape the Third Reich for some years and to find a scientific work place in Prague. Where shall I turn to, if now, due to the occupation of the Sudetes by the German Reich army [Reichsheer], our university in Prague will be destroyed? Can you help me?

Excuse me for adding myself to the never ending queue of scientific petitioners, who have sought help with you in recent years and are continuing to do so, but I think it quite natural that one would like to continue the work to which one has devoted so much time and energy already.[5]

Devotedly, with the friendliest greetings

M. Pinl [signature]

(My address is presently identical with the one of Dr. Freudenthal, a former Berlin fellow student, who is presumably known to you from the redaction of "Compositio Mathematica," and also from other contexts. Dr. Freudenthal will be trying to forward mail to me, in case I should return, against all odds, to Prague at some point.)

---

[4]The so-called Sudetes were heavily populated by ethnic Germans but belonged to Czechoslovakia.

[5]On this Weyl remarks in the same file in a note, dated January 27, 1939: "Since he is still young, perhaps a change of vocation advisable?" OVP, cont. 32, f. Pinl, Max, 1938–39.

A Letter by Emmy Noether of January 1935 to the Emergency Committee in New York Regarding Her Scientific and Political Interests during Emigration[1]

Bryn Mawr College
Bryn Mawr, Pennsylvania
Department of Mathematics

Jan. 30, 1935

Dr. E. R. Murrow
Assistant Secretary of the Emergency Committee i. Aid of Germ. Scholars
New York City

My dear Dr. Murrow:

I thank you for your letter of Jan. 21, asking me to participate in a symposium on American education. It is a great honor to me, but I am very sorry to say that I cannot do it.

I always went my own way, in teaching and research work, here and abroad, and I really could not speak with any authority on the questions you are interested in.

I hope you will find some of the mathematicians, more fit to write the article. I should think that Prof. Courant of New York University would perhaps like to do it. He already has been over here some years ago, and he has seen a good part of the educational work done here.

Yours very sincerely

(signed) Emmy Noether

---

[1]Emergency Committee in Aid of Displaced Foreign Scholars (EC), Box 84, f. Emmy Noether, New York Public Library, Rare Books and Manuscripts Division. Handwritten, English original. This document is published first in Siegmund-Schultze (1998), p. 316.

Richard Courant's Resignation from the German Mathematicians' Association DMV in 1935[1]

[La Rochelle, New York], February 19, 1939

Prof. Dr. W. Blaschke, Mathematical Institute of the University
Rothenbaumchaussee 21
Hamburg 13

Dear Mister Blaschke:

attached I send you my declaration of resignation from the German Mathematicians' Association [Deutsche Mathematiker-Vereinigung] with the request to forward the letter to the person in charge. I need not say to you that I find it hard to take this step, after hearing how you and others were trying to get the Mathematiker-Vereinigung back into reasonable procedures [sachliches Fahrwasser]. My hope that the ill-fated decision of Pyrmont would be tacitly canceled by not publishing it has unfortunately not come true. I can therefore only cancel my membership.

With cordial greetings to you and your family and the other colleagues in Hamburg.

Yours, R. Courant

To the board of the German Mathematicians' Association

The last issue of the Jahresberichte publishes an excerpt from the general meeting on September 11–13. It articulates the "sharpest condemnation" of Harald Bohr because of his public statement, made as a private person in the interest of the honor of our science.

I do not misjudge the tactical context from which this decision may have originated for some colleagues. But the fact remains that from now on the Mathematicians' Association publicly claims a right to control its members, which has nothing to do with the tasks of a purely scientific

---

[1]Typewritten copy, Courant Private Papers (CPP), now at the New York University Archives, Bobst Library. For the meeting of the DMV in Bad Pyrmont in September 1934 and its decisions see Mehrtens (1989). Translation by the editor. The German original is published in Siegmund-Schultze (1998), p. 317.

association. I hope that the Mathematicians' Association will quickly renounce this claim in public and give Herrn Bohr the necessary public satisfaction. Until then I do not want to be a member of the Association and ask therefore to be deleted from the membership list.

My bonds with German mathematics and my German colleagues remain unaffected by this.

R. Courant

Von Mises in His Diary about His Second Emigration, from Turkey to the USA, in 1939[1]

Some background for this document:

Von Mises had resigned his professorship at Istanbul out of protest that the contract for his long-term assistant and personal friend, Hilda Geiringer, was not renewed, and because he feared about his future in Turkey after the death of Atatürk (1938) and under the growing German pressure.[2] He had received an invitation to the Harvard Engineering Department from the dean, Harald M. Westergaard, but without promise of pay. He traveled to an insecure future through Marseille, Paris, Le Havre to New York, and from there on to Boston. Here he met long-time friends such as the physicist Philipp Frank, colleagues like the philosophers Carl Gustav Hempel, Jørgen Jørgensen, and Otto Neurath, as well as his former assistant, Stefan Bergmann. At Harvard, von Mises took part in the Fifth International Congress for the Unity of Science, from September 3 to 9, 1939. Von Mises was of course eager to obtain a permanent position. First hopes raised by the aerodynamicist J. P. Den Hartog, of MIT, proved futile. His scientific orientation remained therefore unclear for some time, in particular the decision between aerodynamics and probability, two of his main research areas.

The diary shows von Mises's steady concern about the global political situation at the time of the outbreak of war in Europe. Occasional reflections on his personal financial conditions are included in his diaries as well, not unexpectedly given his dependence on his financial savings.

What worried von Mises most during the first months of his stay in American exile was the fate of Hilda Geiringer, who, unlike von Mises, did not have an invitation to the United States. To save her life, he took "energetic steps." He traveled to New York City and to Princeton and Philadelphia, where he finally secured a temporary position for Geiringer at Bryn Mawr College. Oswald Veblen (Princeton) and William Graustein (Harvard) were very helpful, while von Mises was personally disappointed with Norbert Wiener (MIT), who, however, had helped many other immigrants.

---

[1]Excerpts from the original German shorthand Gabelsberger, HUG 4574.2 Diaries 1903–52, commentary in square brackets and translation by the editor.

[2]See von Mises's letter to von Kármán March 28, 1939, as quoted in chapter 6.

The fate of Geiringer, who, together with her seventeen-year-old daughter Magda, was trapped in Lisbon "without permission to leave or to stay," and thus threatened with deportation to a Nazi camp, provoked von Mises, who otherwise occasionally showed a certain aristocratic restraint in human relationships, to immediate and emotional action.[3] In the fight for Geiringer's life von Mises even had at times a "terrible vision of the untenability of all these luxury institutions," a reference to the abundance of means in American colleges, which apparently seemed to him meaningless compared to the plight of the refugees.

The excerpts from the diary also relate to von Mises's efforts to learn English, and to adapt to American students. The diary shows von Mises's unabated concern for mathematical and engineering research even in emotionally disturbing times.

In was only in 1943 that von Mises married Geiringer, and in 1945 that he received a full professorship at Harvard University. These events are not reflected in the present excerpts from von Mises's diaries, but they are probably covered in later parts of it, which still await transcription.

1. VIII. 1939. First day on board of the *Theophile Gauthier*. [From Istanbul to Marseille] . . . Learned some English. Only occasionally thoughts about the future. Thinking about previous developments. Huge inner rest. . . .

4. [VIII.39] Begun to work somewhat. Some English. . . . Mainly thoughts about stability in elasticity theory. Designed an existence proof for collectives[4] . . .

7. [VIII.39] Early in the morning in Marseille . . .

8. [VIII.39] . . . Arrived in Paris late in the evening . . . Strong impression by the familiar environment in the *quartier latin* [Latin Quarter]. . . .

9. [VIII.39] . . . In the afternoon visited many book shops, in the evening Montparnasse. Strongly impressed by the need to say goodbye.

10. [VIII.39] . . . Late in the evening in Le Havre. The ship, the equipment and the personal accommodation excellent. Much superior to what I had known earlier. . . .

16. [VIII.39] . . . Much thought about own situation, the justification to leave Istanb[ul]. . . . the insecure future, the possibility of a total debacle, the relationship with Hilda and much else . . .

17. [VIII.39] . . . Arrived in New York in the evening. Received by . . . Phil. Frank and wife. . . .

[3] Reading this passage, Geiringer's daughter, Magda Tisza, wrote to me on August 12, 2008, as a balancing commentary: "It was not the sense of the time in Portugal. There was uncertainty, and, at its most extreme, a threat of statelessness." In hindsight, however, one knows where refugees without visas ended up, even if they were able to survive for a few years in camps in southern France, such a Kurt Grelling did.

[4] This is von Mises's fundamental notion in his theory of probability of 1919.

19. [VIII.39] To library in the morning. . . . Some English. The political situation is getting increasingly tense and disquieting . . .

20. [VIII.39] At the World Fair in the morning [Flushing Meadows, 1939 and 1940], stayed there over noon, great impression. Saw and learned much. . . .

22. [VIII.39] . . . Political situation increasingly threatening.

24. [VIII.39] Started in the library to write an English presentation on probability theory. Also wrote on it in the afternoon. Otherwise very depressed by the horrible political development. This in spite of a certain satisfaction because my prophecies have come true. . . .

25. [VIII.39] In the morning continued work in the library. Letter to Hilda. The fall of the English pound. Very much disturbed and angered. . . .

27. [VIII.39] In the morning at the exhibit [World Fair] and stayed almost the entire day, in particular aviation. Then in the evening a girl show. Then on the Broadway, met Hempel and Joergensen, who live in the same hotel. . . .

30. [VIII.39] Letter from Istanb[ul]. Learn to my delight that Hilda has left. But bad news about the money. Later somewhat better news about the pound. . . .

31. [VIII.39] . . . Left for Boston. On the train read much English. In Cambridge immediately together with Frank . . . later also Bergmann. . . . Impression of the city and environment very pleasant and reassuring. . . .

1. [IX.39] At the faculty in the morning, nice room prepared for me. With Westergaard, who is very kind and pleasant. . . . Also a younger man who is building computers [Rechenmaschinen] . . . Satisfied completely by friendly reception. But still helpless und undecided about my own work. . . .

2. [IX.39] In the afternoon visit by [the noted American logician W.] Quine, then with Fr[ank] in the evening invited by Nissen.[5] Picked up by him with car. He lives in wonderful Eur[opean?] conditions and has prospects to stay. . . .

3. [IX.39] Evening with reception, Neurath, Copel [and] etc. Introduced to the president [of Harvard] by Birkhoff. Rather well entertained. Then with Neur. at Franks. etc. Outbreak of war between England, France, and Germany . . .

4. [IX.39] In the morning beginning of the Congress. Listen to talks. Neur[ath] rather interesting . . . Presentation by myself in the afternoon. Not fully finished. But by and large apparently no bad success. . . .

8. [IX.39] . . . Called on Den Hartog. Learned about a planned course for training in aviation technology, which means that prospects are changing fundamentally. . . .

---

[5] Surgeon Rudolf Nissen (1896–1981), who was a friend of von Mises since their stay in Turkish emigration in Istanbul and was in American emigration since 1938. He practiced mainly in New York City but had apparently a home in Boston or vicinity.

13. [IX.39] . . . again with Den Hartog. Now strong deterioration of situation, no aerodynamics, but rather statistics [!] for engineers in the first semester. In the afternoon in the institute, worked on air-wing theory. . . .

16. [IX.39] . . . Received telegram that Hilda cannot leave Lisbon but cannot stay either. Immediately decided to undertake energetic steps. Talk to Frank, later with his wife. Decision to go to New York. In the evening in very bad mood . . .

17. [IX.39] Continued to work with Fr[ank] on the hydrod[ynamic] problem. In the afternoon Bergmann with me. Later also Norb[ert] Wiener, an insufferable blabber but apparently honest. Then packed. Shortly with Fr[ank] then departed during the night. Earlier wrote vita etc. for Hilda. . . .

18. [IX.39] Early in the morning in N.Y. First errand to Coordin. Com. [??? Jewish Committee?], Mrs. Razowski. After strong efforts and long waiting finally talked to her. Not much success. Afterward also at the Emergency Comm[ittee]. Otherwise little [accomplished], basically in vain. The whole day adventurous, almost unreal. Worn out and almost desperate. . . .

19. [IX.39] In the morning again to the City-Coll[ege ?]. Again futile attempt. . . . Left in the afternoon. In Boston at night . . .

20. [IX.39] In the morning in the university. [Found] many letters. By Hilda from the time before the decision [apparently to leave Istanbul]. . . . In the afternoon long letter to Ladenburg[6] on Hilda's behalf. . . .

22. [IX.39] In the morning in the institute, telegram from Ernst Geir[inger, Hilda's brother] that Hilda in acute emergency, no immigration to England, no stay in Portugal. Irritated to the utmost. Almost incapable to act. Had Mrs. Str[uik] drive me to the MIT, there talk to the former American ambassador to Portugal. Talked to Wiener who is dismissive to the utmost. . . .

23. [IX.39] In the morning to Graustein, who is extremely friendly and obliging. First tangible prospects for Br. M. [Bryn Mawr] and for Smith College. In the afternoon began to write documents for Hilda, vita etc. . . .

25. [IX.39] In the morning with the papers to Graust[ein], who helps very effectively. Other connections do not function or only weakly. Meanwhile extremely worried because of lack of response from Hilda. . . . Tried to work in the afternoon, without success. One of the most disturbed days of my life [Einer der gestörtesten Tage meines Lebens].

26. [IX.39] Another terrible day because of lack of any response from Hilda. . . . . .

27. [IX.39] At last reply from Hilda who has not cabled but written airmail. Somewhat relieved. . . . Departed for New York. . . .

---

[6] Rudolf Ladenburg, German physicist in Princeton.

28. [IX.39] In the morning without stop to Princeton. Here at first Veblen, who is extremely helpful. Happy coincidence that Mrs. Wheeler[7] has come to Princeton. In between lunch with Ladenb[ur]g. and [John von] Neumann. In the afternoon with Wheeler, who is won over immediately. [She] had been already informed by Veblen before. Then with Einstein, who is very kind, writes a letter to the president ["Präsidentin," i.e. the female president of Bryn Mawr College], talks to me about his problems. Then left for Philad[elphia]. . . .

29. [IX.39] In the morning to Bryn Mawr. Walk through the College grounds. Terrible vision of the untenability of all these luxury institutions. At lunch with Mrs. Wheeler, very boring. But she is won for the case. Then with her to president Miss Park. The latter also easily won. All essential points granted. Very relieved, departed at 3 o'clock. Without stop through Philad. and N.Y. Arrived late. Telegram to Hilda and Ernst. Then found refusal from Smith Coll. . . .

30. [IX.39] In the morning more good news, 30 days residence permit in Portug.

3. [X.39] . . . Permanently disturbed by lack of response from Hilda . . .

4. [X.39] . . . In the evening, once again cabled to Hilda . . .

6. [X.39] . . . In the morning at last reply from Hilda to my second telegram. Very angry about the content. . . . Then again cabled to Hilda. Then worked with some more composure.

11. [X.39] . . . At the bank about the money for Hilda. . . .

12. [X.39] . . . At last news from Hilda that she has received the papers [apparently for immigration]. Somewhat relieved over that. . . .

14. [X.39] Lecture on Aerod[ynamics], rather good. Better contact to people. . . . Later decisive progress in the theory of thin air-wings . . .

18. [X.39] Telegram from Hilda, rather good situation. . . .

24. [X.39] In the morning unpacked boxes and cases. . . . English lesson. . . . All clothes and linen put into the closet by myself. . . .

25. [X.39] In the morning telegram from Hilda, has received visa, arrives Thursday. . . .

27. [X.39] . . . Departure late in the night . . .

28. [X.39] In the morning in New York. . . .

30. [X.39] Learned that Hilda is not coming today. . . . Left in the evening. Very tired and worn out. . . .

2. [XI.39] . . . In the evening telephone call from Hilda from New York. At last this means the end of a long anxiety.

4. [XI.39] In the morning lecture. Some errands. Then to the train, picked up Hilda. With her to her hotel and then to my place, drinking tea. In the evening to the Fr[anks].

[7] Dean for mathematics at Bryn Mawr College.

Hermann Weyl to Harlow Shapley on June 5, 1943, Concerning the Problems of the Immigrant from Göttingen, Felix Bernstein[1]

Another case is that of Felix Bernstein, who has been connected with the Dental Department of New York University for the last years, doing work on biometrics, but now he has reached the retiring age of 65. He has only a microscopic pension, and the conditions in his family are not good. But he is still in a pretty active state and I have no doubt that he could do very useful work indeed, in particular in connection with the war effort. He has a brilliant and exceptionally versatile mind, teeming with original ideas (of which probably not more than 20 per cent. ever mature, but even that is plenty). Bernstein started out as a mathematician. Every mathematician knows his equivalence theorem in abstract set theory. Since he proved that, as a young student under Georg Cantor, he has covered a great variety of fields including celestial mechanics. I used to know Bernstein intimately in the years 1910–1913 when we were both Privatdozents at Göttingen, and remember vividly how conversations with him started me on the theory of mean motion and equidistribution modulo 1. In later years he became successor to W. Lexis as Director of the Institute for Statistics in Göttingen, and during the last ten or fifteen years he has concentrated on biometrics and has done quite important work on blood groups. He happened to be on a visit to this country when the Nazi storm broke, and never returned. He is married and now an American citizen. He is exceptional in that he combines so many different fields: mathematics, statistics, economics, biology. Wishing to be frank, I feel I ought to add a word of warning that his personality is not too pleasant, mainly because he seems to feel the necessity of convincing himself at every moment of his own superiority. But I always succeeded in getting on with him quite all right. Einstein, with whom I have discussed both cases [the other was physicist Erich A. Marx; R. S.] does not know Marx, but knows Bernstein pretty well and thinks very highly of his talents and usefulness.

---

[1]Source: EC, box 2, f. Bernstein, F. (1936–45), copy, typewritten, English.

Perhaps the most rational solution of these and similar problems is to overthrow Hitler as quickly as possible, and have the authorities who fill the vacuum in France and Germany recognize the property, pension, etc, rights of refugees that have been infringed illegally by the Nazis. But it may be that this is a Quixotic dream.*

* Not the downfall of Hitler, but the resurrection of these rights!

Richard Courant in October 1945 to the American Authorities Who Were Responsible for German Scientific Reparation[1]

Mr. R. H. Scannell, Chief
Economic Intelligence Division
Foreign Economic Administration
61 Broadway
New York, N.Y.   October 24, 1945

Dear Mr. Scannell:

Following up our conversation of this morning, I shall attempt to summarize some of the points touched upon:

The general question was: What could or should be attempted by the Reparations Commission in the scientific field? Of course, all specific actions have to stem from a basic policy which I shall not go into in this letter. Moreover, the specific recommendations I make below refer only to a rather limited range of objectives concerning which I feel competent to give an opinion offhand.

The point of departure should be a decision whether the German scientific research and development should be (a) impeded or (b) considered with a detached attitude or (c) encouraged. The possibility (a) and in the long run even the possibility (b) seem to be excluded by the friendly attitude of influential American scientists, an attitude which I have observed, for example, in members of American scientific intelligence missions. Therefore, I do not think that much repression of the German scientific potential is feasible even if it should be considered desirable. For example, I do not think that reparations in kind, such as shipping of German libraries or shipping of German scientific laboratory equipment to the United States should be attempted, except, perhaps in the case of equipment belonging to industrial companies (I. G. Farben, Krupp, Kohlenforschungsinstitute in Mülheim, etc.) and, of course, scientific institutions that were expanded or developed for the German armed forces.

---

[1]Richard Courant Papers, New York University Archives (Institute Papers, CIP), Elmer Bobst Library, file Germany, 1935–57, typewritten copy of the English original, 3 pages.

I can see the following productive ways of reparations in the scientific field:

1.) Let German mechanics and workers duplicate such scientific equipment available in German universities and other institutions as may specifically be designated by a scientific commission.

2.) Let the German printing presses reprint German scientific books published or available in manuscript and other scientific material that now exists only in a limited number of copies and which may be of use for research in Allied countries. The surplus of these editions can be left to the Germans.

3.) It is to be expected that during the next years a tremendous overflow of scientific literary production will swamp American and British journals and other outlets of publications. The facilities existing in the United States and probably in England will not be able to handle such volume with the speed desirable for the prompt dissemination of scientific information. In Germany, printing plants, paper mills and a large number of very highly skilled workers are available to do the job. Let it be done under Allied supervision as a reparations measure.

4.) There are many types of activities of a scientific character such as compiling tables, surveys, catalogues, etc., which could easily be handled by groups of German scholars in Germany as a reparations service and which would be exceedingly valuable.

5.) Let German scientific teams be organized in Germany to help digest and evaluate much of the tremendous material that has been produced during the war and of which only the surface has been scratched so far by our Intelligence missions.

6.) While scientific reparations services should be mainly organized within Germany, one must seriously consider bringing over individuals or teams to this country—however, only after careful scrutiny of each individual case.

In my own field, that of mathematics and mathematical physics, I would know of quite a few items where fruitful projects of the described types could be worked out. By and large, I think that a small commission of active scientists of high rank should visit German universities and institutions to prepare concrete proposals. Since most such projects would be at the same time in the interest of German scientists, it should be easy to obtain active cooperation of German scholars or other authoritative persons. In this respect I would suggest that Professor Franz Rellich, director of the Mathematics Institute, University of Goettingen, be contacted. Another person of potentially great value is Professor von Laue, former deputy director of the Kaiser Wilhelm Institute for Physics in Berlin, Dahlem, last known address Hechingen in Württemberg. Laue is a first-rate physicist, has been a passionate hater and active opponent of the Nazis, one of the few really courageous outstanding German scholars. On the other hand, Laue is a German nationalist presumably with anti-Russian feelings. Finally I would mention an old acquaintance of mine, Dr. Carl Still, a leading man in the German heavy industry but profoundly interested in science and connected with scientists. His home is Recklinghausen in Westphalen and he owned a big estate in Rogaetz near Magdeborg [*sic*], where to my knowledge Professor Max Planck lived as Still's guest during the period of Germany's collapse. Still is about seventy years old, a very independent, dynamic, self-made man who as long as I had contact with him was very much against the Nazis and had a very broad view on things. How far he went, as a leading industrialist, with the Nazis I do not know. As a person he is very honest, quite patriotic and may be nationalist but basically fair and unbiased.

Hoping that my remarks may be of some help to you and perhaps to Mr. Angell, I am

Sincerely yours R. Courant

Max Dehn's Refusal to Rejoin the German Mathematicians' Association DMV in 1948[1]

Preliminary remark:

The victim of the Nazi regime Erich Kamke, who had taken over the lead in the new DMV after the war, had written to America to the émigré Max Dehn on June 15, 1948, asking him to rejoin the DMV, alluding to the circumstances under which Dehn had been expelled:

> We assume that it was, in your case as well, one of the illegitimate expulsions, which were performed by the then acting chairman of the DMV in transgression of his rights.

Dehn replied on August 13, 1948:

> Dear Mister Kamke,
> many thanks for your invitation to join the newly founded German Mathematicians' Association [Deutsche Mathematiker-Vereinigung].
>
> I bear no grudge of any kind. As you may know, I am again in close contact with several mathematicians in Germany, of course primarily with those whom I was particularly close to.
>
> But I cannot rejoin the Deutsche Mathematiker-Vereinigung. I have lost the confidence that such an association would act differently in the future than in 1935. I fear it would, once again, not resist an unjust demand coming from outside. The D.M.V. did not have to take care of very important values. That it did not voluntarily dissolve itself in 1935, and that not even a considerable number of members left the association, leads me to this negative attitude. I am not afraid that the new D.M.V. will again expel Jews, but maybe next time it will be so-called communists, anarchists or "colored people."
>
> The contact with Germany, specially also with German mathematicians, is dear to my heart.
>
> With the best greetings . . . I remain.
> Very faithfully yours
> Max Dehn

[1] From Max Dehn Papers, AAM, Austin, box 2, no. 55. Typewritten, Dehn's letter as copy, translation by editor. Original German version published in Siegmund-Schultze (1998), p. 318.

Memoirs for My Children (1933/1988)[1]
Peter Thullen

Introduction by the Editor

The mathematician Peter Thullen (born in Trier, Germany, on August 27, 1907 and died in Lonay on the shores of Lake Geneva, Switzerland, on June 24, 1996) began his scientific career in pure mathematics at the University of Münster (Westphalia, Germany) as a student of Heinrich Behnke (1898–1979), a leading mathematician in complex analysis. Already well known by 1933 through his publications, Thullen decided in 1934, for political reasons and because of his close links with the Catholic youth movement, not to return to Germany after a stay abroad. Instead, he chose to go into exile in Ecuador. He left Latin America in 1952 to take up a position in the social security department at the International Labor Office (ILO) in Geneva, Switzerland, first as chief mathematician and subsequently as director of the department. On retirement from the ILO in 1967, Thullen became professor at the University of Fribourg in Switzerland, where he stayed until 1977. Thereafter, he continued as consultant to many governments in social security matters. Although Thullen had been offered an assignment in the Labor Ministry of the Federal Republic of Germany in 1952, his hopes of being offered a chair at a university in Germany never materialized. His disappointment at being unable to return to academia and the difficulties he encountered in at least having his German citizenship reinstated, can be clearly inferred from the second part of his autobiographical notes, written in 1988.

Thullen's notes were originally intended for his five children. The first, more extensive, part is based on unedited diary entries and correspondence dating back to 1933. The significance of the notes lies in the fact that, on the one hand, they reflect an authentic picture of the atmosphere prevailing at a German university in that year, and, on the other hand, they provide an illustration of how the vicissitudes of life and political factors could lead to new scientific endeavors as a result of emigration. The manuscript, consisting of twenty-eight typewritten pages, was completed

[1]English translation by George Thullen of the German edition (Thullen 2000).

in March 1988. All unpublished source materials that have not been specifically quoted are in Peter Thullen's estate, currently in the hands of his eldest son, Dr. George Thullen (Genthod near Geneva, Switzerland). The scientific papers of the estate will be deposited in the library of the Swiss Federal Polytechnic University in Zurich and the remainder, of more general and historical interest, will be donated to the Exile Archives (1933–45) of the German National Library in Frankfurt, Germany.

Comments and footnotes within square brackets and set in italics are those of the editor. Parenthetical remarks in curly brackets and footnotes not set in italics are Peter Thullen's own. The gaps marked with . . . are in the original March 1988 manuscript. Small corrections, especially of typing errors, in Thullen's own hand in December 1988, have been taken into account.

Finally, it is thanks to the initiative of Madame Jeanne Peiffer (Paris), a former student of Thullen's at the University of Fribourg, Switzerland, that it has been possible to publish these memoirs. Thullen's son, Dr. George Thullen, translated the text from German into English.

Thullen's memoirs have been published in German in the biannual journal *Exil*, 20, Nr. 1, 2000, Frankfurt am Main, pp. 44–57. More extensive comments may be found in a complementary article by the editor in the same issue of *Exil*, "Die autobiographischen Aufzeichnungen Peter Thullens," pp. 58–66. The article also contains additional biographical information.

1933: The Year Hitler Seized Power Notes for My Children

As I went through some of my old papers I came across notes I had jotted down in 1933, the year in which the National Socialists seized power in Germany. I wish to share these with you here, as well as to let you know about events directly pertaining to our emigration. I do not want to go into the complex historical factors that made possible and facilitated Hitler's rise to power, such as the Versailles Peace Treaty, the unbelievably high rates of unemployment at the end of the 1920s, the passive attitude of the Allies in the face of Hitler's growing influence and the political concessions they made to him which they had refused to concede to the Weimar Republic. There are many books in which you can read up on all this in greater detail and in a better-informed way.

These notes were taken during a period when I was totally absorbed by my mathematical research. It must have been a premonition of something unusual, threatening, that drove me to record facts and events as they unfolded, often in a fragmentary way.

Before that, however, let me explain the following: In October 1932 I was to take up a research fellowship at the University of Rome.[2] Instead, I lay in hospital in Münster. What at first seemed to be a minor operation resulted in generalized sepsis. After three days hovering between life and death, I was saved, yet I could not go to Rome until a year later. Thus I was able to witness the takeover of power by the Nazis and their way of exercising it during the first eight months of their regime.

Our opposition, your mother's and mine, to Nazism arose out of our deep involvement in the German youth movement. We rebelled against deceitful authorities and against the hypocrisy of society. We aspired to "truthfulness" and longed for beauty and a healthy lifestyle. Independent youth associations as well as Protestant, Catholic and Socialist youth groups formed a vast fraternal community that firmly believed it could change the world. Educators, authors, theologians and scientists emerged from these movements. Health-food shops, country schools and youth hostels sprung up everywhere.

Then came Hitler. Youth groups were banned. The young were integrated into the Hitler Youth [*Hitlerjugend*]. Even among our own members there were fellow travelers. Others—especially from the religiously oriented and Socialist groups—did not let themselves be deceived. Later on, three of my closest friends from the Catholic youth movement paid for their convictions by incarceration in Dachau (Father Maurus, whom you know, was one of them). We were aware of the brutality of the Nazis long before they reached power, and we knew about the bloody street battles during which they beat up their opponents. In addition, one could read in Hitler's *Mein Kampf* details of his plans (I still have a copy of the unexpurgated 1932 edition). We took these plans seriously, especially because of the frame of mind they reflected. Others found it more convenient just to ignore them.

An additional factor was that we in the Rhineland resented being considered "Prussian." Up until 1919, Prussia treated the Catholic Rhineland[3] almost like a colony. Senior, and even middle-ranking, civil service posts were filled by Prussians, as if we subordinates could not be trusted. Prussia was not *our* State, one with which we could identify. For us, Germany was not so much a "State" as the country of our great poets and musicians—most of whom were Rhinelanders. It was everything that made up German cul-

---

[2][*Thullen's notes from the end of 1990 state:* "I was meant to go to Rome in October 1932 to Prof. Severi charged with the task of seeing how algebraic geometry, in which Italy excelled, could be integrated into the theory of functions of several complex variables."]

[3][*The former Rhine province was absorbed by Prussia in 1824. There had been some aborted attempts after World War I (1919, 1923), to create a "Rhenish Republic" with French support.*]

ture, not to mention its scenery, of which the Rhine and Mosel river valleys were perhaps the most beautiful part. This may sound romantic, but it was the reality of our youth. We hated the Nazis because we loved Germany.

My notes relate to the period of freely consented or forced alignment with the Party (known as "Gleichschaltung") by organizations, people, literature, art and thought. Right from the start, a wave of conformism swept across the universities and affected faculty members and students. The few who refused to toe the line were increasingly overcome by a sense of powerlessness in the face of the demise of the Germany we had loved. The feeling of being left out, of cutting oneself off from a generalized "national upsurge"—along with all the frenzy and rhythms of parades, flags and fanfares—was not easy to bear, and, in the end, many gave up their resistance.

The notes also bear witness to the fact that in 1933, if one really wanted to, one could know, indeed did know, enough about what was already happening. To deny this reality is the great German delusion. Just look at the list I drew up of publicly known facts and events up to the time of my departure in October [see the final footnote].

Unfortunately, I have not found any notes on the first two months following Hitler's rise to power (30 January 1933). Perhaps I was too engrossed with mathematics and myself. I am reproducing my notes just as I wrote them, without any stylistic changes.

Notes from four months after Hitler's seizure of power (1933)[4]

29.4.1933 {On the train to Münster}
In major railway stations one can see SA *[Sturm-Abteilung]*—auxiliary police. Many platforms and locomotives are decorated with green colors and flags for the day after tomorrow (1st of May). Otherwise everything is quiet; only the many brown shirts and swastika flags are a reminder of the great upheaval.

30.4.1933
This afternoon I was at Behnke's. I was surprised at his relative composure and cheerfulness, which I had not expected of him.[5] His fearfulness is, however, often strange—he is afraid of any politically charged conversation and avoids speaking openly. I will have to get used to this a lot more.

---

[4]New explanations have been inserted within curly brackets { }. Words I have been unable to decipher are replaced with . . . ? . . . In some cases they may have concerned fleeting thoughts or ideas intended for letters to your mother. My notes begin with my return from a cure sojourn in the Black Forest.

[5]Behnke's first wife, who died when giving birth to his son, was Jewish. Their son, as a half-Jew, was at risk. This explains much about Behnke's attitudes.

1 May 1933—National Labor Day

This noon was the celebration of Labor Day in the overcrowded assembly hall of the University. Vice-Rector Herrmann delivered the main speech. I only remember that it was pathetic, with no clear ideas, no clear convictions. Next out was the new student leader (Dixweiler), young and spirited, often overly sharp, but sincere. As usual, the singing of the Horst-Wessel song followed. In the corner where I stood I was just about the only one who did not sing or raise my arm in salute. How shameful that many who not long ago were emphatically cursing the "Nazis" and had been either black or red, are now ready to join in, just to avoid being suspected as "nationally unreliable" and being excluded from the race to the spoils. {The [lyrics of the; R. S.] Horst-Wessel song created by himself who was killed in 1930 in a street battle and subsequently proclaimed a martyr. This song became the second national anthem; when singing it one had to raise one's right arm in the "German salute."}

In front of me I saw Rehmann (?) in a SA uniform. He used to declare himself a Communist when that was still interesting and safe. I heard that the Münster SA refused to accept him in their ranks, but that he managed to become a member in a roundabout way in Gelsenkirchen. I also noticed Kentrup proudly raising his hand, he who used to be a docile member of the Center Party and a true petit bourgeois. I wonder whether he will derive any benefit from that turnaround.

The big parade took place in the afternoon—a huge crowd of people. The best looking were the Catholic youth—especially the "Sturmschar" . . . the only group not to display the swastika. The shabbiest were the professors, like a flock of sheep, where only the dog was missing. {I did not take part in the parade; but I watched it out of curiosity.}

3 May

The rift between Hugenberg and Hitler keeps widening ever more. {H[ugenberg] was the chairman of the German National Party—Deutschnationalen—and a leading industrialist who helped Hitler rise to power and was a government minister at the start of the Nazi regime.}[6] Hugenberg is already totally isolated. Following . . . the arrest by the SA of leaders of the "Stahlhelm" in the Palatinate, and the SA's attempt to break up the "Stahlhelm" even in Brunswick, one could see what was coming. Now Seldte has defected to the NSDAP (Nazi Party) and thus placed the "Stahlhelm" under Hitler's command. As a result Hugenberg

---

[6][*Alfred Hugenberg (1865–1951) was chairperson of the board of directors of Krupp AG (1909–1918) as well as a media magnate.*]

has lost what little popular support he had.[7] Düsterberg,[8] a half-Jew, though very popular among his "Stahlhelm" comrades, also had to leave. Amazing how many are now branded as non-Aryans. Even dear old Professor Münster is a Jew. Of the mathematicians, Courant, Reidemeister (although Aryan), have been suspended and subsequently S. Cohn-Vossen and Schur as well. {World-renowned Courant had been an officer during World War I; you met him during one of our visits to New York.}[9]

At noon the new student code was solemnly proclaimed in the city hall. Again a horrible, boring speech by the Vice-Rector—noncommittal, leaving all options open. Then Dixweiler {students' leader} spoke—he feels he is in command of the situation, as if to show that he has more power than the Rector. There was much in his speech one could agree with.

5 May

Kentrup turned up with an enormous swastika insignia. Usually slow on the uptake, he realized how much even Party members looked down upon him with contempt. At first proudly puffing up his chest with pride, he became noticeably embarrassed.

What struck me today at the university was a call by the German students' association titled, "No to the un-German spirit."[10] It states, among other things:

"A Jew cannot write in German. If he writes in German, he is a liar."

"We demand that Jews write only in Hebrew; German language editions should be considered as translations from Hebrew and labeled as such."

The first sentence is an insult to all the Jews who voluntarily served in the armed forces during the First World War, to . . . Jewish scientists and

---

[7][*The federation of war veterans known as "Stahlhelm" ("Steel Helmet") was founded in 1918 by Franz Seldte (1882–1947). Numbering nearly 500,000 members it was the biggest anti-Republican organization from the Weimar Republic period. Forced to align itself with the Nazis in 1933 and officially dissolved in 1935.*]

[8][*Theodor Duesterberg (1875–1950) was the deputy leader of the "Stahlhelm". Already when presidential candidate of the German National People's Party (DNVP) in 1932, he was decried by the Nazis as "half-Jew" despite the fact that he had played a prominent role in the creation in 1931 of the "Harzburger Front" comprising the "Stahlhelm," the DNVP and the NSDAP.*]

[9][*Richard Courant, Stefan Cohn-Vossen, Kurt Reidemeister, Issai Schur. For further details about these mathematicians see above in the present book.*]

[10][*The central office for press and propaganda of the Nazi students' association organized an "enlightenment campaign" between 12 April and 10 May (the day of the book burning) under that slogan. The "Twelve Propositions of the Students," from which Thullen has summarily quoted, were, for instance, reproduced in L. Poliakov and J. Wulf, Das Dritte Reich und seine Denker, Berlin, Arani, 1959, pp. 117–18.*]

intellectuals who reason and think entirely in German. To deny this is idiotic. Both sentences—the second one is a grotesque absurdity—are a disgrace for German students . . . presumably they were written by "turncoat patriots" . . . Precisely such preposterous and nasty statements are bound to hamper any legitimate struggle against the "Jewish spirit." None of the professors, except Spranger,[11] seriously dares to take an opposing stand.

{I have copied the penultimate sentence with some shame. It has to be seen in context. Moralizing Catholicism in those days held the view that one of the principal causes of the alleged "decline" in art and morals in the "Twenties" could be traced to the "Jewish spirit"—without ever defining this term. This had nothing in common with the anti-Semitism propagated by the Nazis. Many of our favorite authors were Jews: for example, Stefan Zweig, Alfred Döblin, Franz Werfel.}

6 May

Today, at long last, I can work again. We will be able to complete our book {"Funktionen mehrer komplexer Veränderlichen"}[12] by the end of the semester.

Rudi Hölker has joined the Party {this "betrayal" by a good friend of mine, who used to belong to my youth group, has stunned me. Later on, as one of Wernher von Braun's collaborators and an important rocket specialist, he was evacuated to the USA. Lives near Boston. The Nazi past of this large group of German scientists and specialists has been blotted out.}[13]

A pillory was set up this morning; a whole range of book covers and periodicals are hanging from it . . . ? . . . The solemn burning of the "un-German" books will take place on Wednesday.

7 May

I find the Behnke atmosphere repugnant. I try to avoid meeting him. Hopefully I shall soon be independent of him. His apprehensiveness borders on cowardice. Whereas others talk openly about the excessive agitation against

[11][*Spranger, Eduard (1882–1963), German philosopher, educator, and leading representative of modern cultural philosophy and pedagogy. Spranger was born in Berlin. He was a professor at Leipzig from 1911 on, subsequently, from 1920 in Berlin and in Tübingen as of 1946. Student of Wilhelm Dilthey. For further information on Spranger's conflicts with the Nazi student bodies and educational authorities see Poliakov/Wulf,* Das Dritte Reich und seine Denker, *pp. 89–94.]*

[12][*H. Behnke and P. Thullen,* Theorie der Funktionen mehrerer komplexer Veränderlichen *(Theory of functions of several complex variables), Berlin, Springer, 1934.]*

[13][*Cf. the "paperclip" campaign of the Americans described in L. Hunt,* Secret Agenda, *New York, 1991.]*

Jews, he dares not utter a word—not even in our presence. How strange that even the most feudalistic student associations are joining the (Nazi) Party in closed ranks, . . . having turned into "socialists." . . .

Brüning has become the "Führer" of the Zentrumspartei.[14] Is this the prelude to a brown-black alliance? Strasser and Brüning might make a good combination that could benefit Germany. {Gregor Strasser had stood in for Hitler during the latter's detention, but left the Nazi Party in 1932 (the same year, by the way, that my eldest sister, Bella, and her husband had also left the Party). Gregor Strasser was murdered by Hitler in 1934. His brother, Otto, also a former leading member of the Party, immigrated to the USA in 1933.}[15]

It is quite obvious from press reports that Hugenberg's position has become very shaky. Twice I read in a National-Socialist newspaper complaints about "outrageous meddling" and "accusations" by the Deutschnationalen [*German National party*].

9 May

Lehrmann-Hartleben is also in danger. He took part in the War, but only as an interpreter. Nobody dares speak up for him in public . . . Not even Party members want to put in a good word for him, for fear of being accused of befriending Jews.

10 May

Burning of "trashy literature" in the Hindenburg Square. All student associations were represented as well as the SA, the Hitler youth and a huge crowd.

Stormy session of the city council. The SA expel the only Social Democratic member.

20 May

Now, the Münster rabbit-breeding club has toed the line, too, so it can meet the challenge of the great tasks ahead for the people and the Fatherland.

---

[14]*[Heinrich Brüning (1885–1970) was a leading politician during the Weimar period of the Catholic "Zentrumspartei" ("Center Party"), which had been founded in 1870, and a Chancellor of the Reich between 1930 and 1932. The use of the term "Führer" ("leader") was symptomatic of a tendency for organizations under the Nazi regime to align themselves with the latter's terminology. Brüning fled to the USA in 1934.]*

[15]*[Gregor Strasser (1892–1934) was murdered during the so-called Röhm Putsch (Röhm coup). His brother, Otto (1897–1974), only emigrated from Europe to Canada in 1943.]*

Of course, the German chess players' association will not want to be seen lagging behind the rabbits.

22 May

The cowardice of the professors really is pathetic. Kratzer (my physics teacher) refuses to sign a petition for Courant.[16]

Scholz (philosopher, my teacher and friend)[17] invites the new student potentates. Only a few professors remain who keep their composure. Hardly anyone greets Goldschmidt (?), Pieper or Lehmann-Hartleben; indeed, they are being fearfully avoided {see further on about Scholz}.

26 May

The future looks bleak. The Disarmament Conference[18] is deadlocked. No one is prepared to make any concessions.

On the home front things do not look any better. Even if the number of unemployed is said to have dropped by 80,000 over the past month, this does not take into account the large number of those employed in the labor service [Arbeitsdienst].

It should not take too long for people to become tired of being fed with the scandals of the "Novemberites" ["Novemberlinge"],[19] and with patriotic celebrations and speeches. Once disillusionment sets in along with the realization that things have not actually improved, the day of reckoning will come. If National Socialism manages to survive this critical period, it will have won.

Why are Germans being forced to act cowardly? Is it really better to have cowardly fellow travelers rather than work with honest upright Germans? Not to speak of all the big and little leaders [Führer], whose success has gone to their head. "Mini-dictators"—they hardly care about the

---

[16][Indeed, the mathematically inclined physicist Adolf Kratzer (1893–1983) is not listed among the twenty-eight friends and students of Richard Courant who signed a petition addressed to the Minister of Education of Prussia in May 1933. Cf. Exodus Professorum, Göttingen, Vandenhoeck & Ruprecht, 1989, pp. 22–24. The list did not include any scientist from Münster.]

[17][Heinrich Scholz (1884–1956), a leading mathematical logician and philosopher of religion.]

[18][The General Disarmament Conference, held in Geneva between 2 February and 14 October 1933, failed to achieve anything substantial.]

[19][The term is more or less equivalent to "November criminals," used as an insult against those who had taken part in the November 1918 revolution on the grounds that they had brought about the defeat of Germany in World War I.]

real needs of the people, . . . let themselves be driven by base instincts, hatred and envy. It is said that Göring has pictures taken every day of himself wearing his uniform with broad general's stripes, made by the best tailor.

Hitler is isolated from the people and seems barely aware of what is going on.

27 May

One could read in the papers today that a bank employee had been punished because he had refused to do the Hitler salute while the Horst-Wessel song was being sung. I guess this will happen to me sooner or later, too. Obliged to watch what is happening, yet refusing to go along, seeing the way noble ideals and all that is good are increasingly being replaced by brutality, meanness, vacuity and the cowardice of petty bourgeois, all this drains one's energy and generates a feeling of impotent rage.

1 June

Today, in a bookstore, I found out how much Hitler enjoys reading Karl May's novels and that these occupy a place of honor in his library. It suddenly dawned on me: Hitler=Old Shatterhand=Kara Ben Nemsi. Idealists and boasters. The dream of a new noble humanity in which he, Hitler, plays the leading role.

14 June

Just at its climax, the conference of Catholic journeymen[20] in Munich has been disbanded. As a first step, the journeymen were forbidden to wear their "uniform"; some Nazis felt "provoked" at the sight of the uniforms, but since the journeymen could hardly be expected to go around naked, the conference was broken up. . . . The enthusiasm of large crowds for ideals other than those advocated by the Nazis is not appreciated.

{I am in Berlin. [*An unnamed*][21] Party member (who, if alive today, would still be an unwavering supporter) gave me an account of internal

[20][*Germany had a long-standing tradition of church-affiliated associations of nonindependent tradesmen such as the journeymen. The most important Catholic journeymen's association was founded in 1846 and was soon influenced by Rev. Adolf Kolping. The so-called Kolpingwerk continues as a worldwide Catholic lay organization.*]

[21][*At the request of Thullen's son, George, the editor has deleted the name of a family member referred to in this passage.*]

Party affairs; I was aghast.} Such a mixture of hypocrisy, pettiness, spiritual narrowness, egoism, political opportunism was not perceptible in the SPD. The latter took 6–10 years to become "bigwigs," whereas the Nazis manage the feat in 6–10 days.

22 June

Proscriptions follow one another rapidly. Now that the campaign against Jews has been suspended temporarily (?),[22] persecution of Christians and Germans has begun.

Within ten days the following have been banned: the "Grossdeutsche Bund" (All German Youth Federation) (comprises "Wandervögel"—ramblers—, "Freischar"—volunteer corps—, and scouts) {all part of the youth movement}, the SPD, the German national battle units [deutschnationalen Kampfstaffeln], the "Steel Helmet" ["*Stahlhelm*"] in its original make-up (they have been fully absorbed into the NSDAP); the Christian trade unions have been declared enemies of the state. The struggle is becoming increasingly brutal and vicious.

I have just read that the Mainz president has decreed that the flags flown by militarized units, i.e., the SA and SS, are to be saluted {with the arm raised, the "German salute"}. Gessler-style hats everywhere, but no William Tell to liberate us!

29 June

The German National Front has disbanded, and Hugenberg tendered his resignation. The followers of the German National People's Party are given assurances that henceforth they would be treated as "authentic" Germans and as equal partners in the new Germany. Civil servants would remain in their posts. New German idealism!

In a few days the Bavarian People's Party will be brought to trial for high treason—because of alleged links with Dollfuß! {Federal Chancellor of Austria; assassinated by the Nazis in 1934.}[23] Attempts are being made to deal with several Catholic associations in the same way. (The pretext: these could "become a vehicle for a counterrevolution," especially the journeymen's associations). Nobody is allowed to contest these fallacious accusations.

---

[22][*Thullen himself puts a question mark to this assertion. In any event, following the anti-Semitic boycott of April 1, 1933, there were tactically motivated periods of intensified anti-Semitic measures by the Nazis, alternating with a relative relaxation of such measures.*]

[23][*Engelbert Dollfuß (1892–1934), Austrian politician, Federal Chancellor of Austria (1932–1934). He represented the Austrian version of fascism (Austrofaschismus).*]

Incidentally, court proceedings in connection with the burning of the Reichstag (Parliament) are continually being dragged out. According to a Berlin joke, it is the Sass brothers[24] who set fire to the Reichstag.

A decree was issued yesterday against "killjoys." Whoever maintains that things are not any better now than under Brüning is accused of being a "disguised Marxist" (Goebbels).

Scholz advised Behnke not to speak with me about politics, on the grounds that I had at one time openly expressed support for Brüning.

I just heard from . . . ? . . . that he had been forced to join the National Socialist Students' Federation—"or else he had better see how he passes the exams and obtains credits for the semester."

Yesterday our seminar almost finished in a political brawl. A long-serving National Socialist student reproached Behnke, "If we had not gone out into the countryside earlier and had our skulls bashed in. . . ." With the best will in the world no one could detect any outer harm done to his skull. It is from such "heroism and martyrdom" that all kinds of rights are being derived.

1 July

The "Sturmschar" dissolved. {Part of the Catholic youth movement; I myself was a leader of the Trier "Sturmschar."}[25] All assets—the hard-earned pennies of young workers and the unemployed—were confiscated. This is how B. Schirach kept his promise {Schirach was the national leader of the Hitler Youth}.[26]

Of course, the "Windthorst" Federation and the Catholic Young Men's Association were also dissolved.[27]

---

[24][*The Sass brothers were notorious burglars, who had carried out spectacular bank robberies in Berlin.*]

[25][*In a later version of his manuscript, dated December 1988, Thullen adds: "The 'Sturmschar' was an important component of the Catholic youth movement, mostly comprising young workers and salaried employees who were refreshingly spontaneous and lively. I owe some of my best friends to the 'Sturmschar.' " In a version dated end of 1990, he noted (p. 22) that during secondary school in Trier he was a leader of the "Waldläufer" ("Forest ramblers") and that it was only during his university years that he joined the leadership of the "Sturmschar".*]

[26][*Baldur von Schirach (1907–1974).*]

[27][*The "Windthorstbund" was named after a former leading Center Party politician, Ludwig Windthorst (1812–1891), who had been Bismarck's main opponent during the socalled Kulturkampf. Thullen's apprehension about dissolution of the Catholic Young Men's Association only materialized in 1939.*]

2 July

Just as well that I am fully absorbed by mathematical problems. My innermost feelings of restlessness and indignation would probably have driven me to prison!

4 July

It is becoming ever clearer how . . . National Socialist leaders are taking advantage of their power to vent their rage on Catholics. In Württemberg they have gone so far as to dissolve the Catholic young women's associations [*Jungfrauenvereine*]. The minister of education declared that only one ideal should prevail in schools, that of National Socialism. {See press cuttings dated 3 July}.[28]

20 July

Some professors who had celebrated the "national revolution" ["*nationale Erhebung*"] with great enthusiasm, have "awakened":[29] H. Scholz, Stählin,[30] Sa.?, even Herrmann. Now they realize in which direction the new spirit, or rather, the evil spirit blows. On top of that comes the suppression of the established Protestant Church by the "German Christians."[31]

Stählin held a very pointed sermon last Sunday: ". . . there is absolutely no question of the church being brought to heel." He is expecting to be jailed. At the close of a public gathering he stated: "The Protestant Christians stand wholeheartedly by the new {church?}. However, the Kingdom

[28][*Thullen's press cuttings have been omitted here, as they could neither be identified nor reproduced properly.*]

[29]This is a reference to the Nazi slogan, "Germany awake!" [*In contrast to the Nazi use of the slogan, Thullen refers to it in the sense of awakening to the crimes being perpetrated by the Nazis.*]

[30][*Wilhelm Stählin (1883–1975), Protestant theologian, professor of practical theology in Münster from 1926 to 1958, one of the leading figures in the German youth movement. Cofounder of the "Berneuchener Circle" in 1923.*]

[31][*This movement within the Protestant church goes back to 1927. After 1933, it set itself the goal of conforming to the Nazi regime, particularly with regard to enforcement of the race laws against "non-Aryan" Christians. Supported by the Nazi Party, the German Christians won a majority in the Church elections of July 23, 1933 in almost all organs of the Protestant Church. The election of "German Christian" Ludwig Müller as "Reich Bishop" on September 27, 1933 and the founding in the same month of an opposing "Pfarrernotbund" by Martin Niemöller mark the starting point of the "Kirchenkampf" (battle within the church) in Germany. Cf. K. Scholder,* Die Kirchen und das Dritte Reich, *Vol. 1, Frankfurt a.M., Ullstein, 1986.*]

of God is still above the new (. . . ? . . .)."[32] No newspapers have dared to publish this sentence!! {Stählin was for the Protestant youth movement what R. Guardini[33] had been for us in the Catholic youth movement. We highly respected Stählin and often attended his sermons.}

23 July

Today is Election Day in the Protestant church. In most cases it is not even necessary to proceed with a ballot because "German Christians" are the only candidates. Tomorrow's newspapers will report on the surprise victory of the German Christians, and "that the entire Protestant Church fully supports the new movement." But what happened in reality? Military chaplain Müller[34] managed to secure the withdrawal of the "Protestant Christians" ["*Evangelische Christen*"] list—on the grounds that this term offended the German Christians. This decision was announced over the radio at 13:00 hours; all lists had to be filed by 15:00 hours at the latest. Hence, only in exceptional instances was it possible to submit new lists.

Newspapers were obliged to publish the German Christians' proclamation every day, whereas others were deprived of any possibility of making their views known.[35]

If the Protestant Church cannot put up resistance, broader Protestant circles will probably be drawn closer to the Catholic Church.

The "Reichskonkordat" text was published today. For the [*Catholic*] Church this is a success, as long as the Nazis do not attempt to achieve their avowed ends through the back door by resorting to fraudulent measures.

It is still not known whether the "Sturmschar" will stay dissolved. It almost looks like it will.

---

[32][*Perhaps "Reiche." It is possible that the word within the preceding parenthesis originally was "Reiche," which resembles "Kirche," and that Thullen may have misread his own notes.*]

[33][*Romano Guardini (1885–1968), German Catholic philosopher of religion. Born in Verona, he acquired German citizenship in 1911. He was appointed to the newly created chair of philosophy of religion and Catholic theology in Berlin in 1923, which he retained until 1939 when the Nazis banned him from teaching. Guardini was also renowned as head of the Catholic youth movement "Quickborn" ("Living Spring") based in the castle of Rothenfels.*]

[34][*The newly elected Protestant Reich Bishop von Bodelschwingh resigned in protest against the appointment of a state inspector for the Prussian regional churches. It was because of this that military chaplain Ludwig Müller took charge of the Protestant Churches of the Old-Prussian Union; later he became "Reich Bishop" (cf. note 31 above).*]

[35][*The manuscript reproduces a number of press cuttings, of unidentified origin, dated July 23, 1933, reporting majorities between 75 percent and 100 percent won by the "German Christians" in the church elections.*]

[What follows are Peter Thullen's notes dated March 1988]

Here is where my diary entries end. Still two months to go before my departure for Rome.

I felt isolated at the university in Münster. Only at home and among my good friends in Trier could I talk freely.

I must admit that in Münster even I was tempted to give up resistance, at least outwardly. However, I just could not bring myself even to raise my arm, which felt as if it were weighed down by lead, for the German salute or—worse still—to end my letters with the obligatory "Heil Hitler," in the name of the man who ruined our Germany. The research grant I obtained in 1932 to go to Rome now seemed providential. It would allow me to gain some distance and to observe events from outside. It was clear to me that I would not return to Germany as long as Hitler remained in power. I told this to Behnke. He would not have any of that; even much later in life he really did not quite forgive me for having deserted the "German university."[36] Behnke, in his book, *Ein Leben an deutschen Universitäten im Wandel der Zeit* (Göttingen, 1978, p. 129), wrote the following, when discussing the expulsion of Jewish professors:

> Thus were we deprived of the dignity of office and of human dignity. Were not those better off who were forced to resign from their posts? This was often said, and yet most colleagues defended their position. Of course it looks impressive if one leaves head high before one is caught and perhaps driven out by the brown bureaucracy. In certain cases, such a voluntary disappearance does not cost much; for example, if one is a successful young mathematician. Such freedom gave me the creeps. I like the people better who had qualms about leaving their German university. If they took their calling seriously, they were bound to feel remorse at leaving behind, as it were, a substantial portion of their own lives at their former place.

The reference to a "successful young mathematician" could only have meant me (at that time I was an "untenured scientific assistant").[37] Behnke never fully understood the true nature of National Socialism; at most he saw its anti-Semitic traits, and he seemed not to have understood that the young mathematician in question preferred the insecurity of life in exile to both the "loss of human dignity" and the loss of the German university, which had capitulated to Hitler. *[End of digression.]*

---

[36]*[The following digression was a footnote in the original manuscript. Because of its significance, the editor decided to include it in the body of the memoirs.]*

[37]*[According to information in his papers, Thullen's monthly income in that position amounted to 140 Reichsmark.]*

In Rome I devoured whatever news came from Germany. Only once, in June 1934, when the "Röhm Putsch"[38] took us all by surprise, did I see any hope of returning to Germany. I figured that if they started killing each other off the regime would collapse. It was a vain hope. The frightening aspect of this incident was that the extrajudicial execution, ordered from above, of Hitler's erstwhile companions, Röhm and his SA entourage, only served to strengthen Hitler's position. The middle class greeted the elimination of the "SA hooligans"[39] with a sigh of relief, even if this had been at the price of the murder of prominent political opponents (such as Klausener[40] and General Schleicher). By the way, Rudi Salat, Secretary General of the international Catholic students' organization, Pax Romana[41] (he was Ursula's godfather) was also on the hit list, but thankfully, he was away from home in Munich on the night of the killings. He managed to flee abroad.

My one-year grant came to an end. Mathematically speaking it had been very productive. I sent an account of the results obtained to the "Notgemeinschaft der Deutschen Wissenschaft" [*Emergency Society for German Scholars*],[42] but turned down offers from Germany. I just could not see myself serving Hitler's totalitarian State in which individual resistance had become impossible, unless one was deliberately seeking death. Instead I stayed in Rome. Again good fortune came to my rescue. Severi's[43] senior assistant moved to Paris with a Rockefeller grant; Severi offered me the vacant post on a fixed-term basis.

---

[38][*Ernst Röhm (1887–1934), chief of staff of the SA and one of Hitler's earliest followers, was liquidated by him because the SA stood in the way of the Wehrmacht and the SS, on whose support Hitler was counting.*]

[39]Quoted from the memoirs (1986) of a former pupil of our (Catholic) secondary school, who even today has not recognized the full extent of the horrendous illegality of these murders.

[40][*Erich Klausener (25.1.1885 Düsseldorf–30.6.1934 Berlin). Central Party politician, in 1924 director at the Social Welfare Ministry, in 1926 chief of the police department within the Ministry of the Interior of Prussia, head of the "Katholische Aktion" ("Catholic Action") in the Diocese of Berlin and dismissed in 1933. At the Berlin Catholic Congress on June 24, 1934 he stood up against the racist policies of the Nazis; subsequently he was placed on the "hit list" by Göring and executed during the "Röhm Putsch."*]

[41][*Pax Romana, founded in 1921, has since 1947 included academic associations. Headquartered originally in Fribourg and more recently in Geneva (Switzerland). It is perhaps no coincidence that Fribourg was also Thullen's last stop in his professional life. When he wrote a letter on June 30, 1971 to Rudi Salat, he reiterated his desire to continue maintaining his "human contacts."*]

[42][*A copy of the report to the "Notgemeinschaft der Deutschen Wissenschaft" ("Emergency Society for German Scholars," today's "Deutsche Forschungsgemeinschaft") is kept in the Thullen estate.*]

[43][*Francesco Severi (1879–1961), a leading Italian geometer.*]

I took advantage of this opportunity to look around for other possibilities outside of Germany and Italy, such as in the USA, Switzerland, Austria and even Peking [*Beijing*]. This was not an easy undertaking, even with the help of a few friends. At long last came an unexpected offer from Ecuador to join a group of seven professors who were to found a "Polytechnic School." I lost no time in accepting, without really knowing where Quito was located. Two or three days later Severi informed me that as a "Tedesco" [*German*] I was no longer allowed to give lectures. Hitler and Mussolini had fallen out with each other. Severi was greatly relieved when informed of my good news. Some other time I will tell you about the wonderful one and a half years spent in ancient and modern Rome.

This is how your mother and I, at last united, ended up in Ecuador. Before our departure we met up in Germany. I had the benefit of a "German abroad" passport ["*Auslanddeutscher Pass*"] with a conspicuous visa issued by the Ecuadorian embassy in Geneva (Switzerland). After she obtained a university degree (equivalent to a master's degree) your mother also refused to teach in Nazi schools and tided herself over as a private tutor. We were married in the Benedictine abbey of Maria Laach and bid farewell to our families and friends. A few days before we left, Behnke wrote to us that "someone" had reproached him for allowing his best student to go abroad. How did this "someone" find that out? We boarded the train in Trier for Amsterdam with some anxiety. We reached the Dutch border—no one hauled us out of the train—we were free!

Your mother has described our life in Quito, where all five of you were born, in the family chronicle. The following are just a few supplementary comments of a political nature.

There were two groups of German expatriates in Quito: the first included well-established merchants and businessmen and the embassy staff; the second was the new, rapidly expanding, group of émigrés, nearly all "non-Aryan" or married to them, except ourselves. German-speaking Jews from Prague later joined this second group. Until 1939 both groups lived more or less peacefully side-by-side, particularly since the ambassador, Dr. Klee, had been anything but a Nazi (indeed, rumor had it that he was "transferred for disciplinary reasons" to insignificant Ecuador). When Dr. Klee amicably tried to persuade us at least to take part in "patriotic" celebrations, however, we made it plain to him that we would feel very ill at ease among so many Nazis and that we would surely be the only ones not to raise our hands in the "German salute" during the inevitable singing of the German anthem.

Meanwhile, a local National Socialist section was founded in Quito headed by a former sailor (so we were told) as "Ortsgruppenleiter" [*local section leader*]. Ecuador and Germany were still at peace, and nobody prevented the Nazis from pursuing their activities. Most of the long-established businessmen belonged to the local chapter of the Nazi Party; undoubtedly they were unaware of what was really going on in their home country. The zealous section leader managed to lodge himself in the embassy and controlled visitors entering the premises. As the only "Aryans" among the émigrés we probably irritated him. Several times he would lie in wait for me and summon us to participate in the embassy's official events. Finally I lost patience and told him to leave us alone, and that we were not afraid of him or of the Gestapo. Thereupon he immediately forbade the ambassador, with whom we had developed a friendly relationship, to have any further contacts with us.

[A section on Thullen's further problems with the embassy is omitted here.]

Events in Germany accelerated: occupation of the Rhineland [*March 1936*], annexation of Austria to the "Greater German Reich" (March 1938) and then the dismantling of Czechoslovakia and transformation of what was left into a "Protectorate" (1939). On that decisive day we were listening to news from Germany on a radio at a neighbor's home (we did not have a radio of our own). All I can remember is that I cried out in anger: "This means war!" My neighbor—Hitler's portrait hung over his armchair—retorted, "The 'Führer' only wants peace." Since then we stopped listening to the radio at his home. Shortly afterward war broke out in Europe.

Later on, toward the end of 1941, when the USA entered the war against the "Axis powers," Ecuador, like most other Latin American countries, declared war on Germany. The old established German expatriate community and the embassy staff were expelled; a ship repatriated them to Germany. In July 1942 we acquired Ecuadorian citizenship.[44]

The German émigré community had grown considerably in the meantime, to somewhere between a thousand and two thousand persons in Quito alone, among them many academics. A shared fate brought us closer together. A stimulating intellectual life developed. Even a German-language theater group was launched ([. . .]). The focal point was the association of Free Germans [*Vereinigung der Freien Deutschen*]. It was our intention to show the world an image of a "different" Germany. We also sought contacts with other like-minded circles in Latin America.

---

[44][*In a letter dated August 5, 1980 to Gerd Muhr, former vice-chair of the German Confederation of Trade Unions (DGB), Thullen stressed that circumstances in 1942 forced him to acquire Ecuadorian citizenship, without ever having, on his own initiative, renounced his German nationality.*]

[Thullen follows with excerpts from correspondence, speeches, etc. relating to the activities of the "Freien Deutschen," which were mostly concerned with defining their identity and in particular with discussions about the nature of National Socialism. They have been omitted here.]

Even the little I have to tell you about the activities of the Free Germans shows what moved us in those times. Our prime concern was the emergence of a truly democratic system in our native country. In our new home country I was soon faced with new tasks that absorbed all my time and energies. In March 1938 I took charge of the future actuarial department of the Instituto Nacional de Previsión (National Social Security Institute); its scope extended far beyond mathematics alone to cover the most important aspects of social security. This was the beginning of a new, creative and fascinating area of work that became a substitute for my earlier passion for pure mathematics.

Ecuador had turned into our real home country. Many from the cultural and intellectual elite became close friends. Yet, in 1947 we had to go into exile for the second time. From what I told you in the past you are familiar with the complex reasons that led to our departure from Ecuador. It began with my refusal to carry out a reform of the pension scheme (that is, to do the necessary actuarial calculations) as demanded by the then dictator (Velasco Ibarra). I considered such a reform would jeopardize the scheme's unity and principle of equality of treatment. On top came my struggle against the incredibly corrupt president of our institute. The end result of all this was that although my technical advice was heeded and the president was forced to relinquish his post, the latter managed to get his successor to oblige me to leave as well.[45] We moved to Colombia. With that began an international career, which finally led me to the International Labor Office in Geneva and thus back to Europe.

We discovered a German Federal Republic in which—contrary to what we had hoped—democracy had been introduced by decree from *outside*, by the Allies. Immediately after the war we received books and journals, openly critical and reflecting a contagious intellectual vivacity in which

---

[45]A few months later the Institute's board of directors was changed—largely as a result of pressure both from trade unions that remained loyal to me and from our many friends. The new board sent me an apology in Bogotá couched in very moving terms and ordered it to be published in the two leading newspapers in Quito and Guayaquil. However, since I was bound by new contracts (in Colombia) there was no way I could contemplate a return to Ecuador.

we hoped to participate.[46] Alas, the Restoration began already under the first West German chancellor Adenauer (1949). Critical voices became less frequent. Apart from war criminals convicted by the Allies and those who had managed to escape abroad in time, most others retained their posts or returned to them or just kept on as usual in their professions. Among the latter were Hitler's former "bloody judges" ["*Blutrichter*"] and the euthanasia doctors. What counted were "competence" and loyalty to the State and the Law. People like us who, when in doubt, would rather follow their conscience than acquiesce to the dictates of so-called civic virtues, were not welcome.

Back in 1935 when we left Germany for good we had faced a similar situation: why would a young mathematician be so crazy as to renounce a "brilliant career" and vanish to a little known country in Latin America?[47] Only in the immediate post-war period were we so-called Americans welcome, we who converted our savings into Care parcels that were sent to the old country. Once the economic boom set in in Germany, however, attitudes changed: "You were so well off over there, whereas we . . . etc."

I also recall appeals published in newspapers encouraging scientists in exile to return to their home country. These appeals gradually ceased. In fact, I never received an offer to teach at a German university, but instead, in 1955, one from the respected Catholic University in Washington (I declined the offer because at that time I was wholly devoted to problems of the Third World). As soon as I retired from the UN system, I was called to the University of Zurich as associate professor and later to the University of Fribourg as tenured professor. Thanks to Swiss universities I have been able to come around full circle and return to academic life.

[46][*Thullen was probably referring to the* Frankfurter Hefte, *a monthly politico-cultural journal founded in 1946 by Eugen Kogon and Walter Dirks as a forum for independent ("leftist") current of thought in Catholicism. This journal merged in 1985 with the* Die neue Gesellschaft, *which was close to the SPD. At one time Thullen considered the option of publishing these memoirs in the weekly* Die Zeit *in Hamburg, particularly since its editor and publisher, Gräfin Dönhoff, held views very similar to his regarding the Nazis.*]

[47][*These bitter comments probably were not intended to draw a parallel between the Hitler dictatorship and the democratic social order of the Federal Republic of Germany; instead, they are more likely referring to the comparable reactions of ordinary "apolitical" individuals, including scientists.*]

Only of late[48] can one detect a gradual change of mindset in the Federal Republic of Germany. Post-war generations are reacting to the denial tactics of their elders and demanding to know the full truth about "what really happened." May my notes contribute to that process.[49]

Marly, March 1988
Your Father

[48][*In the second half of the 1960s in the wake of student unrest, lectures were already being held on the history of Nazism and its impact on German universities. However, it is indeed true that systematic historical analysis of the role of science during the Third Reich only dates to the beginning of the 1980s. Cf. H. Mehrtens and S. Richter (editors),* Naturwissenschaft, Technik und NS-Ideologie, *Frankfurt a.M., Suhrkamp, 1980, and B. Müller-Hill,* Tödliche Wissenschaft. Die Aussonderung von Juden, Zigeunern und Geisteskranken 1933–1945, *Hamburg, Rowohlt, 1984.*]

[49][*The memoirs end with a two-page annex titled, "Publicly known facts from the period between Hitler's seizure of power and my departure for Rome in 1933." It has not been included on the assumption that the information contained therein is familiar to readers of these memoirs.*]

Archives, Unprinted Sources, and Their Abbreviations

AAM	Archives of American Mathematics; Center for American History; University of Texas at Austin, USA
AAMS	Archives of the American Mathematical Society (AMS) at the John Hay Library, Providence, Rhode Island, USA
Adams Papers	Clarence Raymond Adams Papers, BUA
Bernstein Papers	Papers of Marianne Bernstein, Sarasota, Florida (with Ron Bernstein-Wiener)
Bieberbach Papers	Partial estate Ludwig Bieberbach, Oberaudorf (Germany), possession Ulrich Bieberbach
Birkhoff Papers (GDBP)	George David Birkhoff Papers at HUA (HUG 4213)
Bohr Archives	Niels Bohr Archive Copenhagen, Partly estate Harald Bohr
Brun Papers	Papers of Viggo Brun at the National Library, Oslo, no. 719.
BUA	Brown University Archives, Providence, Rhode Island, USA; contains e.g., papers of Richardson, Tamarkin, Adams, and documents of the Applied Mathematics Division
CIP	Courant Institute Papers = part of CP at Bobst Library, New York City
Courant Papers	Richard Courant Papers at Bobst Library, New York City, New York University Archives, not yet fully catalogued
CP	see Courant Papers
CPP	Courant Private Papers = part of CP at Bobst Library, New York City
Dehn Papers	Max Dehn Papers at AAM, Austin, USA
Dunn Papers	Papers of geneticist L. C. Dunn at the Archives of the American Philosophical Society, Philadelphia (correspondence with E. J. Gumbel and F. Bernstein)

EC	Emergency Committee in Aid of Displaced Germans (since 1939: Foreign) Scholars, files at the New York Public Library, Manuscripts and Archives Division, USA
Einstein Papers	Albert Einstein Archives, Hebrew University Jerusalem, and Mugar Library, Boston (the latter has materials in copy)
EP	see Einstein Papers
ETH	Library of Eidgenössische Technische Hochschule Zürich (Zurich), Wissenschaftshistorische Sammlungen
Franck Papers	James Franck Papers, Regenstein Library, Chicago, USA
Freudenthal Papers	Partly estate Hans Freudenthal at ETH, HS 1183 (Copies from Freudenthal Archives Utrecht)
GSA	Geheimes Staatsarchiv Preußischer Kulturbesitz, Berlin, files of the Prussian Ministry of Culture, Rep. 76 Va
Gumbel Papers	Emil Julius Gumbel Papers, Regenstein Library, Chicago, USA
Hauser Papers	Partly estate Wilhelm Hauser within papers Günter Wirth, Archives of Konrad Adenauer Stiftung, Sankt Augustin, Germany
Hellinger Papers	Ernst Hellinger Papers, NWUA, Evanston, USA
Hopf Papers	Papers of Heinz Hopf at ETH, Hs 621
HSP	see Shapley Papers
HUA	Harvard University Archives; Cambridge, Massachusetts; contains papers of G. D. Birkhoff, R. von Mises, H. Shapley, E. B. Wilson, etc., and records of the Mathematics Department (UAV 561)
IASA	Institute for Advanced Study, Princeton, USA, Institute Archives
IBD-Microfilm	Microfilms with original questionnaires for the IBD (Strauss/Röder), available for instance in the "Zentrum für Antisemitismusforschung," Berlin
Kármán Papers	The Theodore von Kármán Collection at the California Institute of Technology, Pasadena (Microfiche Edition available in various libraries)
Milne Papers	Papers of Edward Arthur Milne, Bodleian Library, Oxford
Mises Papers	Papers of Richard von Mises at HUA, HUG 4574
Mordell Papers	Papers of Louis Joel Mordell, St, John's College Library, Cambridge, UK

MITA	Massachusetts Institute of Technology, Archives: quoted Wiener Papers and AC 4
Neyman Papers	Jerzy Neyman Papers, University of California, Berkeley, USA
NWUA	Northwestern University Archives, Evanston, USA: Hellinger Papers and Department of Mathematics Papers
OVP	Oswald Veblen Papers, Library of Congress, Washington, DC, USA
Pólya Papers Zurich	Estate George Pólya in ETH, Hs 89
Pólya Papers Stanford	Partly estate George Pólya in the Special Collections of the Stanford University Libraries; SC 337, USA
RAC	Rockefeller Archive Center, Sleepy Hollow, New York, USA
Richardson Papers	Papers of Roland George Dwight Richardson, BUA
SBPK	Staatsbibliothek Preußischer Kulturbesitz Berlin; here quoted: Manuscript Division, estate Alfred Landé
Shapley Papers	Harlow Shapley Refugee Files at HUA, HUG 4773.10
SPLS	Files of the Society for the Protection of Science and Learning, formerly Academic Assistance Council, Bodleian Library, Oxford, in particular, boxes 277–86 concerning refugee-mathematicians
Stone Papers	Marshall Harvey Stone Papers, AAMS
Szegö Papers	Gabor Szegö Papers in the Special Collections of the Stanford University Libraries; SC 323, USA
Tamarkin Papers	Jacob David Tamarkin Papers, BUA
Thomas Mann Archives	Thomas Mann-Archiv at ETH, quoted here ms. of F. Behrend "Die Fahrt zu den Vätern" (1961), 19 pp.
Thullen Papers	Peter Thullen Papers, in the possession of Georg Thullen, Genthod, Switzerland
Toeplitz Papers	Otto Toeplitz Papers, University Library Bonn, Manuscript Division
Reissner Papers	Hans Reissner Papers, Mandeville Special Collections Library, University of California, San Diego, MSS 0030
UAB	University Archives, Humboldt University, Berlin (Personal file J. von Neumann, UK 44, and NSDozentenschaft, no. 222 [E. Schmidt])
UAH	University Archives, Ruprecht-Karls-Universität, Heidelberg
Van der Waerden Papers	B. L. van der Waerden Papers, ETH, Hs 652

Veblen Papers	see OVP above
Weyl Papers	Hermann Weyl Papers, ETH, Hs 91
Wiener Papers	Norbert Wiener Papers at MIT, Institute Archives
Wilson Papers	Edmund Bidwell Wilson Papers at HUA, HUG 4878

Letters from Witnesses and Relatives to the Author

Franz L. Alt (July 12, 1993, September 9, 1997); Rafael Artzy (January 11, 1998; February 4, 1998); John Bather (for Stefan Vajda, December 2, 1997); Marianne Bernstein (for Felix Bernstein, February 11, 1998); Patricia Bing (for Kurt Bing, March 6, 1998); Curtis M. Fulton (March 20, 1994); Michael Golomb (July 19, 1993); Kurt Hohenemser (December 6 and 23, 1997); Barbara Jaeckel (for Richard Fuchs: February 13, 1998; March 1, 1998); Walter Ledermann (December 11 and 29, 1997); Henry B. Mann (August 18, 1993); Bernhard H. Neumann (August 22, 1993; February 16, 1998); Lutz Noack (for Walter Jacoby: February 25, 1999), Eric Reissner (March 18, 1994); Hans Samelson (May 2, 1994; March 19, 1998); Hanna Schwerdtfeger (for Hans Schwerdtfeger, July 21, 1993; September 8, 1997); Olga Taussky-Todd (August 11, 1993); Rachel Theilheimer (for Feodor Theilheimer: December 17, 1997); Georg Thullen (for Peter Thullen: several letters between 1999 and 2006), Ros Whiting (for Th. Estermann, July 6, 1993); Henry Zatzkis (March 16, 1994)

Interviews by the Author with Witnesses and Relatives

Marianne Bernstein (repeatedly by phone 1992 through 2003); Ludwig Boll (personally November 29, 1983, in Berlin); Lilly Görke (personally March 19, 1986 in Berlin); Walter Ledermann (phone, January 10, 1998); Curt Siodmak (phone February 12, 1998); Dirk J. Struik (personally December 18, 1991 in Boston, USA); Georg Thullen (repeatedly between 1999 and 2007); Horst Tietz (personally May 19, 1994 in Copenhagen); Magda Tisza (for H. Pollaczek-Geiringer and R. v. Mises, personally January 30, 1992 and November 5, 1997 in Boston, USA).

Further Abbreviations Used, Mostly concerning Institutions and Published Sources, in Particular Journals

AAC	Academic Assistance Council, London, later SPSL
AC	*American Council for Émigrés in the Professions. Records 1941–1974*
AMS	American Mathematical Society

AMWS	*American Men and Women of Science*, New York City
BMG	Berliner Mathematische Gesellschaft
Caltech	California Institute of Technology, Pasadena, USA
CW	*Collected Works*
DMV	Deutsche Mathematiker-Vereinigung
DSB	*Dictionary of Scientific Biography* (with Supplement), 1970–81, NY Scribners
EC	Emergency Committee in Aid of Displaced German (since 1939: Foreign) Scholars, New York City—abbreviation also used for files of the Committee
ETH	Eidgenössische Technische Hochschule Zürich
IAS	Institute for Advanced Study, Princeton, USA
IBD	*International Biographical Dictionary of Central European Émigrés 1933–1945* [= Röder/Strauss, eds. (1983)]
JDMV	*Jahresbericht der DMV*
LDS	*Lists of Displaced Scholars 1936/37*, issued by the SPSL, reprinted in Strauss/Buddensieg/Düwell (1987)
LMS	London Mathematical Society
MIT	Massachusetts Institute of Technology, Cambridge, USA
NDB	*Neue Deutsche Biographie*, 1953–; Berlin, Duncker and Humblot
NÖMG	*Nachrichten der Österreichischen Mathematischen Gesellschaft*
NTM	*Zeitschrift für Geschichte der Wissenschaften, Technik und Medizin*, Birkhäuser, Basel
Poggendorff	*Biographisch-literarisches Handwörterbuch zur Geschichte der exakten Wissenschaften*, Leipzig 1863–2003 (available as CD-ROM 2004), supplement "Ergänzungsband Mathematiker" in two parts, 2004
RF	Rockefeller Foundation, New York City (see also archives at RAC)
R. S.	Initials of the author
SPSL	Society for the Protection of Science and Learning, London, formerly Academic Assistance Council (AAC)
T	Translation into English by the author R. S.
Tob	Tobies (2006)
Toep	Toepell ed. (1991)

References

Abikoff, W. (1995). "Remembering Lipman Bers." *Notices American Mathematical Society* 42, no. 1, 8–18.

Albers, D. J., and G. L. Alexanderson, eds. (1985). *Mathematical People: Profiles and Interviews.* Boston: Birkhäuser.

Albers, D. J., G. L. Alexanderson, and C. Reid, eds. (1990). *More Mathematical People.* Boston: Harcourt Brace Jovanovich.

Alberts, G. (1994). "On Connecting Socialism and Mathematics: Dirk Struik, Jan Burgers, and Jan Tinbergen." *Historia Mathematica* 21, 280–305.

Alexanderson, G. L. (2000). *The Random Walks of George Pólya.* Washington, DC: Mathematical Association of America.

Aleksandrov, P. S. (1935/81). "Obituary of Emmy Noether," in Dick (1981), 153–79.

American Council for Émigrés in the Professions: Records, 1941–1974, 2 vols., mimeographed, 778 pp. (=AC).

American Mathematical Society Semicentennial Publications in Two Volumes (1938). New York: American Mathematical Society.

Anon. (1970). "Øystein Ore (1899–1968)." *Journal of Combinatorial Theory* 8, i–iii.

Ansprachen anläßlich der Feier des 75.Geburtstages von Erhard Schmidt durch seine Fachgenossen (13.1.1951), mimeographed manuscript, typewritten, Berlin, 34 pp.

Archibald, R. C. (1938). *A Semicentennial History of the American Mathematical Society, 1888–1938*; volume 1 of *American Mathematical Society Semicentennial Publications in Two Volumes*, New York: AMS.

Archibald, R. C. (1950). "R. G. D. Richardson, 1878–1949." *Bulletin AMS* 56, 256–65.

Arnold, W. (1986). "Dr. h. c. Ludwig Boll—in Memoriam." *Historische und Philosophische Probleme der Mathematik* 6 (Humboldt Universität Berlin), no. 12, 10–14.

Artin, E. (1950). "The Influence of J. H. M. Wedderburn on the Development of Modern Algebra." *Bulletin AMS* 56, 65–72.

Artzy, R. (1972). "Kurt Reidemeister." *Jahresbericht DMV* 74, 96–104.

Artzy, R. (1994). *Reminiscences of an Albertina Mathematics Student, 1930–1933*, typewritten, 2 pages. Sent in copy to this author.

Ash, M. (1985). "Gestalt Psychology: Origins in Germany and Reception in the United States," in C. E. Buxton, ed., *Points of View in the Modern History of Psychology.* Orlando: Academic Press, 1985, 295–344.

Ash, M. (1996). "Common and Disparate Lemmas of German and American Universities," in Muller, ed. (1996), 37–46.

Ash, M., and A. Söllner, eds. (1996). *Forced Migration and Scientific Change: Émigré German-Speaking Scientists and Scholars after 1933*. Washington, DC: German Historical Institute; Cambridge: Cambridge University Press.

Askey, R. (1982). "Gabor Szegö: A Short Biography," in Szegö (1982), 1–7.

Askey, R., and P. Nevai (1996). "Gabor Szegö: 1895–1985." *Mathematical Intelligencer* 18, no. 3, 10–22.

Aspray, W. (1988). "The Emergence of Princeton as a World Center for Mathematical Research, 1896–1939," in P. Duren, ed. (1988/89), vol. 2, 195–215.

Barneck, A. (1922). "Nachruf auf Eugen Jahnke." *Sitzungsberichte der Berliner Mathematischen Gesellschaft* 21, 30–39.

Barrau, J. A. (1948). "In Memoriam Prof. Dr. J. Wolff." *Nieuw Archief voor Wiskunde* (2) 22, 113–14.

Basch, A. (1950). "Karl Wolf." *Nachrichten der Österreichischen Mathematischen Gesellschaft (NÖMG)* 4, no. 11, 4–6.

Bather J. A. (1996). "Stefan Vajda." *Independent*, Monday, January 1, 1996 (includes photo).

Bayart, D., and P. Crépel. "Statistical Control of Manufacture," in Grattan-Guinness, ed. (1994), 1386–91.

Becker, H. (1987). "List of Displaced Göttingen Scholars," in Becker, Dahms, and Wegeler, eds. (1987), 489–501.

Becker, H., H.-J. Dahms, and C. Wegeler, eds. (1987). *Die Universität Göttingen unter dem Nationalsozialismus. Das verdrängte Kapitel ihrer 250 jährigen Geschichte*. Munich: Saur.

Begehr, H., ed. (1998). *Mathematik in Berlin. Geschichte und Dokumentation*. Aachen: Shaker, 2 vols.

Behnke, H. (1973). *Das Haus Still und seine Freunde aus der Wissenschaft*. Recklinghausen: Aurel Bongers.

Behnke, H. (1978). *Semesterberichte. Ein Leben an deutschen Universitäten im Wandel der Zeit*. Göttingen: Vandenhoeck and Ruprecht.

Behnke, H., and G. Köthe (1963). "Otto Toeplitz zum Gedächtnis." *Jahresbericht DMV* 66, 1–16.

Bell, E. T. (1938). "Fifty Years of Algebra in America, 1888–1938," in *American Mathematical Society Semicentennial Publications in Two Volumes*. New York: AMS, vol. 2, pp. 1–34.

Bernays, P. (1969). "Hertz, Paul." *Neue Deutsche Biographie* 8, 711–12.

Berndt, B. C. (1992). "Hans Rademacher (1892–1969)." *Acta Arithmetica* 61, 209–31.

Bers, L. (1988). "The Migration of European Mathematicians to America," in P. Duren, ed. (1988), vol. 1, 231–43.

Beyerchen, A. (1977). *Scientists under Hitler: Politics and the Physics Community in the Third Reich*. New Haven, CT: Yale University Press.

Bieberbach, L. (1934). "Persönlichkeitsstruktur und mathematisches Schaffen." *Unterrichtsblätter für Mathematik und Naturwissenschaften* 40, 236–43.

Biermann, K.-R. (1988). *Die Mathematik und ihre Dozenten an der Berliner Universität, 1810–1933*. Berlin: Akademieverlag.

Bigalke, H.-G. (1988). *Heinrich Heesch. Kristallgeometrie, Parkettierungen, Vierfarbenforschung*. Basel: Birkhäuser.

Binder, Ch. (1984). "Alfred Tauber (1866–1942)—ein österreichischer Mathematiker." *Jahrbuch Überblicke Mathematik* 17, 151–66.

Binder, Ch. (1992). "Hilda Geiringer: Ihre ersten Jahre in Amerika," in Demidov, S. S. et al., eds. (1992), 25–53.

Birkhoff, G. (1976a). "The Rise of Modern Algebra to 1936," in Tarwater et al., eds. (1976), 41–63.

Birkhoff, G. (1976b). "The Rise of Modern Algebra, 1936–1950," in Tarwater et al., eds. (1976), 65–85.

Birkhoff, G. (1977). "Some Leaders in American Mathematics, 1891–1941," in Tarwater, ed. (1977), 25–78.

Birkhoff, G. (1980). "Computer Developments, 1935–1955, as Seen from Cambridge, U.S.A.," in N. Metropolis et al. (1980), 21–30.

Birkhoff, George D. (1938). "Fifty Years of American Mathematics," in *American Mathematical Society Semicentennial Publications in Two Volumes*. New York: AMS, vol. 2, 270–315.

Birnbaum, Z. W. (1982). "From Pure Mathematics to Applied Statistics," in Gani, ed. (1982), 76–82.

Birnbaum, Z. W., ed. (1970). "William Feller, 1906–1970." *Annals of Mathematical Statistics* 41, iv–xiii (with photo).

Bochner, S. (1976). "Mathematical Americana," in Tarwater et al., eds. (1976), 87–98.

Böhm, J. (1965). "Wilhelm Maier zum 70. Geburtstag." *Wissenschaftliche Zeitschrift Universität Jena*, Math.-naturwiss. Reihe 14, no. 5, 219–20.

Böhne, E., and W. Motzkau-Valeton, eds. (1992). *Die Künste und Wissenschaften im Exil*. Gerlinen: Schneider.

Boll, L. (1977). "Die Bedeutung der russischen Übersetzungen in der Mathematik der DDR." *Mitteilungen der Mathematischen Gesellschaft der DDR* 1977, nos. 2–3, 7–14.

Bolza, O. (1936). *Aus meinem Leben*. Munich: Reinhardt.

Born, G., ed. (2005). *The Born-Einstein Letters*. New York: Macmillan [revision of the English edition of 1971, which was a translation of the original German edition of 1969].

Bos, H. (1992). "In Memoriam: Hans Freudenthal (1905–1990)." *Historia Mathematica* 19, 106–8.

Branges, L. de, I. Gohberg, and J. Rovnyak, eds. (1990). *Topics in Operator Theory: Ernst Hellinger Memorial Volume*. Basel, Boston, Berlin: Birkhäuser (includes photo).

Brauer, A. (1973). "Gedenkrede auf Issai Schur," in Schur (1973), vol. I, v–xiii.

Brauer, R. (1967). "Emil Artin." *Bulletin AMS* 73, 27–43.

Brauer, R., H. Hasse, and E. Noether (1932). "Beweis eines Hauptsatzes in der Theorie der Algebren." *Journal für die reine und angewandte Mathematik* 167, 399–404.

Braun, H. (1990). *Eine Frau und die Mathematik, 1933–1940. Der Beginn einer wissenschaftlichen Laufbahn*, edited by Max Koecher. Berlin: Springer.

Brewer, J. W., and M. K. Smith, eds. (1981). *Emmy Noether: A Tribute to Her Life and Work*. New York, Basel: Dekker.

Briegel, M., and M. Frühwald, eds. (1988). *Die Erfahrung der Fremde. Kolloquium des Schwerpunktprogramms "Exilforschung" der Deutschen Forschungsgemeinschaft.* Weinheim: VCH Verlagsgesellschaft.

Brieskorn, E., ed. (1996). *Felix Hausdorff zum Gedächtnis, Band I. Aspekte seines Werkes.* Opladen: Westdeutscher Verlag.

Brocke, B. vom, ed. (1991). *Wissenschaftsgeschichte und Wissenschaftspolitik im Industriezeitalter. Das "System Althoff" in historischer Perspektive.* Hildesheim: Lax.

Brown, J. D., M. T. Chu, D. C. Ellison, and R. J. Plemmons, eds. (1994). *Proceedings of the Cornelius Lanczos International Centenary Conference.* Philadelphia: SIAM.

Browne, C. A. (1940). "The Role of Refugees in the History of American Science." *Science* 91, 203–8.

Bru, B. (1991). "Doeblin's Life and Work from His Correspondence," in H. Cohn, ed. (1991), 1–64.

Brüning, B., D. Ferus, and R. Siegmund-Schultze (1998). *Terror and Exile: An Exhibition on the Occasion of the International Congress of Mathematicians 1998* (Catalogue). Berlin: Deutsche Mathematiker-Vereinigung, 72 pp.

Butler, L. (1992). *Mathematical Physics and the American Mathematics Community,* unpublished PhD dissertation, University of Chicago.

Butler, L. (1997). "Mathematical Physics and the Planning of American Mathematics: Ideology and Institutions." *Historia Mathematica* 24, 66–85.

Butzer, P., and L. Volkmann (2006). "Otto Blumenthal (1876–1944) in Retrospect." *Journal of Approximation Theory* 138, 1–36.

Buxton, C. E., ed. (1985). *Points of View in the Modern History of Psychology.* Orlando: Academic Press.

Cassels, J. W. S., and A. Fröhlich (1977). "Hans Arnold Heilbronn." *Bulletin London Mathematical Society* 9, 219–32.

Chandler, B., and W. Magnus (1982). *The History of Combinatorial Group Theory: A Case Study in the History of Ideas.* New York: Springer.

Cherry, T. M., and B. H. Neumann. (1964). "Felix Adalbert Behrend." *Journal Australian Mathematical Society* 4, 264–70.

Cohen, I. B. (1983). "Willy Hartner, 22 January 1905–16 May 1981." *ISIS* 74, 86–87.

Cohen, J. W. (1981). "Obituary: Felix Pollaczek." *Journal of Applied Probability* 18, 958–63 (includes photo).

Cohn, H., ed. (1991). *Doeblin and Modern Probability,* Contemporary Mathematics 149. Providence: AMS, 1991.

Coifman, R. R., and R. S. Strichartz (1989). "The School of Antoni Zygmund," in P. Duren, ed. (1989), vol. 3, 343–68.

Coleman, A. J. (1997). "Groups and Physics—Dogmatic Opinions of a Senior Citizen." *Notices AMS* 44, 8–17.

Corry, L. (1996). *Modern Algebra and the Rise of Mathematical Structures.* Basel, Boston, Berlin: Birkhäuser.

Corry, L. (2004). *David Hilbert and the Axiomatization of Physics (1898–1918).* Berlin: Springer.

Coser, L. (1984). *Refugee Scholars in America: Their Impact and Their Experiences*. New Haven, CT: Yale University Press.

Coser, L. (1988). "Die österreichische Emigration als Kulturtransfer Europa-Amerika," in Stadler, ed. (1988), 93–101.

Courant, R. (1943). "Variational Methods for the Solution of Problems of Equilibrium and Vibrations." *Bulletin AMS* 49, 1–23.

Courant, R, and K. Friedrichs (1948). *Supersonic Flow and Shock Waves*. New York: Interscience.

Courant, R., K. Friedrichs, and H. Lewy. (1928). "Über die partiellen Differenzengleichungen der mathematischen Physik." *Mathematische Annalen* 100, 32–74 [English in *IBM Journal of Research and Development* 11 (1967), 215–34].

Courant, R., and D. Hilbert (1953/62). *Methods of Mathematical Physics*, vol. I. New York and London: Interscience Publishers, 1953; Vol. II, *Partial Differential Equations*, by R. Courant. New York and London: Interscience Publishers, 1962.

Crawford, E., T. Shinn, and S. Sörlin, eds. (1993). *Denationalizing Science: The Contexts of International Scientific Practice*. Dordrecht: Kluwer.

Dahan Dalmedico, A. (1996). "L'essor des mathématiques appliquées aux États-Unis: L'impact de la Seconde Guerre Mondiale." *Revue d'histoire des mathématiques* 2, 149–213.

Dahms, H.-J. (1987). "Die Emigration des Wiener Kreises," in Stadler, ed. (1987), 66–122.

Dähnhardt, W., and B. S. Nielsen, eds. (1993). *Exil in Dänemark. Deutschsprachige Wissenschaftler, Künstler und Schriftsteller im dänischen Exil nach 1933*. Heide: Westholsteinische Verlagsanstalt.

Dalen, D. v. (2005). *Mystic, Geometer, and Intuitionist: The Life of L. E. J. Brouwer, 1881–1966*, vol. 2, *Hope and Disillusion*. Oxford: Clarendon Press.

Daniels, R. (1983). "American Refugee Policy in Historical Perspective," in Jackman and Borden, eds. (1983), 61–77.

Danneberg, L., A. Kamlah, and L. Schäfer, eds. (1994). *Hans Reichenbach und die Berliner Gruppe*. Braunschweig, Wiesbaden: Vieweg.

David, H. A. (1982). "Obituary Notice: H. O. Hartley, 1912–1980." *International Statistical Review* 50, 327–30.

Davie, M. R. (1947). *Refugees in America: Report of the Committee for the Study of Recent Immigration from Europe*. New York, London: Harper and Brothers.

Dawson Jr., J. W. (1997). *Logical Dilemmas: The Life and Work of Kurt Gödel*. Wellesley, MA: A. K. Peters.

Dawson Jr., J. W. (2002). "Max Dehn, Kurt Gödel, and the Trans-Siberian Escape Route." *Internationale Mathematische Nachrichten*, no. 189, 1–13.

Demidov, S., M. Folkerts, D. Rowe, and Ch. Scriba, eds. (1992). *Amphora: Festschrift für Hans Wussing zu seinem 65. Geburtstag*. Basel: Birkhäuser.

Diaz, J. B., ed. (1978). *Alexander Weinstein Selecta*. London, San Francisco, Melbourne: Pitman (includes photos and biography).

Dick, A. (1981). *Emmy Noether, 1882–1935*. Boston, Basel, Stuttgart: Birkhäuser.

Dieudonné, J. (1970). "The Work of Nicholas Bourbaki." *American Mathematical Monthly* 77, 134–45.

Diggins, J. P. (1992). *The Rise and Fall of the American Left*. New York, London: Norton (original 1973).

Dinghas, A. (1945). "Erinnerungen aus den letzten Jahren des Mathematischen Instituts der Universität Berlin," in H. Begehr, ed. (1998), vol. 2, 183–203.

Dresden, A. (1942). "The Migration of Mathematicians." *American Mathematical Monthly* 49, 415–29.

Duggan, S. (1943). *A Professor at Large*. New York: Macmillan.

Duggan, S., and B. Drury. (1948). *The Rescue of Science and Learning: The Story of the Emergency Committee in Aid of Displaced Foreign Scholars*. New York: Macmillan.

Dunnington, G. W. (1935). "Ludwig Schlesinger." *Scripta Mathematica* 3, 67–68.

Duren, P., ed. (1988/89). *A Century of Mathematics in America*, 3 vols. Providence: AMS.

Duren, W. L. (1989). "Mathematics in American Society 1888–1988: A Historical Commentary," in P. Duren, ed., vol. 2, 399–447.

Ebbinghaus, H.-D., in cooperation with V. Peckhaus (2007). *Ernst Zermelo: An Approach to His Life and Work*. Berlin, Heidelberg: Springer.

Ebner, H. (1959). "Hans J. Reissner 85 Jahre." *Zeitschrift für Flugwissenschaften* 7, no. 1, 23 (includes photo).

Eckert, M. (1993). *Die Atomphysiker. Eine Geschichte der theoretischen Physik am Beispiel der Sommerfeldschule*. Braunschweig, Wiesbaden: Vieweg.

Eckert, M. (2006). *The Dawn of Fluid Mechanics: A Discipline between Science and Technology*. Weinheim: Wiley.

Ehrlich, H., H. Kneser, and W. Walther (1968). "Erich Kamke zum Gedächtnis." *Jahresbericht DMV* 69, 191–208.

Einhorn, R. (1985). *Vertreter der Mathematik und Geometrie an den Wiener Hochschulen, 1900–1940*, 2 vols. Vienna: VWGÖ.

Einstein, A. (1935). "The Late Emmy Noether." *New York Times*, May 4, p. 12.

Engel, W. (1985). "Mathematik und Mathematiker an der Universität Rostock." *Rostocker Mathematisches Kolloquium* 27, 41–79.

Epple, M. (2004). "Knot Invariants in Vienna and Princeton during the 1920s: Epistemic Configurations of Mathematical Research." *Science in Context* 17, 131–64.

Eppel, P. (1988). "Die Vereinigten Staaten von Amerika." in Stadler, ed. (1988), 986–96.

Erichsen, R. (1994). "Emigrantenhilfe von Emigranten—Die Notgemeinschaft Deutscher Wissenschaftler im Ausland." *Exil* 14, no. 2, 51–69.

Exodus Professorum (1989). Göttingen: Vandenhoeck and Ruprecht.

Feferman, A. B., and S. Feferman (2004). *Alfred Tarski: Life and Logic*. Cambridge: Cambridge University Press.

Feigl, H. (1969). "The Wiener Kreis in Amerika," in Fleming and Bailyn, eds. (1969), 630–73.

Feit, W. (1979). "Richard D. Brauer." *Bulletin AMS* (n.s.) 1, 1–20.

Feller, W. (1937). "Zur Theorie der stochastischen Prozesse (Existenz- und Ein-deutigkeitssätze)." *Mathematische Annalen* 113, 113–60.

Feller, W. (1950/66). *An Introduction to Probability Theory and Its Applications*, vol. 1 (1950), vol. 2 (1966). New York: John Wiley.

Fenster, D. D. (1997). "Role Modeling in Mathematics: The Case of Leonard Eugene Dickson (1874–1954)." *Historia Mathematica* 24, 7–24.

Fischer, G., F. Hirzebruch, W. Scharlau, and W. Toernig, eds. (1990). *Ein Jahrhundert Mathematik 1890–1990; Festschrift zum Jubiläum der DMV*. Braunschweig: Vieweg.

Fischer, K. (1991). "Die Emigration deutschsprachiger Physiker nach 1933: Strukturen und Wirkungen," in Strauss et al., eds. (1991), 25–72.

Fleming, D., and B. Bailyn, eds. (1969). *The Intellectual Migration*. Cambridge, MA: Harvard University Press.

Fletcher, C. R. (1986). "Refugee Mathematicians: A German Crisis and a British Response, 1933–1936." *Historia Mathematica* 13, 13–27.

Forman, P. (1971). "Weimar Culture, Causality, and Quantum Theory, 1918–1927: Adaptation by German Physicists and Mathematicians to a Hostile Intellectual Environment." *Historical Studies in the Physical Sciences* 3, 1–115.

Fosdick, R. B. (1952). *The Story of the Rockefeller Foundation*. New York: Harper and Brothers.

Fox, K. A., J. K. Sengupta, and G. V. L. Narasimham, eds. (1969). *Economic Models, Estimation and Risk Programming: Essays in Honor of Gerhard Tintner*. New York: Springer (=*Lecture Notes in Operations Research and Mathematical Economics* 15, includes photo).

Fraenkel, A. (1938). "Alfred Loewy (1873–1935)." *Scripta Mathematica* 5, 17–22.

Fraenkel, A. A. (1967). *Lebenskreise: Aus den Erinnerungen eines jüdischen Mathematikers*.Stuttgart: Deutsche Verlagsanstalt.

Frank, W. (1988). "Richard von Mises und Hilda Geiringer-Mises: Anmerkungen zu deren Lebenslauf," in Stadler, ed. (1988), 751–55.

Franksen, O. I. (1997). "Boole's Development Process Revisited: From an Array-Theoretic Viewpoint." *Acta historica Leopoldina* 27, 175–88.

Frei, G., and U. Stammbach. (1992). *Hermann Weyl und die Mathematik an der ETH Zürich, 1913–1930*. Basel, Boston: Birkhäuser.

Frewer, M. (1981). "Felix Bernstein." *Jahresbericht DMV* 83, 84–95.

Friedman, S. S. (1973). *No Haven for the Oppressed: United States Policy toward Jewish Refugees, 1938–1945*. Detroit: Wayne State University Press.

Frucht, R. W. (1982). "How I Became Interested in Graphs and Groups." *Journal of Graph Theory* 6, 101–4.

Fry, Th. (1941). "Industrial Mathematics." *Bell System Technical Journal* 20, 255–92 [reprinted in *American Mathematical Monthly* 48 (1941), 1–38].

Fuchs, L., and R. Göbel (1993). "Levi," in L. Fuchs and R. Göbel, eds., *Abelian Groups: Proceedings of the 1991 Curaçao Conference*. New York: Marcel Decker 1993, 1–13.

Fuchs, W. H. J. (1975). "Hermann Kober." *Bulletin London Mathematical Society* 7, 185–90 (includes photo).

Gähler, S., and W. Gähler (1993). "Nachruf auf Josef Naas." *Mathematische Nachrichten* 161, 4–5 (includes photo).

Gaier, D. (1992). "Abraham Ezechiel Plessner (1900–1961): His Work and His Life." *Mathematical Intelligencer* 14, no. 3, 31–36.

Gaines, F. J., and Th. J. Laffey (1985). "The Mathematical Work of Walter Ledermann." *Linear Algebra and Its Applications* 69, 1–8 (includes photo).

Gani, J., ed. (1982). *The Making of Statisticians*. New York, Heidelberg, Berlin: Springer.

Geiger, R. L. (1986). *To Advance Knowledge: The Growth of American Research Universities, 1900–1940*. Oxford, New York: Oxford University Press.

Georgiadou, M. (2004). *Constantin Carathéodory: Mathematics and Politics in Turbulent Times*. Berlin: Springer.

Georgiadou, M. (2007). "Humanism and Human Responsibility." *Mathematical Intelligencer* 29, no. 4, 5–6.

Gluchoff, A. (2005). "Pure Mathematics Applied in Early Twentieth-Century America: The Case of T. H. Gronwall, Consulting Mathematician." *Historia Mathematica* 32, 312–57.

Göbel, S. (1988–1992). "Über die Verwendung der Sprachen in mathematischen Zeitschriften." *Sitzungsberichte Berliner Mathematische Gesellschaft*, 137–75.

Golczewski, F. (1988). *Kölner Universitätslehrer und der Nationalsozialismus: Personengeschichtliche Ansätze*. Köln, Vienna: Böhlau.

Goldstine, H. H. (1972). *The Computer from Pascal to von Neumann*. Princeton, NJ: Princeton University Press.

Golomb, S., Th. Harris, and J. Seberry (1997). "Albert Leon Whiteman (1915–1995)." *Notices AMS* 44, 217–19.

Gottwald, S., and L. Kreiser (1984). "Paul Mahlo—Leben und Werk." *NTM* 21, no. 2, 1–22.

Grattan-Guinness, I., ed. (1994). *Companion Encyclopedia of the History and Philosophy of the Mathematical Sciences*, 2 vols. London, New York: Routledge.

Gray, J., and K. H. Parshall, eds. (2007). *Episodes in the History of Modern Algebra (1800–1950)*. Providence and London: AMS and LMS.

Green, J. A. (1978). "Richard Dagobert Brauer." *Bulletin London Mathematical Society* 10, 317–42.

Greenberg, K. J. (1996). "The Refugee Scholar in America: The Case of Paul Tillich," in Ash and Söllner, eds. (1996), 273–89.

Grimshaw, M. E. (1958). "Hans Ludwig Hamburger." *Journal London Mathematical Society* 33, 377–84.

Grinstein, L. S., and P. J. Campbell, eds. (1987). *Women of Mathematics: A Bio-bibliographic Sourcebook*. New York: Greenwood Press.

Grosswald, E. (1974). "Biographical Sketch," in E. Grosswald, ed. (1974). *Collected Papers of Hans Rademacher*, vol. 1. Cambridge: MIT Press, xiii–ixx.

Gruenberg, K. W. (1981). "Reinhold Baer." *Bulletin London Mathematical Society* 13, 339–61 (includes photo).

Gruenberg, K. W. (1988). "Obituary: Kurt August Hirsch." *Bulletin London Mathematical Society* 20, 350–58 (includes photo).

Grundmann, S. (2004). *The Einstein Dossiers*. Berlin: Springer.

Gumbel, E. J. (1933). "Amerika zerstört Illusionen (1933)," in Jansen (1991), 306–15.

Gumbel, E. J. (1936). "Die Quadratur des Kreises." *Das Neue Tage-Buch* (Paris) 4, no. 9, 213.

Gumbel, E. J., ed. (1938). *Freie Wissenschaft: Ein Sammelbuch aus der deutschen Emigration.* Strasbourg: Sebastian Brant.

Gumbel, E. J. (1991). *Auf der Suche nach Wahrheit: Ausgewählte Schriften, versehen mit einem Essay von Annette Vogt.* Berlin: Dietz.

Haller, R., and F. Stadler, eds. (1993). *Wien—Berlin—Prag: Der Aufstieg der wissenschaftlichen Philosophie.* Vienna: Hölder-Pichler-Tempsky.

Halmos, P. R. (1988). "Some Books of Auld Lang Syne," in P. Duren, ed. (1988/89), vol. 1, 131–74.

Hanle, P. (1982). *Bringing Aerodynamics to America.* Cambridge, MA: MIT Press.

Hannak, J. (1952). *Emanuel Lasker: Biographie eines Schachweltmeisters.* Berlin: Siegfried Engelhardt (contains photos and cartoons, depicting Lasker).

Harnack, A. von (1964). "Mein Abiturientenexamen." *Neue Sammlung. Göttinger Blätter für Kultur und Erziehung* 4, 469–73.

Hartman, P. (1962). "Aurel Wintner." *Journal London Mathematical Society* 37, 483–503.

Hasse, H. (1936). "Kurt Hensel zum 75. Geburtstag." *Forschungen und Fortschritte* 12, 458.

Haupt, O. (1960). "Arthur Rosenthal." *Jahresbericht DMV* 63, 89–96.

Hayman, W. K. (1965). "Werner Wolfgang Rogosinski." *Biographical Memoirs of Fellows of the Royal Society* 11, 135–45 (includes photo).

Hecht, H., and D. Hoffmann (1982). "Die Berufung Hans Reichenbachs an die Berliner Universität." *Deutsche Zeitschrift für Philosophie* 30, 651–62.

Hecht, H., and D. Hoffmann (1991). "Die Berliner Gesellschaft für wissenschaftliche Philosophie." *NTM* 28, no. 1, 43–59.

Heilbronn, H., and G. H. Hardy (1938). "E. Landau (Obituary)." *Journal London Mathematical Society* 13, 310–18.

Hemmer, P. Ch. (2005). "Werner Romberg (1909–2003)." *Aarbok Norges Tekniske Vitenskapsakademi*, 125–26.

Hentschel, K., ed. (1996). *Physics and National Socialism: An Anthology of Primary Sources.* Basel, Boston, Berlin: Birkhäuser.

Hermann, G. (1985). *Die Überwindung des Zufalls.* Hamburg: Felix Meiner (postum with photo of H.).

Hertz, S. (1997). *Emil Julius Gumbel (1891–1966) et la statistique des extrêmes.* Unpublished doctoral thesis, Lyon University.

Higman, G. (1974). "Hanna Neumann." *Bulletin London Mathematical Society* 6, 99–100.

Hildebrandt, S. (1988). "Laudatio (Ehrenpromotion von Fritz John an der Freien Universität Berlin 2.6.1988)." *Sitzungsberichte Berliner Mathematische Gesellschaft* 1988–92, 193–98.

Hille, E. (1980). "In Retrospect." *Mathematical Intelligencer* 3, no. 1, 3–13.

Hintikka, J. (1993). "Carnaps Arbeiten über die Grundlagen der Logik und Mathematik aus historischer Perspektive," in Haller and Stadler, eds. (1993), 73–97.

Hirsch, K. (1986). "Sixty Years of Mathematics." *Mathematical Medley* 14, Singapore Mathematical Society, 35–50.

Hoch, P. K. (1983). "The Reception of Central European Refugee Physicists of the 1930s: U.S.S.R., U.K., U.S.A." *Annals of Science* 40, 217–46.

Hoch, P. K. (1987). "Migration and the Generation of New Scientific Ideas." *Minerva* 25, 209–37.

Hoch, P., and J. Platt (1993). "Migration and the Denationalization of Science," in Crawford et al., eds. (1993), 133–52.

Hochkirchen, Th. (1998). "Wahrscheinlichkeitsrechnung im Spannungsfeld von Mass—und Häufigkeitstheorie—Leben und Werk des 'Deutschen' Mathematikers Erhard Tornier (1894–1982)." *NTM* 6 (n.s.) 22–41.

Hoehnke, H. J. (1986). "66 Jahre Brandtsches Gruppoid." *Wissenschaftliche Beiträge Martin Luther Universität Halle* 47 (M 43), 15–79.

Hoffmann, D. (1993). "Die Berliner 'Gesellschaft für empirische/wissenschaftliche Philosophie,'" in Haller and Stadler, eds. (1993), 386–401.

Hoheisel, G. (1966). "Hamburger, Hans Ludwig." *Neue Deutsche Biographie* 7, 581.

Holl, F. (1996). *Produktion und Distribution wissenschaftlicher Literatur: Der Physiker Max Born und sein Verleger Ferdinand Springer, 1913–1970.* Frankfurt: Buchhändler-Vereinigung.

Holton, G. (1983). "The Migration of Physicists to the United States," in Jackman and Borden, eds. (1983), 169–88.

Hopf, E. (1937). *Ergodentheorie.* Berlin: Springer.

Hornich, H. (1956). "Paul Funk—70 Jahre." *NÖMG* 10, nos. 43/44, 69–70.

Hunter, P. W. (1996). "Drawing the Boundaries: Mathematical Statistics in 20th-Century America." *Historia Mathematica* 23, 7–30.

Høyrup, E. (1987). "Käte Fenchel (1905–1983)," in Grinstein and Campbell, eds. (1987), 30–32.

Jackman, J. C., and C. M. Borden, eds. (1983). *The Muses Flee Hitler: Cultural Transfer and Adaptation, 1930–1945.* Washington, DC: Smithsonian Institution Press.

Jansen, Ch. (1991). *Emil Julius Gumbel: Portrait eines Zivilisten.* Heidelberg: Wunderhorn.

Jarausch, K. H. (1984). *Deutsche Studenten, 1800–1970.* Frankfurt: Suhrkamp.

Jessen, B. (1993). "Mathematiker unter den deutschsprachigen Emigranten," in Dähnhardt and Nielsen, eds. (1993), 127–33.

Johnson, A. (1952). *Pioneer's Progress: An Autobiography.* New York: Viking Press.

Johnson, N. L., and S. Kotz, eds. (1997). *Leading Personalities in Statistical Sciences.* New York: Wiley.

Jones, B. Z. (1984). "To the Rescue of the Learned: The Asylum Fellowship Plan at Harvard, 1938–1940." *Harvard Library Bulletin* 32, 205–38.

Kac, M. (1987). *Enigmas of Chance: An Autobiography.* Berkeley: University of California Press.

Kahane, J.-P., K. Krickeberg, and L. Lorch (1994). "Concerns about Obituaries Published in JDMV." *Notices AMS* 41, 571–72.

Kallen, H. M., and J. Dewey, eds. (1941). *The Bertrand Russell Case.* New York: Viking Press.

Kamlah, A. (1993). "Hans Reichenbach—Leben, Werk und Wirkung," in Haller and Stadler, eds. (1993), 238–83.

Kani, E. J., and R. A. Smith, eds. (1988). *The Collected Papers of Hans Heilbronn*. New York: Wiley.

Karabel, J. (2005). *The Chosen: The Hidden History of Admission and Exclusion at Harvard, Yale, and Princeton*. Boston: Houghton Mifflin.

Kass, S. (1996). "Karl Menger." *Notices AMS* 43, 558–61.

Kegel, O. H., and V. Remmert (2003). "Friedrich Wilhelm Daniel Levi (1888–1966)," in G. Wiemers, ed., *Sächsische Lebensbilder*, vol. 5. Leipzig: Verlag Sächsische Akademie der Wissenschaften, 2003, 395–403.

Kent, D. P. (1953). *The Refugee Intellectual: The Americanization of the Immigrants of 1933–1941*. New York: Columbia University Press.

Kimberling, C. (1981). "Emmy Noether and Her Influence," in Brewer and Smith, eds. (1981), 3–61.

Kline, R. R. (1992). *Steinmetz: Engineer and Socialist*. Baltimore: Johns Hopkins University Press.

Kluge, W. (1983). *Edmund Landau: Sein Werk und sein Einfluß auf die Entwicklung der Mathematik*. Unpublished work for the state exam, University Duisburg (Germany), mimeographed, typewritten, 154 pp.

Knobloch, E. (1998). *Mathematik an der Technischen Hochschule und der Technischen Universität Berlin 1770–1988*. Berlin: Dr. Michael-Engel-Verlag.

Knoche, M. (1990). "Wissenschaftliche Zeitschriften im nationalsozialistischen Deutschland," in M. Estermann and M. Knoche, eds. (1990), *Von Göschen bis Rowohlt: Beiträge zur Geschichte des deutschen Verlagswesens*. Wiesbaden: Otto Harrasowitz, 260–81.

Korn, A. (1945). "Teaching of Mathematics in the United States and in Germany." *Journal of Engineering Education* 35, 407–13.

Kowalewski, G. (1950). *Bestand und Wandel*. Munich: Oldenbourg.

Kracht, M. (1981). "Maximilian Pinl in Memoriam." *Jahresbericht DMV* 83, 119–24.

Krause, E., L. Huber, and H. Fischer, eds. (1991). *Hochschulalltag im "Dritten Reich": Die Hamburger Universität, 1933–1945*, 3 parts. Berlin, Hamburg: Dietrich.

Kröner, H.-P. (1989). "Die Emigration deutschsprachiger Mediziner im Nationalsozialismus," *Berichte zur Wissenschaftsgeschichte* 12, special issue, 1–44.

Krohn, C.-D. (1987). *Wissenschaft im Exil: Deutsche Sozial—und Wirtschaftswissenschaftler in den USA und die New School for Social Research*. Frankfurt, New York: Campus.

Krohn, C.-D., P. von zur Mühlen, G. Paul, and L. Winckler, eds. (1998). *Handbuch der deutschsprachigen Emigration, 1933–1945*. Darmstadt: Primus-Verlag.

Kühnau, R. (1997). "Herbert Grötzsch zum Gedächtnis." *Jahresbericht DMV* 99, 122–45.

Lausch, H. (1997). "Felix Adalbert Behrend and Mathematics in Camp 7, Hay, 1940–41." *Australian Jewish Historical Society Journal* 14, part I, 110–19.

Lax, P. (1989). "The Flowering of Applied Mathematics in America," in P. Duren, ed., (1988/89), vol. 2, 455–66.

Ledermann, W. (1983). "Issai Schur and His School in Berlin." *Bulletin London Mathematical Society* 15, 97–106.

Lemmermeyer, F., and P. Roquette, eds. (2006). *Helmut Hasse und Emmy Noether: Die Korrespondenz, 1925–1935*. Göttingen: Universitätsverlag [also available free through http://univerlag.uni-goettingen.de/hasse-noether/].

Lenz, H., M. Aigner, and W. Deuber (1991). "Richard Rado, 1906–1989." *Jahresbericht DMV* 93, 127–45.

Litten, F. (1993). " 'Vielleicht hilft uns Professor Röntgen mit der Zeit?': Die Korn-Röntgen-Affäre." *Kultur and Technik*, no. 4, 43–49.

Litten, F. (1996). "Ernst Mohr—das Schicksal eines Mathematikers." *Jahresbericht DMV* 98, 192–212.

Lorenz, E. (1992). *Exil in Norwegen: Lebensbedingungen und Arbeit deutschsprachiger Flüchtlinge, 1933–1943*. Baden-Baden: Nomos.

Luchins, A. S., and E. H. Luchins. (2000). "Kurt Grelling: Steadfast Scholar in a Time of Madness." *Gestalt Theory* 22 (2000), 228–81 [extended as http://gestalttheory.net/archive/kgbio.html].

Luchins, E. H. (1987). "Olga Taussky-Todd (1906–)," in Grinstein and Campbell, eds. (1987), 225–35.

Lukacs, E. (1982). "From Riemannian Spaces to Characteristic Functions: The Evolution of a Statistician," in J. Gani, ed. (1982), 14–19.

Lüneburg, R. (1964). *Mathematical Theory of Optics*. Berkeley: University of California Press.

Maas, Ch. (1991). "Das Mathematische Seminar der Hamburger Universität in der Zeit des Nationalsozialismus," in Krause, Huber, and Fischer, eds. (1991), part 3, 1075–95.

Macrakis, K. (1993). *Surviving the Swastika: Scientific Research in Nazi Germany*. New York: Oxford University Press.

Mader, K. (1950). "Hofrat Dr. F. Hopfner." *NÖMG* 4, no. 10, 5–7.

Magnus, W. (1978). "Max Dehn." *Mathematical Intelligencer* 1, no. 3, 132–43.

Mehrtens, H. (1979). *Die Entstehung der Verbandstheorie*. Hildesheim: Gerstenberg.

Mehrtens, H. (1989). "The Gleichschaltung of Mathematical Societies in Nazi Germany." *Mathematical Intelligencer* 11, no. 3, 48–60.

Mehrtens, H. (1990a). "Verantwortungslose Reinheit: Thesen zur politischen und moralischen Struktur mathematischer Wissenschaften am Beispiel des NS-Staates," in G. Fülgraff and A. Falter, eds. (1990). *Wissenschaft in der Verantwortung: Möglichkeiten der institutionellen Steuerung*, Frankfurt, New York: Campus, 37–54.

Mehrtens, H. (1990b). *Moderne, Sprache, Mathematik*. Frankfurt: Suhrkamp.

Menger, K. (1994). *Reminiscences of the Vienna Circle and the Mathematical Colloquium*, edited by L. Golland, B. McGuinness, and A. Sklar. Dordrecht: Kluwer.

Menzler-Trott, E. (2007). *Logic's Lost Genius: The Life of Gerhard Gentzen*. Providence and London: AMS and LMS [translation of the German original of 2001, Birkhäuser Verlag].

Merzbach, U. (1992). "Robert Remak and the Estimation of Units and Regulators," in S. Demidov et al., eds. (1992), 481–522.

Metropolis, N., J. Howlett, and G.-C. Rota, eds. (1980). *A History of Computing in the Twentieth Century: A Collection of Essays*. New York, London: Academic Press.

Mises, R. v. (1927). "Pflege der angewandten Mathematik in Deutschland." *Die Naturwissenschaften* 15, 473.

Mises, R. v. (1930). "Über das naturwissenschaftliche Weltbild der Gegenwart." *Die Naturwissenschaften* 18, 885–93.

Mises, R. v. (1934). "Problème de deux races." *Matematicheskij Sbornik* 41, 359–74.

Möller, H. (1992). "Die Remigration von Wissenschaftlern nach 1945," in Böhne and Motzkau-Valeton, eds. (1992), 601–14.

Moore, W. (1989). *Schrödinger: Life and Thought*. Cambridge: Cambridge University Press.

Morse, M. (1946). "George David Birkhoff and His Mathematical Work." *Bulletin AMS* 52, 357–91.

Müller, C. (1986). "Zum 100. Geburtstag von Hermann Weyl." *Jahresbericht DMV* 88, 159–89.

Müller, G. H. (1978). "Paul Bernays." *Mathematical Intelligencer* 1, no. 1, 27–28 (includes photo).

Muller, S., ed. (1996). *Universities in the Twenty-First Century*. Providence, Oxford: Berghahn.

Mußgnug, D. (1988). *Die vertriebenen Heidelberger Dozenten: Zur Geschichte der Ruprecht-Karls-Universität nach 1933*. Heidelberg: Carl-Winter-Universitätsverlag.

Neuenschwander, E. (1996). "Felix Hausdorffs letzte Lebensjahre nach Dokumenten aus dem Bessel-Hagen-Nachlass," in Brieskorn, ed. (1996), 253–70.

Neumark, F. (1980). *Zuflucht am Bosporus: Deutsche Gelehrte, Politiker und Künstler in der Emigration 1933–1953*. Frankfurt: Josef Knecht.

Newman, M. F., and G. E. Wall (1974). "Hanna Neumann." *Journal Australian Mathematical Society* 17, 1–28 (includes photo).

Nicholas, P., and C. Benson (2002). "Files Reveal How FBI Hounded Chess King." *Philadelphia Inquirer*, November 17, A01

Niven, I. (1988). "The Threadbare Thirties," in P. Duren, ed. (1988/89), vol. 1, 209–29.

Olff-Nathan, J., ed. (1993). *La science sous le Troisième Reich*. Paris: Seuil 1993.

O'Malley, R. E. (1995). "Wolfgang W. Wasow." *Results in Mathematics* 28, 12–14.

Ortiz, E. L., and A. Pinkus. (2005). "Herman Müntz: A Mathematician's Odyssey." *Mathematical Intelligencer* 27, no. 1, 22–31.

Oswald, M. C. (1983). "Prof. Dr. Dr. h. c. Wilhelm Hauser." *Novocastrian News, Newcastle*, no. 25 (Autumn), 42–47.

Owens, L. (1989). "Mathematics and War: Warren Weaver and the Applied Mathematics Panel, 1942–1945," in Rowe and McCleary, eds. (1989), vol. 2, 287–305.

Pach, J. (1997). "Two Places at Once: A Remembrance of Paul Erdös." *Mathematical Intelligencer* 19, no. 2, 38–48.

Pais, A. (1982). *"Subtle Is the Lord . . .": The Science and the Life of Einstein*. Oxford: Oxford University Press.

Papcke, S. (1988). "Fragen an die Exilforschung heute." *Exilforschung* (Jahrbuch) 6, 13–27.

Parikh, C. (1991). *The Unreal Life of Oscar Zariski*. Boston: Academic Press.

Parshall, K. H. (1991). "Mathematics in National Contexts (1875–1900): An International Overview." *Proceedings of the International Congress of Mathematicians, Zürich 1994*. Basel: Birkhäuser 1995, 1581–91.

Parshall, K. H., and D. E. Rowe (1994). *The Emergence of the American Mathematical Research Community, 1876–1900: J. J. Sylvester, Felix Klein, and E. H. Moore*. Providence and London: AMS and LMS.

Peckhaus, V. (1990a). *Hilbertprogramm und Kritische Philosophie*. Göttingen: Vandenhoeck and Ruprecht.

Peckhaus, V. (1990b). " 'Ich habe mich wohl gehütet, alle Patronen auf eimal zu verschießen': Ernst Zermelo in Göttingen." *History and Philosophy of Logic* 11, 19–58.

Peckhaus, V. (1994). "Von Nelson zu Reichenbach: Kurt Grelling in Göttingen und Berlin," in Danneberg et al., eds. (1994), 53–86.

Perron, O. (1953). "Alfred Pringsheim." *Jahresbericht DMV 56*, 1–6.

Pietzsch, G. (1992). "Prof. Dr. Lilly Görke." *Mathematik in der Schule* 30, no. 9, 501.

Pinl, M. (1965). "In Memory of Ludwig Berwald." *Scripta Mathematica* 27, no.3, 193–203.

Pinl, M. (1969–1972). "Kollegen in einer dunklen Zeit," part 1: *Jahresbericht DMV* 71 (1969), 167–228; part 2: *JDMV* 72 (1971), 165–89; part 3: *JDMV* 73 (1971/72), 153–208.

Pinl, M., and A. Dick (1974–76). "Kollegen in einer dunklen Zeit. Schluß." *JDMV* 75 (1974), 166–208; "Nachtrag und Berichtigung." *JDMV* 77 (1976), 161–64.

Pinl, M., and L. Furtmüller (1973). "Mathematicians under Hitler." *Yearbook Leo Baeck Institute* 18, 129–82.

Polanyi, M. (1966). *The Tacit Dimension*. Garden City, NY: Doubleday.

Poorten, A. J. van der (1991). "Obituary Kurt Mahler, 1903–1988." *Journal Australian Mathematical Society* (series A) 51, 343–80.

Porter, L. S. (1988). *From Intellectual Sanctuary to Social Responsibility: The Founding of the Institute for Advanced Study, 1930–1933*. PhD dissertation, Princeton University. manuscript, 479 pp. (kind communication by S. Sigurdsson).

Prager, W. (1972). "Introductory Remarks." *Quarterly of Applied Mathematics* 30, no. 1, 1–9 (special issue).

Pross, H. (1955). *Die deutsche akademische Emigration nach den Vereinigten Staaten, 1933–1941*. Berlin: Duncker and Humblot.

Pyenson, L. (1979). "Mathematics, Education, and the Göttingen Approach to Physical Reality, 1890–1914." *Europa, A Journal of Interdisciplinary Studies* 2, 91–127.

Quaisser, E. (1984). "Zur 'Deutschen' Mathematik." *Wissenschaftliche Zeitschrift der Ernst-Moritz-Arndt-Universität Greifswald, Mathematisch-naturwissenschaftliche Reihe* 32, nos. 1–2, 35–39.

Rammer, G. (2002). "Der Aerodynamiker Kurt Hohenemser." *NTM* (n.s.) 10, 78–101.

Ramskov, K. (1995). *Matematikeren Harald Bohr*. Aarhus: Institut for de eksakte videnskabers historie, Aarhus Universitet, viii+451 pp.

Reich, K., and A. Kreuzer, eds. (2007). *Emil Artin (1898–1962): Beiträge zu Leben, Werk und Persönlichkeit*. Augsburg: Rauner Verlag.

Reichenbach, H. (1949). *The Theory of Probability: An Inquiry into the Logical and Mathematical Foundations of the Calculus of Probability*, English translation by Ernest H. Hutten and Maria Reichenbach, 2nd ed. Berkeley: University of California Press.

Reid, C. (1976). *Courant in Göttingen and New York: The Story of an Improbable Mathematician*. New York: Springer.

Reid, C. (1982). *Neyman—from Life*. Berlin: Springer.

Reid, C. (1983). "K. O. Friedrichs, 1901–1982." *Mathematical Intelligencer 5*, no. 3, 23–30.

Reingold, N. (1981). "Refugee Mathematicians in the United States of America 1933–1941." *Annals of Science 38*, 313–38.

Reisch, G. (2005). *How the Cold War Transformed Philosophy of Science: To the Icy Slopes of Logic*. Cambridge, New York: Cambridge University Press.

Reisman, A. (2006). *Turkey's Modernization: Refugees from Nazis and Atatürk's Vision*. Washington, DC: New Academia Publishing.

Reissner, E. (1977). "Hans Reissner, Engineer, Physicist and Engineering Scientist." *Engineering Science Perspective 2*, no. 4, 97–105.

Reiter, W. L. (2001). "Die Vertreibung der jüdischen Intelligenz: Verdoppelung eines Verlustes—1938/1945." *Internationale Mathematische Nachrichten*, no. 187, 1–20.

Remmert, V. (1995). "Zur Mathematikgeschichte in Freiburg: Alfred Loewy (1873–1935): Jähes Ende späten Glanzes." *Freiburger Universitätsblätter 129*, 81–102.

Remmert, V. (2004). "Die Deutsche Mathematiker-Vereinigung im 'Dritten Reich.'" *Mitteilungen der Deutschen Mathematiker-Vereinigung 12*, 159–77, 223–45.

Renewing U.S. Mathematics: A Plan for the 1990s, Excerpts (1990), *Notices AMS 37*, 542–46.

Rescher, N. (1997). "H_2O: Hempel-Helmer-Oppenheim: An Episode in the History of Scientific Philosophy in the 20th Century." *Philosophy of Science 64*, 334–60.

Rescher, N., ed. (1969). *Essays in Honor of Carl G. Hempel*. Dordrecht: Reidel.

Richards, J. L. (1987). "Hilda Geiringer von Mises," in Grinstein and Campbell, eds. (1987), 41–46.

Richards, P. S. (1994). *Scientific Information in Wartime: The Allied-German Rivalry, 1939–1945*. Westport, CT, London: Greenwood Press.

Richardson, R. G. D. (1936). "The Ph.D. Degree and Mathematical Research." *American Mathematical Monthly 43*, 199–215, reprinted in P. Duren, ed. (1988/89), vol. 2, 361–78.

Richardson, R. G. D. (1943). "Applied Mathematics and the Present Crisis." *American Mathematical Monthly 50*, 415–23.

Rider, R. (1984). "Alarm and Opportunity: Emigration of Mathematicians and Physicists to Britain and the United States, 1933–1945." *Historical Studies in the Physical Sciences 15*, 107–76.

Röder, W., and H. Strauss, eds. (1983). *International Biographical Dictionary of Central European Émigrés, 1933–1945*, vol. 2 (in two parts), *The Arts, Sciences, and Literature*. Munich, New York: Sauer [=IBD].

Rogers, C. A. (1991). "Richard Rado." *Biographical Memoirs of Fellows of the Royal Society of London* 37, 413–26 (includes photo).

Rohrbach, H. (1929/1998). "Die mathematisch-physikalische Arbeitsgemeinschaft (Mapha)," in H. Begehr, ed. (1998) [Publikation of the mimeographed manuscript of 1929], 121–31.

Rohrbach, H. (1981). "Richard Brauer zum Gedächtnis." *Jahresbericht DMV* 83, 125–34.

Rohrbach, H. (1988). "Alfred Brauer zum Gedächtnis." *Jahresbericht DMV* 90, 145–54.

Rohsa, E. (1983). "Nelli Neumann fand die Mathematik einfacher als den Haushalt." *Göttinger Monatsblätter*, November 1983, 5–7.

Roquette, P. (2004). "The Brauer-Hasse-Noether Theorem in Historical Perspective." *Schriften der Mathematisch-Physikalischen Klasse der Heidelberger Akademie der Wissenschaften* no. 15, 92 pp. [updated version at http://www.rzuser.uni-heidelberg.de/~ci3/brhano.pdf].

Rota, G.-C. (1989). "Fine Hall in Its Golden Age: Remembrances of Princeton in the Early Fifties," in P. Duren, ed. (1988/89), vol. 2, 223–36.

Roth, Ph. (2004). *The Plot against America*. New York: Random House.

Rovnyak, J. (1990). "Ernst David Hellinger, 1883–1950: Göttingen, Frankfurt Idyll, and the New World," in Branges, Gohberg, and Rovnyak, eds. (1990), 1–41.

Rowe, D. E. (1989). "Klein, Hilbert, and the Göttingen Tradition." *Osiris* (2) 5, 186–213.

Rowe, D., and J. McCleary, eds. (1989). *The History of Modern Mathematics*, vol. 2, *Institutions and Applications*. Boston: Academic Press.

Rudin, W. (1997). *The Way I Remember It*. Providence and London: AMS and LMS.

Rürup, R., ed. (1979). *Wissenschaft und Gesellschaft: Beiträge zur Geschichte der Technischen Universität Berlin 1879–1979*, 2 vols. Berlin: Springer.

Rybarz, J. (1957). "Adalbert Duschek zum Gedächtnis." *NÖMG* 11, nos. 51/52, 61/62.

Sauer, R., and H. Heinrich (1960). "Friedrich Adolf Willers. Sein Leben und Wirken," *Zeitschrift für Angewandte Mathematik und Mechanik* 40, 1–8.

Scaife, B. K. P., ed. (1974). *Studies in Numerical Analysis: Papers in Honor of Cornelius Lanczos*. London, New York: Academic Press.

Scanlan, M. (1991). Who Were the American Postulate Theorists?" *Journal of Symbolic Logic* 56, 981–1002.

Schaefer, C. (1954). "Wolfgang Sternberg." *Physikalische Blätter* 10, 30.

Schaper, R. (1992). "Mathematiker im Exil," in Böhne and Motzkau-Valeton, eds. (1992), 547–68.

Schappacher, N. (1987). "Das Mathematische Institut der Universität Göttingen, 1929–1950," in Becker, Dahms, and Wegeler, eds. (1987), 345–73.

Schappacher, N. (1993). "Questions politiques dans la vie des mathématiques en Allemagne (1918–1935)," in Olff-Nathan, ed. (1993), 51–89.

Schappacher, N. (2006). "Seventy Years Ago: The Bourbaki Congress at El Escorial and Other Mathematical (Non) Events of 1936," in *Madrid Intelligencer 2006*, edited by F. Chamizo and A. Quirós, Springer, 8–15.

Schappacher, N. (2007). "A Historical Sketch of B. L. Van der Waerden's Work on Algebraic Geometry, 1926–1946," in J. Gray and K. H. Parshall, eds. (2007), 245–84.

Schappacher, N., and M. Kneser (1990). "Fachverband-Institut-Staat," in Fischer et al., eds. (1990), 1–82.

Schappacher, N., and E. Scholz, eds. (1992). "Oswald Teichmüller—Leben und Werk." *Jahresbericht DMV* 94, 1–39.

Scharlau, W., ed. (1990). *Mathematische Institute in Deutschland 1800–1945.* Braunschweig: Vieweg.

Schiffer, M. M. (1986/1998). "Issai Schur: Some Personal Reminiscences," in H. Begehr, ed. (1998), vol. 2, 177–81.

Schiffer, M, and H. Samelson (1979). "Dedicated to the Memory of Stefan Bergman." *Applicable Analysis* 8, 195–99.

Schlote, K.-H. (1991). "Fritz Noether—Opfer zweier Diktaturen." *NTM* 28, no. 1, 33–41.

Schmolze, G. (1983). "Eine jüdische Familiengeschichte" (review of Wirth [1982]). *Kirche im Sozialismus* (Berlin) 1983, no. 4, 52–55.

Schreiber, F., and P. Le Gall (1993). "In Memoriam Félix Pollaczek (1892–1981)." *Archiv für Elektrotechnik und Übertragungstechnik* 47, 275–81 (includes photo).

Schreiber, P. (1996). "Clemens Thaer (1883–1974)—ein Mathematikhistoriker im Widerstand gegen den Nationalsozialismus." *Sudhoffs Archiv* 80, 78–85.

Schur, I. (1973). *Gesammelte Abhandlungen*, 3 vols., edited by Alfred Brauer and Hans Rohrbach. Berlin: Springer.

Schuster, C. A. (1976). "Hans Reichenbach." *Dictionary of Scientific Biography*, vol. 11, 355–59.

Schwarz, W., and J. Wolfart (1988). *Zur Geschichte des Mathematischen Seminars der Universität Frankfurt am Main von 1914 bis 1945.* Unpublished preprint, 21 pp.

Schweber, S. S. (1986). "The Empiricist Temper Regnant: Theoretical Physics in the United States, 1920–1950." *Historical Studies in the Physical and Biological Sciences* 17, 55–98.

Schwerdtfeger, P., R. Potts, and T. Wall (1991). "Hans Schwerdtfeger, 1902–1990." *Australian Mathematical Society Gazette* 18, no. 3, 65–70 (includes photo).

Scriba, Ch. J. (2004). "In Memoriam Adolf Prag (1906–2004)." *Historia Mathematica* 31, 409–13.

Segal, S. L. (1980). "Helmut Hasse in 1934." *Historia Mathematica* 7, 46–56.

Segal, S. L. (2003). *Mathematicians under the Nazis.* Princeton, NJ: Princeton University Press.

Selberg, S. (1965). "Ernst Jacobsthal." *Det Kongelige Norske Videnskabers Selskabs Forhandlinger* (Trondheim) 38, no. 16, 70–73 (includes photo).

Sher, R. B. (1994). "Max Dehn and Black Mountain College." *Mathematical Intelligencer* 16, no. 1, 54–55.

Sheynin, O. (2003). "Mises on Mathematics in Nazi Germany." *Historia Scientiarum* 13, 134–46.

Siebauer, U. (2000). *Leo Perutz: Ich kenne alles. Alles nur nicht mich.* Gerlingen: Bleicher.

Siegel, C. L. (1956). *Vorlesungen über Himmelsmechanik.* Berlin, Göttingen, Heidelberg: Springer.

Siegel, C. L. (1965). *Zur Geschichte des Frankfurter Mathematischen Seminars.* Frankfurt: Klostermann.

Siegel, C. L. (1968). "Zu den Beweisen des Vorbereitungssatzes von Weierstraß," in C. L. Siegel, *Gesammelte Abhandlungen*, vol. 4. Berlin: Springer, 1979, 1–8.

Siegmund-Schultze, R. (1984). "Theodor Vahlen-zum Schuldanteil eines deutschen Mathematikers am faschistischen Mißbrauch der Wissenschaft." *NTM* 21, no. 1, 17–32.

Siegmund-Schultze, R. (1992). "Über das Interesse von Mathematikern an der Geschichte ihrer Wissenschaft," in S. Demidov et al., eds. (1992), pp. 705–35.

Siegmund-Schultze, R. (1993a). *Mathematische Berichterstattung in Hitlerdeutschland. Der Niedergang des Jahrbuchs über die Fortschritte der Mathematik (1869–1945).* Göttingen: Vandenhoeck and Ruprecht.

Siegmund-Schultze, R. (1993b). "Hilda Geiringer-von Mises, Charlier Series, Ideology, and the Human Side of the Emancipation of Applied Mathematics at the University of Berlin during the 1920s." *Historia Mathematica* 20, 364–81.

Siegmund-Schultze, R. (1993c). Dealing with the Political Past of East German Mathematics." *Mathematical Intelligencer* 15, no. 4, 27–36.

Siegmund-Schultze, R. (1994). " 'Scientific Control' in Mathematical Reviewing and German-U.S.-American Relations between the Two World Wars." *Historia Mathematica* 21, 306–29.

Siegmund-Schultze, R. (1996). "Zu den ost-westdeutschen mathematischen Beziehungen bis zur Gründung der Mathematischen Gesellschaft der DDR 1962." *Hochschule Ost* 5, no. 3, 55–63.

Siegmund-Schultze, R. (1997). "The Emancipation of Mathematical Research Publishing in the United States from German Dominance (1878–1945)." *Historia Mathematica* 24, 135–66.

Siegmund-Schultze, R. (1998). *Mathematiker auf der Flucht vor Hitler: Quellen und Studien zur Emigration einer Wissenschaft.* Braunschweig, Wiesbaden: Vieweg.

Siegmund-Schultze, R. (1998a). "Eliakim Hastings Moore's 'General Analysis'." *Archive for History of Exact Sciences* 52, 51–89.

Siegmund-Schultze, R. (1999). "The Shadow of National Socialism," in K. Macrakis and D. Hoffmann, eds. (1999), *Science under Socialism: East Germany in Comparative Perspective.* Cambridge, MA: Harvard University Press, 64–81, 315–17.

Siegmund-Schultze, R. (2000). "Die autobiographischen Aufzeichnungen Peter Thullens." *Exil* 20, no. 1, 58–66.

Siegmund-Schultze, R. (2001): *Rockefeller and the Internationalization of Mathematics between the Two World Wars: Documents and Studies for the Social History of Mathematics in the 20th Century.* Basel: Birkhäuser.

Siegmund-Schultze, R. (2002). "The Effects of Nazi Rule on the International Participation of German Mathematicians: An Overview and Two Case Studies," in K. H. Parshall and A. Rice, eds., *Mathematics Unbound: The Evolution of an International Mathematical Research Community, 1800–1945.* Providence and London: AMS and LMS, 335–57.

Siegmund-Schultze, R. (2003a). "The Late Arrival of Academic Applied Mathematics in the United States: A Paradox, Theses, and Literature." *NTM* (n.s.) 11, 116–27.

Siegmund-Schultze, R. (2003b). "Military Work in Mathematics, 1914–1945: An Attempt at an International Perspective," in B. Booß-Bavnbek and J. Høyrup, eds. (2003), *Mathematics and War.* Basel: Birkhäuser, 2003, 23–82.

Siegmund-Schultze, R. (2004a). "Helmut Grunsky (1904–1986) in the Third Reich: A Mathematician Torn between Conformity and Dissent," in Oliver Roth and Stephan Ruscheweyh, eds. (2004), *Helmut Grunsky: Collected Papers.* Lemgo: Heldermann, 2004, XXXI–L.

Siegmund-Schultze, R. (2004b). "A Non-Conformist Longing for Unity in the Fractures of Modernity: Towards a Scientific Biography of Richard von Mises (1883–1953)." *Science in Context* 17, 333–70.

Siegmund-Schultze, R. (2007). "Einsteins Nachruf auf Emmy Noether in der New York Times 1935." *Mitteilungen der Deutschen Mathematiker-Vereinigung* 15, no. 4.

Siegmund-Schultze, R., and S. Zabell (2007). "Richard von Mises and the 'Problem of Two Races': A Statistical Satire in 1934." *Historia Mathematica* 34, 206–20.

Sigmund, K. (2001). *"Kühler Abschied von Europa"—Wien 1938 und der Exodus der Mathematik* (catalogue to an exhibition, Vienna, September 2001). Vienna: Österreichische Mathematische Gesellschaft.

Sigmund, K. (2004). "Failing Phoenix: Tauber, Helly, and Viennese Life Insurance." *Mathematical Intelligencer* 26, no. 2, 21–33.

Sigmund, K., Dawson, J., and K. Mühlberger (2006). *Kurt Gödel: The Album.* Wiesbaden: Vieweg.

Sigurdsson, S. (1991). *Hermann Weyl, Mathematics and Physics, 1900–1927.* Unpublished PhD thesis, Harvard University, 307 pp. typewritten.

Sigurdsson, S. (1996). "Physics, Life, and Contingency: Born, Schrödinger, and Weyl in Exile," in Ash and Söllner, eds. (1996), 48–70.

Slembek, S. (2002). *Oscar Zariski und die Entstehung der modernen algebraischen Geometrie.* Unpublished PhD dissertation, University of Mainz, Germany.

Slembek, S. (2007). "On the Arithmetization of Algebraic Geometry," in Gray and Parshall, eds. (2007), 285–300.

Soderberg, C. R. (1982). "Stephen P. Timoshenko (1878–1972)." *Biographical Memoirs of the National Academy of Sciences* 53, 322–49.

Soifer, A. (2004/5). "In Search of Van der Waerden." *Geombinatorics* 14, 21–40, 72–102, 124–61.

Sommerfeld, A., and F. Seewald (1952/53). "Ludwig Hopf zum Gedächtnis." *Jahrbuch der Rheinisch-Westfälischen Technischen Hochschule Aachen* 5, 24–26.

Spalek, J. M. (1978). *Guide to the Archival Materials of the German-Speaking Emigration to the United States after 1933.* Charlottesville: University Press of Virginia.

Srinivasan, B. (1984). "Ruth Moufang." *Mathematical Intelligencer* 6, no. 2, 51–55.

Srinivasan. B., and J. D. Sally, eds. (1983). *Emmy Noether in Bryn Mawr*. New York: Springer.

Srubar, I., ed. (1988). *Exil, Wissenschaft, Identität; Die Emigration deutscher Sozialwissenschaftler, 1933–1945*. Frankfurt: Suhrkamp.

Stachel, J. (1994). "Lanczos's Early Contributions to Relativity and His Relationship with Einstein," in J. D. Brown et al., eds. (1994), 201–21.

Stadler, F. (1988). "Philosophie-Wissenschaftstheorie-Mathematik-Logik: Prolog zu Emigranten-Wissenschaften," in Stadler, ed. (1988), 118–24.

Stadler, F. (2001). *The Vienna Circle: Studies in the Origins, Development, and Influence of Logical Empiricism*. New York: Springer.

Stadler, F., ed. (1987). *Vertriebene Vernunft I. Emigration und Exil österreichischer Wissenschaft, 1930–1940*. Vienna, Munich: Jugend und Volk.

Stadler, F., ed. (1988). *Vertriebene Vernunft II. Emigration und Exil österreichischer Wissenschaft*. Vienna, Munich: Jugend und Volk.

Stein, E. (1986). *The Collected Works of Edith Stein—Sister Teresa Benedicta of the Cross Discalced Carmelite (1891–1942)*, vol. 1, edited by L. Gelber and R. Leuven. Washington, DC: ICS Publications.

Strassmann, W. P. (2006). *Die Strassmanns: Schicksal einer deutsch-jüdischen Familie über zwei Jahrhunderte*. Frankfurt, New York: Campus Verlag.

Strauss, H. (1991). "Wissenschaftsemigration als Forschungsproblem," in Strauss et al., eds. (1991), 7–23.

Strauss, H., T. Buddensieg, and K. Düwell, eds. (1987). *Emigration: Deutsche Wissenschaft nach 1933. Entlassung und Vertreibung*. Berlin: Technical University.

Strauss, H. A., K. Fischer, Ch. Hoffmann, and A. Söllner, eds. (1991). *Die Emigration der Wissenschaften nach 1933. Disziplingeschichtliche Studien*. Munich: K. G. Saur.

Stürzbecher, M. (1997). "Dr. med Albert Fleck und die Suche nach seiner Fermat-Klinik." *Acta Historica Leopoldina* 27, 339–46.

Süss, I. (1967). *Entstehung des Mathematischen Forschungsinstituts in Oberwolfach im Lorenzenhof*. Oberwolfach; privately printed, 59 pp.

Swerdlow, N. M. (1993). "Otto E. Neugebauer." *Proceedings of the American Philosophical Society* 137, 137–65.

Synott, M. G. (1979). *The Half-Opened Door: Discrimination and Admissions at Harvard, Yale, and Princeton, 1900–1970*. Westport, CT: Greenwood.

Szaniawski, K., ed. (1989). *The Vienna Circle and the Lvov-Warsaw School*. Dordrecht, Boston, London: Kluwer.

Szegö, G. (1954). "Otto Szász." *Bulletin AMS* 60, 261–63.

Szegö, G. (1982). *Collected Papers*, vol. 1, edited by R. Askey. Boston, Basel: Birkhäuser.

Tamari, D. (2007). *Moritz Pasch (1843–1930): Vater der modernen Axiomatik. Seine Zeit mit Klein und Hilbert und seine Nachwelt. Eine Richtigstellung*. Aachen: Shaker.

Tarwater, J. D., J. T. White, and J. D. Miller, eds. (1976). *Men and Institutions in American Mathematics*. Lubbock: Graduate Studies, Texas Tech University.

Tarwater, D., ed. (1977). *The Bicentennial Tribute to American Mathematics, 1776–1976*. Washington, DC: Mathematical Association of America.

Taussky-Todd, O. (1985). "An Autobiographical Essay," in Albers and Alexanderson, eds. (1985): *Mathematical People*, 309–36 (includes photos).

Taussky-Todd, O. (1988a). "Zeitzeugin," in F. Stadler, ed. (1988), 132–34.

Taussky-Todd, O. (1988b). "My Personal Recollections of Hans Heilbronn," in Kani and Smith, eds. (1988), 27–37.

Thiel, Ch. (1984). "Folgen der Emigration deutscher und österreichischer Wissenschaftstheoretiker und Logiker zwischen 1933 und 1945." *Berichte zur Wissenschaftsgeschichte 7*, 227–56.

Thullen, P. (2000). "Erinnerungsbericht für meine Kinder." *Exil 20*, no. 1, 44–57 [edited by R. Siegmund-Schultze, English translation in this book as Appendix 6].

Tietz, H. (1980). "Fundstellen für biographische und bibliographische Angaben über deutsche Mathematiker, die nach 1933 verstorben sind." *Jahresbericht DMV 82*, 181–92.

Tietz, H. (1998). *Erlebte Geschichte: mein Studium—meine Lehrer*. Stuttgart: printed manuscript, 23 pp.

Titze, H. (1990). *Der Akademikerzyklus. Historische Untersuchungen über die Wiederkehr von Überfüllung und Mangel in akademischen Karrieren*. Göttingen: Vandenhoeck and Ruprecht.

Tobies, R. (1991). "Wissenschaftliche Schwerpunktbildung: Der Ausbau Göttingens zum Zentrum der Mathematik und Naturwissenschaften," in Brocke, ed. (1991), 87–108.

Tobies, R. (1994). "Mathematik als Bestandteil der Kultur—Zur Geschichte des Unternehmens 'Encyklopädie der Mathematischen Wissenschaften mit Einschluß ihrer Anwendungen'." *Mitteilungen Österreichische Gesellschaft für Wissenschaftsgeschichte 14*, 1–90.

Tobies, R. (2006). *Biographisches Lexikon in Mathematik promovierter Personen an deutschen Universitäten und Technischen Hochschulen WS 1907/08 bis WS 1944/45*. Augsburg: Dr. Erwin Rauner Verlag.

Tobies, R. (2007). "Zur Position von Mathematik und Mathematiker/innen in der Industrieforschung, am Beispiel früher Anwendung von mathematischer Statistik in der Osram G.m.b.H." *NTM* (n.s.) 15, 241–70.

Tobies, R., ed. (1997, 2nd ed. 2008). *"Aller Männerkultur zum Trotz": Frauen in Mathematik und Naturwissenschaften*. Frankfurt: Campus.

Toepell, M. (1996). *Mathematiker und Mathematik an der Universität München. 500 Jahre Lehre und Forschung*. Munich: Institut für Geschichte der Naturwissenschaften.

Toepell, M., ed. (1991). *Mitgliedergesamtverzeichnis der Deutschen Mathematikervereinigung 1890–1990*. Munich: Institut für Geschichte der Naturwissenschaften.

Tollmien, C. (1991). "Die Habilitation von Emmy Noether an der Universität Göttingen." *NTM* 28, no. 1, 13–32.

Ulam, S. (1958). "John von Neumann, 1903–57." *Bulletin AMS* 64, 1–49 [special issue devoted to von Neumann, May 1958, 1–129].

Ulam, S. (1976). *Adventures of a Mathematician*. New York: Scribners.

Van der Waerden, B. L. (1930/31). *Moderne Algebra*, 2 vols. Berlin: Springer.

Van der Waerden, B. L. (1935). "Nachruf auf Emmy Noether." *Mathematische Annalen* 111, 469–76.

Van der Waerden, B. L. (1975). "On the Sources of My Book *Moderne Algebra*." *Historia Mathematica* 2, 31–40.

Voigt, K. (1988). "Die jüdische Emigration in Italien. Ein Überblick," in Briegel and Frühwald, eds. (1988), 13–32.

Vogt, A. (1998). "Die Berliner Familie Remak—eine deutsch-jüdische Geschichte im 19. und 20. Jahrhundert," in M. Toepell, ed., *Mathematik im Wandel*, vol. 1, *Mathematikgeschichte und Unterricht*. Hildesheim: Franzbecker, pp. 331–42.

Wallis, W. A. (1980). "The Statistical Research Group, 1942–1945." *Journal of the American Statistical Association* 75, 320–30.

Wasow, W. (1986). *Memories of Seventy Years: 1909–1979*. Madison: typewritten, 426 pp. [a copy in the possession of Prof. H. Wefelscheid, Essen, has been used by this author].

Weaver, W. (1972). "Richard Courant (1888–1972)." *Yearbook American Philosophical Society*, 142–49.

Weiher, S. von (1983). *Männer der Funktechnik*. Berlin: VDE-Verlag.

Weil, A. (1992). *The Apprenticeship of a Mathematician*. Basel: Birkhäuser.

Weiner, Ch. (1969). "A New Site for the Seminar: The Refugees and American Physics in the Thirties," in Fleming amd Bailyn, eds. (1969), 190–234.

Weyl, H. (1935). "Emmy Noether." *Scripta Mathematica* 3, 201–20.

Weyl, H. (1938). "Courant and Hilbert on Partial Differential Equations (Review)." *Bulletin AMS* 44, 602–4.

Weyl, H. (1944). "David Hilbert and His Mathematical Work." *Bulletin AMS* 50, 612–54.

Widmann, H. (1973). *Exil und Bildungshilfe; Die deutschsprachige akademische Emigration in die Türkei nach 1933*. Frankfurt: Lang.

Wiener, N. (1935). "Once More . . . the Refugee Problem Abroad." *Jewish Advocate*, February 5, 2.

Wiener, N. (1956). *I Am a Mathematician: The Later Life of a Prodigy*. Garden City, NY: Doubleday.

Wilder, R. L. (1989). "Reminiscences of Mathematics at Michigan," in P. Duren, ed. (1988/89), vol. 3, 191–204.

Williamson, F. (1980). "Richard Courant and the Finite Element Method: A Further Look." *Historia Mathematica* 7, 369–78.

Wirth, G. (1982). *Die Hauser-Chronik*. Berlin: Der Morgen.

Woyczynski, W. A. (2001). "Seeking Birnbaum, or Nine Lives of a Mathematician." *Mathematical Intelligencer* 23, no.2, 36–46.

Wunderlich, W. (1948). "Gedenken an Ludwig Eckhart." *Nachrichten der Mathematischen Gesellschaft in Wien* (later *NÖMG*) 2, no. 4, 16–18.

Wunderlich, W. (1959). "Hofrat Alfred Basch und Magnifizenz Franz Magyar zum Gedächtnis." *NÖMG* 13, nos. 59/60, 60–62.

Young, L. C. (1981). *Mathematicians and Their Times*. Amsterdam, New York, Oxford: North Holland Publishing.

Zassenhaus, H. ed. (1977). *Number Theory and Algebra: Collected Papers Dedicated to Henry B. Mann, Arnold E. Ross, and Olga Taussky-Todd.* New York: Academic Press (photos of Mann, Taussky, and Ross).

Zienkiewicz, O. C. (2000). "Achievements and Some Unsolved Problems of the Finite Element Method." *International Journal for Numerical Methods in Engineering* 47, 9–28.

Photographs Index and Credits

Figure 1. Berlin Exhibit
P. **xxvi**. Title page of Brüning et al. (1998).

Figure 2. Richard Courant
P. **xxviii**. From the collection Natasha Artin-Brunswick (Princeton), kind mediation by Liliane Beaulieu (Nancy).

Figure 3. Hans Freudenthal
P. **10**. Private possession and courtesy Hans Freudenthal.

Figure 4. Wilhelm Hauser
P. **19**. In the possession of the Konrad-Adenauer-Foundation, St. Augustin, kind mediation by Manfred Agethen.

Figure 5. Hanna and Bernhard Neumann
P. **21**. Reproduced from Ledermann (1983), p. 104. Courtesy W. Ledermann (London).

Figure 6. Käte Sperling-Fenchel
P. **22**. Private possession and courtesy of Bernhelm Booss-Bavnbek (Roskilde, Denmark).

Figure 7. Theodor von Kármán
P. **39**. From the collection of portraits at the California Institute of Technology, Pasadena, kind mediation by Judith Goodstein and Bonny Ludt.

Figure 8. Oswald Veblen
P. **49**. Reproduced from Archibald (1938), pp. 206–7, published by AMS.

Figure 9. John von Neumann
P. **50**. Reproduced from John von Neumann, *Collected Works*, vol. 1, 1961, ed. A. H. Taub, p. iii, courtesy Pergamon Press.

Figure 10. Hermann Weyl
P. **57**. From the collection of Natasha Artin-Brunswick (Princeton), kind mediation by Liliane Beaulieu (Nancy).

Figure 11. Veblen, Landau, and Bohr
P. **64**. Courtesy Archives Institute for Advanced Science, Princeton.

Figure 12. Karl Löwner
P. **67**. From Charles Loewner, *Collected Papers*, ed. L. Bers, 1988, p. iii, courtesy Birkhäuser, Boston.

Figure 13. Petition by Wilhelm Blaschke
P. **74**. From Geheimes Staatsarchiv Berlin-Dahlem, Rep. 76 Va, Sekt. 11, Tit. IV, Nr. 37, fol. 49.

Figure 14. Circular of the Weierstrass-Commission
P. **76**. Archives Berlin-Brandenburger Akademie der Wissenschaften, II–VII, 17, fol. 18.

Figure 30. Hilda Geiringer
P. **145**. Private possession and courtesy of Geiringer's daughter, Magda Tisza (Boston).

Figure 31. Tombstone of Issai Schur
P. **154**. Cemetery Little Tel Aviv, private possession and courtesy Leo Corry (Tel Aviv).

Figure 32. Otto Neugebauer
P. **164**. From the collection of Natasha Artin-Brunswick (Princeton), kind mediation by Liliane Beaulieu (Nancy).

Figure 33. Richard von Mises
P. **172**. Private possession and courtesy Magda Tisza (Boston).

Figure 34. Von Kármán to W. Tollmien in June 1933
P. **174**. Kármán Papers Pasadena, 20.37. Courtesy Archives California Institute of Technology.

Figure 35. *Free Science*
P. **179**. Title page of *Free Science*, edited by Emil Julius Gumbel in Strasbourg.

Figure 36. Rudolf Lüneburg
P. **181**. Title page, Lüneburg, *Mathematical Theory of Optics*, Berkeley 1964, courtesy University of California Press.

Figure 37. Herbert Busemann
P. **183**. Collection of portraits, Mathematical Research Institute, Oberwolfach, kind mediation by Mrs. Disch.

Figure 38. Stephen Duggan
P. **193**. From Duggan (1943), dust cover, courtesy Macmillan publishers.

Figure 39. Roland G. D. Richardson
P. **196**. Reproduced from Archibald (1938), pp. 104–5, published by AMS.

Figure 40. Harlow Shapley
P. **198**. Reproduced from Jones (1984), title page (p. 204), courtesy Harvard University Press.

Figure 41. Max Dehn
P. **211**. Collection of portraits, Mathematical Research Institute, Oberwolfach, kind mediation by Mrs. Disch.

Figure 42. Einstein on Noether in the *New York Times*.
P. **215**. "The Late Emmy Noether," *New York Times*, May 4, 1935, p. 12. Courtesy Albert Einstein Archives, Hebrew University, Jerusalem.

Figure 43. Norbert Wiener
P. **220**. Collection of portraits, Mathematical Research Institute, Oberwolfach, kind mediation by Mrs. Disch.

Figure 44. Wolfgang Wasow
P. **221**. From *Results in Mathematics* 28 (1995), nos. 1–2, p. 3, courtesy Heinrich Wefelscheid (Essen).

Figure 45. George David Birkhoff
P. 224. Collection of portraits, Mathematical Research Institute, Oberwolfach, kind mediation by Mrs. Disch.

Figure 46. Carl Ludwig Siegel
P. 239. Reproduced from C. L. Siegel, *Gesammelte Abhandlungen*, vol. 1 (1966), title page, courtesy Springer publishers.

Figure 47. Felix Bernstein
P. 263. Private possession and courtesy Ronald Wiener (Sarasota, Florida), Bernstein's grandson.

Figure 48. Rademacher-Tree
P. 286. Reproduced from Golomb, Harris, and Seberry (1997), p. 218. Courtesy AMS.

Figure 49. Walter Ledermann
P. 289. From the private possession and courtesy of Walter Ledermann, London.

Figure 50. Emmy Noether
P. 292. Reproduced from Srinivasan and Sally, eds. (1983), title page, courtesy Bryn Mawr College, Philadelphia.

Figure 51. Franz Alt
P. 302. Private possession and courtesy of Franz Alt, New York City.

Figure 52. Feodor Theilheimer
P. 304. Private possession and courtesy of Rachel Theilheimer, Las Cruces, New Mexico.

Figure 53. Richard Brauer
P. 311. Reproduced from R. Brauer, *Collected Papers*, vol. 1, 1980, title page, courtesy MIT Press, Cambridge.

Figure 54. Bartel Leendert van der Waerden
P. 317. Collection of portraits, Mathematical Research Institute, Oberwolfach, kind mediation by Mrs. Disch.

Figure 55. Carl Gustav Gustav Hempel
P. 326. Private possession and courtesy of Volker Peckhaus (Erlangen).

Figure 56. Kurt Hohenemser
P. 332. Private possession and courtesy of Kurt Hohenemser, St. Louis, Missouri, now deceased.

Figure 57. Ernst Jacobsthal
P. 335. Reproduced from Selberg (1965), an obituary in the *Communications of the Norwegian Academy of Sciences* in 1965.

Figure 58. Vienna Exhibit, September 2001
P. 338. Private possession and courtesy of Karl Sigmund (Vienna).

Subject Index

This subject index has been generated with respect to the main body of the text only, excluding appendixes and literature. However, subjects repeated in the latter can be located in the index, too. The index is restricted to notions of specific historical or scientific/mathematical meaning, excluding those that occur very often such as acculturation, algebra, AMS, Austria, Berlin, Boston, Brown University, California, Cambridge, DMV, Emergency Committee, emigration, England, ETH, Europe, France, Frankfurt, Germany, Göttingen, Harvard, immigration, Institute for Advanced Study, Jewish, Jews, mathematician, Nazi, New York, OVP, physics, political, Princeton, Rockefeller, science, SPSL, United States, Vienna, victims, war, and Zurich (Zürich).

Name Index

This name index has been generated with respect to the main body of the text only, excluding appendixes and literature. However, names repeated in the latter can be located in the index, too. The index does not as a rule contain persons thanked in the preface or authors of literature. Also the Polish victims mentioned in footnote 15, page 4, are not listed again.